Flight Strategies of Migrating Hawks

Flight Strategies of Migrating

Paul Kerlinger

The University of Chicago Press

Chicago and London

Hawks

Paul Kerlinger is director of the Cape May Bird Observatory at Cape May Point, New Jersey, and director of research for the New Jersey Audubon Society.

The University of Chicago Press, Chicago 60637
The University of Chicago Press, Ltd., London

© 1989 by The University of Chicago
All rights reserved. Published 1989
Printed in the United States of America

98 97 96 95 94 93 92 91 90 89   54321

Library of Congress Cataloging-in-Publication Data
Kerlinger, Paul.
    Flight strategies of migrating hawks / Paul Kerlinger.
        p. cm.
    Bibliography: p.
    Includes index.
    ISBN 0-226-43166-5. — ISBN 0-226-43167-3 (pbk.)
    1. Hawks—Migration. 2. Birds—Migration. I. Title.
QL696.F3K45   1989
598'.916—dc19   89-30037
                                           CIP

To Nina

Contents

Preface

This book is the product of nearly ten years of research. It focuses on the empirical aspects of flight behavior of migrating hawks, mostly through my own studies using radar and direct visual observations. It is more about behavior than about hawks. For this reason I have included many references and examples from the literature dealing with the migration of passerines, waterfowl, shorebirds, and nonraptorial soaring birds. I hope that readers who are interested primarily in raptors and secondarily in migration will not be disappointed. Because I have attempted to integrate information about the general ecology, physiology, behavior, and evolution of raptors with their migratory flight behavior, raptor biologists may gain new insight about the natural history of the animals they study. Furthermore, students interested in the migratory flight behavior of nonraptorial species may be introduced to new ways of conceptualizing or analyzing important questions in migration research.

I was faced with the dilemma of whether to write for the layperson or the scientist. Because of the keen awareness of the nonscientific public, this is becoming less of a problem. My decision was to write for people who will take the time to think about the material. Scientists and others familiar with graphs, equations, statistical analyses, and migration and raptor jargon should find the book straightforward. Those who are not well versed on technical topics will need to devote more time in places or may wish to skip some sections. I have attempted to explain some of the more complex material (or give references) so that familiarity with the technical approaches is not really necessary. Had I omitted the technical aspects of my research, I would have cheated readers.

I have chosen to ignore much of the hawk migration literature that is based on counts of hawks at particular locations except where researchers have made conclusions about flight behavior. The reason will become clear in chapter 2 and elsewhere in the book. In addition, I have avoided conservation/management-oriented material. These omissions may deter some readers, but I believe they strengthen the book for several reasons. Including hawk migration count literature and conservation/management related material would "dilute" the material I consider most important. This book is concerned with basic research—the first step in the conservation/management process.

The format I have adopted seems to be unique among works on migration. The book is divided into two parts that should be read in the order presented. The first four chapters provide an introduction to hawk migration. I begin by defining migration and describing the types of migrations hawks undertake. Next are chapters detailing the methods used to study migrating hawks, the methodological and philosophical approach used to study the flight behavior of migrants, and atmospheric structure as it relates to soaring flight. The second part of the book includes nine chapters addressing aspects of the flight behavior of migrating raptors. These include flight mechanics (aerodynamics, energetics, and morphology), flight direction, altitude of flight, flocking, water crossing behavior, flight-speed selection, daily distance of migration, and flight strategies. These nine chapters are based on published and unpublished field data, the literature, and theory.

In many ways this book is a quantitative natural history of raptor migration. I realize that natural history is a forbidden concept among "modern" ecologists and behaviorists (evolutionary ecologists and behavioral ecologists included), but it need not be taboo when properly done. Before my studies there were few measurements of altitude, flock size, flight direction, speed, climb rate, glide ratio, and other aspects of the flight behavior of migrating hawks. Thus quantitative descriptions were a necessary first step. These descriptions are the superstructure of many chapters. By testing specific hypotheses regarding various aspects of flight behavior, I have addressed questions pertaining to how individual migrants make "decisions." Some of these tests used a priori hypotheses, whereas others were post hoc. These tests potentially can help us to determine how natural selection has shaped the flight strategies of different raptor species. These behavioral strategies, presumed to be the result of natural selection, are examined in the last chapter of the book along with questions and directions for future research.

Two manuscript reviewers criticized some sections of this book for lacking a stronger theoretical (mathematical) perspective and for not conducting more sophisticated analyses of the existing data. I am guilty of not incorporating sophisticated models in some places. The strength of this book is the empirical nature of my research. I systematically describe several aspects of the migratory flight behavior of numerous species of raptors, an approach never before used in a monograph on migration. As for more sophisticated analyses of existing data, I believe I have been conservative in that I have refrained from conducting elegant analyses on rough data, nor have I used theory that is not well founded empirically. For example, one reviewer suggested that more thorough analysis was necessary to investigate the differences in water crossing tendency among species. To do this would require a much better knowledge of real animals' ability to use continuous powered flight. At this time the empirical studies of avian flight energetics show so much variance that the results of such an analysis would be tenuous.

Finally, this book is not complete. Instead, it is the midpoint in a three-step process. Heintzelman's 1975 volume was the first step, summarizing what was known at the time about hawk migration. This book presents a framework for studying migration and details various aspects of the flight behavior of migrants. But it is the third step that I view as crucial. Future research should build on the research presented here, showing how the various aspects of migratory flight behavior are integrated into an overall flight strategy. An understanding of flight strategies will provide new insight on how natural selection has shaped the behavior of animals. Lack of time, funding, and energy have prevented me from following this course of study thus far.

Acknowledgments

During my fieldwork and writing I was assisted in many ways. Financial support for fieldwork was provided by the Frank Chapman Memorial Fund of the American Museum of Natural History, the Electric Power Research Institute (Palo Alto, California), the National Science Foundation (graduate assistantship), and the Society of Sigma Xi. Clemson University, the State University of New York at Albany, the Cape May Bird Observatory (New Jersey Audubon Society), and the Santa Ana National Wildlife Refuge contributed logistical support during my fieldwork. The University of Calgary and the University of Southern Mississippi provided me with computer time and office space while I wrote several chapters. Most of this book was written while I was unemployed, and many of the expenses were paid out of my own pocket (with assistance from Nina Tumosa).

Kenneth P. Able, Verner P. Bingman, and Sidney A. Gauthreaux, Jr., spent many hours in the field with me as well as in discussing migration. Many of the data in this book are the result of a collaborative effort. Able and Gauthreaux also introduced me to radar ornithology, maintained the equipment during fieldwork, and paid me a salary during portions of my research. I am very grateful to them.

Ramon Cipriano of the Atmospheric Sciences Research Center, State University of New York at Albany, was my initial tutor on atmospheric structure, soaring flight, and aerodynamic performance. Several sailplane flights with Ramon, including one to 3,200 meters above sea level, provided me with new insights on flight and how difficult it is to stay aloft. Roland B. Stull and Edwin Eloranta spent hours discussing the structure of the atmosphere and provided some of the photographs.

While I was a postdoctoral fellow at the University of Calgary and Southern Mississippi, my supervisors, M. Ross Lein and Frank R. Moore, were extremely tolerant of the time I devoted to initiating this work.

I wish that John G. New could see this volume. It was with his help that I undertook my first hawk migration project as an undergraduate at the State University College of New York at Oneonta.

Some of the photographs were graciously provided by Sidney A. Gauthreaux, Kenneth Hardy, Frank Schleicher, Clay Sutton, and Edwin Eloranta.

Several government agencies and private sources (the National Science Foundation, the United States Fish and Wildlife Service, and the Smithsonian Tropical Research Institute) declined to fund some of the projects included here or other studies I proposed that would have made the book more complete. I continue to be baffled about how and why research funds are awarded. Funding of "sexy" research or trendy tests of hypotheses by the National Science Foundation using mechanistic approaches is important, but it is difficult to study evolutionary mechanisms when the basic biology of an organism or a phenomenon is unknown. Programmatic research on sound biological questions is often ignored, while flashy, sometimes ill-conceived projects are favored. Projects funded by the United States Fish and Wildlife Service tend to be conservation/management-oriented, but too often they suffer from a dearth of knowledge regarding the basic biology (quantitative natural history) of the organisms in question. Before conservation research or tests of mechanistic hypotheses are conducted, a sound knowledge of the natural history of organisms is needed. That natural history information is not available now and will not be until rigorous, quantitative field studies are conducted.

I thank the following people for assisting in the field or with other aspects of my research and writing efforts: James Bednarz, Michael Bennett, William S. Clark, Lorne Curren, Peter Dunne, Ronald Enck, William Gergits, Thomas Gilmore, Patricia A. Gowaty, Ronald Gregg, Robert Jaeger, Frank Iwen, Stephen Kerlinger, M. Ross Lein, H. Elliott McClure, Timothy Moermond, Frank Moore, John G. New, Joseph Rasmussen, Paul Roberts, Myron Schulman, Nancy Schulman, David A. Sibley, Robert Snell, Clay Sutton, Patricia Sutton, David Willard, and Floyd Wolfarth.

Paul Roberts made invaluable comments on earlier drafts of several chapters. Patient and thoughtful reviews by Colin J. Pennycuick and Thomas Alerstam prevented me from making several errors in the text and were helpful in explaining important details about flight, the atmosphere, and migratory behavior. I am grateful to them.

Fred Kerlinger and Betty Kerlinger provided financial and analytical assistance throughout my research and writing.

Nina Tumosa provided emotional, financial, and editorial assistance throughout the writing of this book and during much of the fieldwork. Without her encouragement and level head this project could not have been completed.

1

The Ecology and Geography of Hawk Migration

Migration occurs throughout the animal kingdom (Baker 1978). A variety of movements have been interpreted as migration, probably as a result of the varying definitions proposed (see Dingle 1980 for a review). Thompson (1926) stated that migration included "changes in habitat, periodically recurring and alternating in direction." Such a definition describes the return migrations of some birds, fish, mammals, and zooplankton but excludes many movements that have more recently been considered migration. To remedy this narrow view, Baker (1978) expanded the definition to include "the act of moving from one spatial unit to another." Unfortunately, Baker's definition is too general to be practical (Dingle 1980). A definition less general than Baker's yet not as narrow as Thompson's is that of Dingle (1980), who defined migration as "a specialized behavior, *especially evolved* for the displacement of an individual in space" (italics mine).

Although Dingle's definition is general, it is useful because it emphasizes the role of natural selection and specialized behaviors. That natural selection has produced such a diversity of movements may explain why migration is so hard to define. Migration is, after all, a construct used by humans who seek to understand animal behavior.

To eliminate some of the confusion regarding migration and definitions, avian biologists recognize at least six types of bird movements that can be considered migration. Complete migrations include seasonal movements from a breeding range to a non-breeding range and back again, with little or no overlap between the ranges (fig. 1.1). The movement after the breeding season usually takes place during autumn and the movement before the breeding season takes place during spring. Partial migration includes seasonal movements to and from a non-breeding range by some members of a population but not by all. An exodus of individuals from a portion of the breeding range is also included in this type of migration (fig. 1.1). Thus, there is a varying degree of overlap between breeding and non-breeding ranges for par-

1

tial migrants. Complete and partial migrations qualify as migrations according to Thompson's (1926) definition.

Several other movements fit Dingle's definition of migration: natal dispersal, and irruptive, nomadic, and local movements. Natal dispersal is "the movement an individual makes from its point of origin to the place it reproduces or would have reproduced had it survived" (Gauthreaux 1982a). Irruptive movements and invasions are those made "irregularly at intervals of a few years," often "triggered by high population levels coupled with reduced food resources" (Gauthreaux 1982a). Some authors consider nomadic movements a special case of irruptive movements. Nomadism, where birds continually move from place to place, usually in response to fluctuating resources, can occur during the breeding or non-breeding season. Finally, the term "local movements" is found frequently in the literature (Brown and Amadon 1968) but is rarely defined. These types of movements may fall into one of the other categories. Raptors undertake all the types of movements defined above. This volume deals mainly with partial and com-

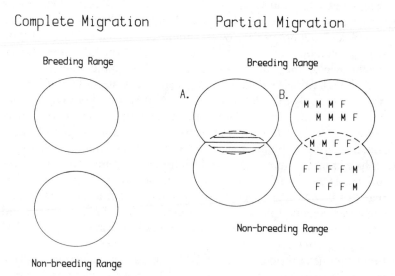

Figure 1.1 Schematic representation of complete migration, where all individuals leave the breeding range, and of two of the many patterns of partial migration. (A) A partial migration in which only a portion (horizontal lined area) of the breeding range is inhabited during the non-breeding seasons; (B) a partial migration in which most females leave the breeding range, creating skewed sex ratios in both breeding and non-breeding portions of the range during the non-breeding season.

plete migrants, although a few irruptive and nomadic movements are also examined.

Migration Tendencies of the Falconiformes

With the exception of Mindell's (1985) review of New World raptors, no general overview of the migratory tendencies of any taxon of birds has been attempted. I shall review the migratory tendencies of the 285 species of the order Falconiformes, based on information gleaned from the literature. I have used the listing of falconiform species given by Brown and Amadon (1968) with some modifications (extinct species omitted and some species lumped; e.g., *Falco kreyenborgi* = *F. peregrinus*). Each species is categorized as a nonmigratory resident, partial migrant, or complete migrant (operationally defined as more than 90% of the individuals of a species leaving the breeding range during the non-breeding season). In addition, all species known to undertake local, irruptive, or nomadic movements have been put in one category. Species for which no information on migration could be found have been placed in the resident or nonmigratory category. For each species that is known to migrate, the following characteristics of migration are noted:

1. Distance traveled during migration: <300 km one way = short distance; 300–1,500 km one way = medium distance; and >1,500 km one way = long distance.

2. Water crossing tendency and approximate distance of crossings: not known to make crossings; <25 km = short crossing; 25–100 km = moderate crossing; >100 km = long crossing.

3. Flocking tendency and approximate size of flocks formed: flocking categorized as not known to fly in flocks; irregularly or sometimes flocking; and regularly flocking.

4. Whether the species is known to follow insects during the non-breeding season.

Each of these behaviors will be discussed in detail. The classification scheme and subsequent analyses are preliminary and will need to be updated as more is learned about falconiform migration.

A majority (*N* = 152, 53.3%; fig. 1.2) of the 285 falconiform species are not migratory, at least based on information in the literature. Of the remaining 133 species, 24 (8.4% of all falconiforms) are irruptives, 18 (6.3%) are complete migrants, and 91 (31.9%) are partial migrants (appendix 1). Thus, partial migration is the most common form of migratory movement among the raptors and complete migration is the least common.

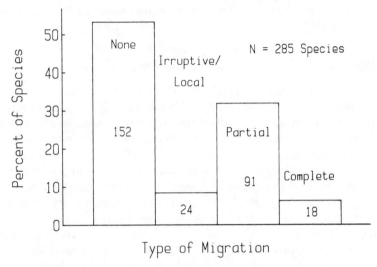

Figure 1.2 Summary of migratory tendency among the 285 species of falconiforms.

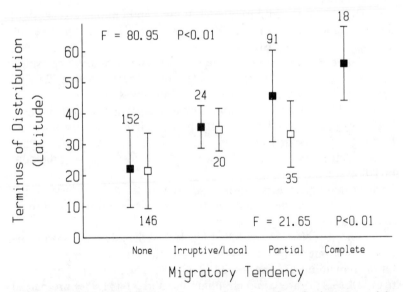

Figure 1.3 Relation of migratory tendency to the extreme latitudinal terminus of the distribution of species (solid squares = all species, $r^2 = 0.46$; open squares = species with latitudes extending north of 30° N, which were deleted from the analysis, $r^2 = 0.15$).

Among falconiform species the tendency to migrate is related strongly to the extreme latitudinal terminus of a species' distribution (maximum latitude in either the Northern or Southern hemisphere; fig. 1.3), which accounts for 46.3% of the variance. When all species distributions exceeding 30°S latitude are deleted from the analysis, the amount of variance accounted for by the relation increases nearly 13% to 58.9%. A decline of nearly 44% in the relation occurs when species with ranges extending north of 30° N are excluded (from 58.9% down to 15.2% of the variance). Therefore, it seems that species in the Southern Hemisphere are less migratory. An alternative explanation is that the number of species that migrate in the Southern Hemisphere is actually greater than described in the published literature. It is difficult to understand how the Gray Eagle-Buzzard, Red-tailed Buzzard, White-throated Caracara, Forster's Caracara, and Common Caracara do not migrate despite living in inhospitable climes at latitudes south of 50° S.

The coefficients of determination (percentage of the variance accounted for by an independent variable) given above are preliminary approximations. Some readers will argue that regression statistics are inappropriate with these data. The coefficients that emerge show a meaningful biological relation, but the relations in figure 1.3 do not answer the question whether migration is influenced directly by climate or indirectly through trophic (Newton 1979) or social relations (Gauthreaux 1978a, 1982a). For an introduction to the evolutionary and ecological underpinnings of migration in birds and other animals, see Cox (1968, 1985), Baker (1978), Gauthreaux (1982a), and Dingle (1980).

Among the 152 nonmigratory species of falconiforms, 33 (21.7%) belong to five genera (*Leucopternis, Spizaetus, Phalcoboenus, Micrastur,* and *Microhierax*). These and several other genera with fewer species (*Henicopernis, Harpagus, Spilornis, Harpyhaliaetus, Daptrius,* and *Poliohierax*) are mostly nonmigratory or at most make very short movements that remain undetected. The genus *Accipiter* has the largest number of nonmigratory species (37 of 47 species within the genus, 78.7%), although some of the other species are highly migratory. Together these genera constitute about half (83 of 152 species, 54.6%) of the nonmigratory species. The remaining 69 nonmigratory species are scattered among various taxonomic groups.

As shown in the analysis above, the distributional termini of nonmigratory species lie closer to the equator, on average, than do those of migratory species. Furthermore, many nonmigratory species are insular, being restricted to a few islands, or even one. Most of these 152 species (117, 77.0%) are tropical or subtropical, being restricted to the region from 30° N to 30° S latitude. Thus, only 6 of 152 (3.9%) species with distributions in the Northern Hemisphere exceeding 30° N are not known to migrate, and only 10 species (6.6%) range north of 50° N or south of 50° S.

Complete Migrants

At the other end of the continuum from the nonmigratory species are 18 complete migrants. This group includes 6 species from the genus *Falco*, 3 from *Buteo*, 2 from *Aquila*, and 1 each from seven other genera (appendix 1). All undertake round-trip flights exceeding 3,000 km. Although the Amur Falcon flies more than 15,000 km round trip, species that make return flights of 6,000–8,000 km are more typical. These migrations take one month to six weeks (Brown and Amadon 1968; Moreau 1972; Smith 1980; Cade 1982). Distance traveled during migration is highly dependent upon the type of migration undertaken by a species (log ratio $\chi^2 = 177$, df $= 4$, $p << 0.001$); complete migrants tend to travel farther than partial migrants, and partial migrants travel farther than irruptive/local migrants.

The breeding range for all complete migrants extends north of 31° N latitude, and for 10 species (55.6%) the breeding range exceeds 50° N (fig. 1.3). None of the complete migrants has a breeding range that is predominantly within the Southern Hemisphere. The Osprey is the only complete migrant that breeds in both hemispheres (although some southern populations or individuals may be nonmigratory, perhaps making the species a partial migrant). Only the Sooty Falcon breeds predominantly in the subtropics and tropics (northern extension to about 31° N). Although these species mostly exploit vertebrate prey (exceptions are the Honey Buzzard, Lesser Kestrel, and perhaps Red-footed and Amur falcons) during the breeding season, at least 9 of 18 (50.0%) become insectivorous during the non-breeding season. Of these, 7 or 8 species form nomadic flocks that seek locust and grasshopper "plagues," emerging termites, and ants. These odd bits of information demonstrate that the long-distance migrations of some falconiforms rival those of shorebirds, passerines, and waterfowl, although the subject is poorly studied.

Partial Migrants

The vast majority of falconiforms for which migration has been reported are partial migrants. At least 91 species are known to undertake partial migrations. This number will certainly increase as more is learned, especially about tropical species. Approximately one-half (47 of 91, 51.6%) of all partial migrants are from four genera: *Falco*, 18 species (19.8% of partial migrants); *Buteo*, 11 species (12.1%); *Accipiter*, 10 species (11.0%); and *Circus*, 8 species (8.9%). The remaining 46 species are taxonomically mixed. The breeding range of partial migrants extends from the Arctic (>70° N) through the tropics and southward to about 55° S. This range is far greater than the range of complete migrants, irruptives, or nonmigratory species. Similarly, the distance these species travel during migration ranges from very short distances (<100 km) to distances rivaling those traveled by many complete migrants. Most partial migrants however, undertake migrations that are usually less than 3,000 km round trip. Thus, the average distance traveled by partial migrants is shorter than that traveled by complete migrants (appendix 1).

The migration patterns of partial migrants are more difficult to characterize than those of complete migrants. Great diversity and variability of migration patterns occur among species, among age and sex classes, and among populations of a species. Some of this variability is discussed in the sections on geographic and temporal patterns of migration.

"Local" and Irruptive Movements

The 24 species in the "local" and irruptive movement category are from a mixed taxonomic background. At present, this category is a "dustbin" into which species not considered resident or "truly" migratory are classified. Movements of individuals within this group seldom if ever exceed 3,000 km round trip. One-half (12 of 24) of the species within this group are from five genera: (*Falco*, 3 species; *Buteo*, 3; *Elanus*, 2; *Hieraaetus*, 2; and *Circaetus*, 2).

Although "local" or irruptive migrants do not seem to make round-trip migrations between breeding and non-breeding ranges as partial and complete migrants do, they do undertake seemingly adaptive movements (sensu Dingle 1980). Whether these are migrations may be a semantic argument. The movements of these animals are often correlated with unpredictable fluctuations of the environment, analogous to regular migrations that occur in response to seasonal environmental fluctuations. The similarity of selective pressures in true migrants (partial and complete migrants) and in the irruptive species is striking; it shows that geographic movements are a fundamental adaptation of this taxonomic group.

Geographic and Temporal Patterns of Raptor Migration

In the previous section I presented a classification scheme that placed raptor species in one of four categories ranging from nonmigratory to highly migratory. Like most taxonomies, the categories are artificial. Errors of classification within my system are attributable to a paucity of knowledge, primarily regarding tropical species, or from variance of migratory tendency within species at the level of the population (deme), age and sex classes, and eventually the individual. The cause(s) of this variance will interest evolutionary ecologists, behaviorists, and wildlife managers. Within species, variance has been studied in only a few species of hawks, although it is a "hot" topic in the current migration literature (Ketterson and Nolan 1976, 1983; Gauthreaux 1978a, 1982a; Myers 1981; Terrill and Ohmart 1984; Rabenold and Rabenold 1985; Kerlinger and Lein 1986). The remainder of this chapter summarizes the geographic and temporal aspects of variability among and within falconiform species.

A popular idea among those who inhabit north temperate latitudes is that migratory hawks (as well as other birds) move south in autumn and return north the following spring. Although this pattern is probably the most common, it is only one of many. Variations in geographic patterns of migration include movements across latitudinal, longitudinal, and elevational gradients. Migratory movements of most raptors are combinations of these three patterns. Within these three categories are several types of movements that have been referenced in the literature: intratropical migration, habitat changes, movement to offshore islands, coastward movements, and so on.

Latitudinal Migrations

Migration across latitudinal gradients, as stated above, is the most common geographic pattern among hawks. Most species that breed in north temperate latitudes move south during autumn and north in spring. The distance of these movements can vary among species breeding at similar latitudes. For instance, some Gyrfalcons that breed at above 70° N move only a few degrees southward (although every winter some individuals move south to 45–50° N), whereas others remain north of 65° N (Platt 1976; National Audubon Society Christmas Bird Counts). Rough-legged Hawks and Peregrine Falcons (*Falco peregrinus tundrius*) that breed at the same latitudes as Gyrfalcons migrate farther south. Rough-legged Hawks winter from southern Canada to the central United States (AOU 1983), and tundra Peregrine Falcons winter from the southern United States into Central and South America (Brown and Amadon 1968; White 1968; Hunt, Rogers, and Stowe 1975; Cade 1982).

A similar situation occurs among the buteos inhabiting the prairies of western North America. Many Red-tailed Hawks breeding in Oklahoma are hardly migratory, whereas Swainson's Hawks from the same locale migrate to central South America across nearly 70° of latitude.

In southern South America and Australia several species of hawks are known to undertake latitudinal movements northward toward the equator during the austral autumn. This migration is generally not pronounced, but in some species the movements involve more than 1,000 km, similar to migrations in the Northern Hemisphere. Most of these species are from South America (Black-shouldered Kite, Plumbeous Kite, Snail Kite, Long-winged Kite, White-tailed Hawk, and Chimango, to name a few), whereas the remainder are from Australia and Tasmania (Australian Black-shouldered Kite, Marsh Harrier, Australian Kestrel, Brown Hawk, and Little Falcon). The absence of African species in this group may result because Africa extends only to about 34° S latitude, whereas Australia (Tasmania) extends beyond 43° S and South America beyond 55° S. There are several species in southern South America (Patagonia and Tierra del Fuego) that may migrate latitudinally, but little is known of their movements (Turkey Vulture, Gray Eagle-Buzzard, Red-backed Buzzard, Red-tailed Buzzard, Aplomado Falcon, and Peregrine Falcon [*F. p. cassini*]). A minimal amount of fieldwork might reveal that some populations or individuals of these species engage in latitudinal or other types of movements each year.

The latitudinal movements described above include migration toward or across the equator from temperate or even subtropical locations. Latitudinal movements are also known to occur solely within the tropics, although the distances involved are less than in temperate areas. The African Red-tailed Buzzard and African Steppe Eagle (Tawny Eagle) move across less than 10° of latitude between breeding and non-breeding sites (appendix 1). Very little is known about intratropical migrations, although Moreau (1972), Brown (1970), Thiollay (1978), and Curry-Lindahl (1981) discuss the topic. Migrations within the tropics are gaining the attention of more field biologists, as is evident from a recent volume concerned with migrants in the Neotropics (Keast and Morton 1980).

Longitudinal Movements

Migrations are less prevalent along longitudinal gradients than along latitudinal gradients. To my knowledge there are no species in which migratory movements are strictly along longitudinal gradients; yet there are several in which longitudinal distance equals or exceeds latitudinal distance. The best known of the longitudinal migrants is the Prairie Falcon of North America (Enderson 1964). Falcons from the western Great Plains and the eastern edge of the Rocky Mountains radiate eastward and southward into the eastern plains and Midwest. This eastward movement is characteristic of the migration of many hawks and owls that breed on the western prairies and in the western forests of North America, including Marsh Hawks (Houston 1968), Red-tailed Hawks (Houston 1967), Great Horned Owls (*Bubo virginianus;* Houston 1978), Bald Eagles (Gerrard et al. 1978), and Red-tailed Hawks (Brinker and Erdman 1985).

Raptors moving from Asia to Africa by way of the Middle East land bridge also traverse extensive longitudinal gradients. This eastward movement is in part a result of the Indian Ocean, which is a barrier to southward migration. Of the dozens of species that move through the Middle East (Christensen et al. 1981; Leshem 1985), the Amur Falcon makes the most extensive longitudinal movements. Thousands of these falcons fly from northeastern China (Amur Province) to east-central Africa and back each year, and many cross portions of the Indian Ocean. Similar longitudinal movements (although rarely over large bodies of water) are an integral part of the migrations of the Oriental Hobby into western India from China (Ali and Ripley 1978) and the Merlin into Scotland and England from Iceland (Williamson 1954).

Probably the most unusual of the longitudinal migrations is that of the Letter-winged Kite, which has been interpreted as an irruptive species. These kites radiate centrifugally from central Australia toward the coasts (Brown and Amadon 1968). Return movements probably occur, but when or how often is not known. These radiative movements from the hot, arid interior may be linked to the unpredictable resources and rainfall upon which the breeding of so many Australian birds depends.

Elevational Movements

The migration of some species involves movements from high to low elevations in the mountains. Elevational migrations probably are more common than noted in the literature, especially in temperate areas where high-elevation climates are more unsuitable during winter than those in the tropics. Elevational movements are particularly difficult to document because movements are relatively short, some individuals (or populations) may make both latitudinal and elevational migrations, and populations undertaking latitudinal migrations may overlap with those making altitudinal movements during the nonbreeding season, obscuring elevational patterns. The best documentation of elevational migration is that of Rabenold and Rabenold

(1985), who marked more than 1,000 Carolina Juncos (*Junco hyemalis carolinensis*) in the mountains of the southeastern United States. Rabenold and Rabenold found that the winter range of Carolina Juncos overlapped with that of a different subspecies of junco (the Northern Junco, *J. h. hyemalis*) that is a latitudinal migrant. Dozens of raptors may undertake such migrations. If similar patterns occur among them, they would be difficult to detect.

Seasonal Timing of Migration

At north temperate latitudes, where a large proportion of species are migrants, seasonal patterns of migration vary. Autumn migration can begin as early as August or as late as December and January. For young of the year the time to migrate should be a compromise among the seasonal decline of prey, weather, and developmental state of the bird (including molt and feather growth). For adults that can feed themselves adequately, weather and a decline of prey should determine when migration occurs. Geller and Temple (1983) showed that immature Red-tailed Hawks migrating from northern latitudes had less body fat and migrated earlier than immatures from farther south. They suggested that the shorter breeding season of northerly populations necessitated migration at an earlier stage of life. Do individuals from northern populations experience greater mortality than those from southern latitudes because of poorer body condition when migration begins (Greenberg 1980)?

Timing of spring migration also varies. Some species such as Gyrfalcons return to nest sites as early as January/February even though they nest at arctic latitudes (Platt 1976). Tundra-breeding Peregrine Falcons, however, do not return until late May or early June (Cade 1960). An analogous situation occurs with Red-tailed and Swainson's hawks that breed on the prairies of southern Alberta. Although both rely on Richardson's ground squirrels (*Spermophilus richardsonii*) as the predominant prey species, Red-tailed Hawks return in late March and early April whereas most Swainson's Hawks do not return until early May. Selective pressures that probably select for timing of arrival on the breeding grounds include food availability, climate, and competition for nest sites. Early arrival may be rewarded by a choice of the best nesting sites (but see Bedard and LaPointe 1984), although harsh weather may make prey unavailable, so that birds who arrive too early starve. Starvation need not kill the bird to influence reproductive success (fitness). A slightly starved bird may not be able to defend a territory, attract a mate, build a nest, or yolk up eggs. Any of these will decrease the number of offspring produced in that year.

Until now I have discussed migration during autumn and spring. In the tropics seasons are not usually termed spring and autumn but are often divided into rainy and dry seasons that can be as predictable as seasonal fluctuations at temperate latitudes. The onset of the rainy season in many parts of the tropics is distinct. One day it is dry, the next the rains have arrived. The onset of the rainy season can vary from year to year by several weeks (Brown 1970; Oke 1978), as can the onset of tropical migrations and breeding seasons.

The migration of birds within tropical environments with rainy and dry seasons is less noticeable than migration at temperate latitudes because of the short distances migrants travel. Moreover, these movements have rarely been studied. Among the falconiforms several species undertake rainy or dry season migrations within the tropics. Brown (1970) presents a readable introduction to these movements within Africa, and the works by Moreau (1972), Thiollay (1978), and Curry-Lindahl (1981) are helpful. Brown states that for species such as the large vultures the rains prohibit soaring flight, but that the convection (thermals) associated with the rains promotes migration to drier areas. Some species that undertake rainy/dry season migrations are noted in appendix 1.

Differential Migration: Variation in Seasonal Timing and Geography within Species

In the previous sections the focus was primarily on the species as the unit of observation and analysis. This view of species specific migratory tendencies can be misleading. Researchers are beginning to document variability within many migrant species with respect to seasonal timing of migration and geographic (including habitat) distribution during the non-breeding season. This variance is manifested at the level of the population (or perhaps subspecies), age and sex classes, and individuals. This variation within a species is termed "differential migration" (Gauthreaux 1978a, 1982a; Ketterson and Nolan 1983). It is attracting much attention among ornithologists studying the behavioral ecology of migrants.

Differential Migration by Population

Differences in seasonal timing and geographic pattern of migration have been documented among populations of the same species. The most commonly cited example is "leapfrog" migration where individuals from northern populations migrate earlier and fly farther than individuals from southern populations. The most cited example of leapfrog migration is that of the Peregrine Falcon in North America (Beebe 1960; Enderson 1965; Herbert and Herbert 1965; White 1968) and Eurasia (Vaurie 1965). In North America, tundra-breeding falcons (*F. p. tundrius*) migrate farther south than the slightly migratory temperate falcons of the *anatum* and *pealei* subspecies. In Eurasia a parallel phenomenon is known where the falcons of this species from farthest north leapfrog "over" those from England and other nonmigrating populations. Similar, but less dramatic, examples have been documented for Turkey Vultures in eastern North America (Stewart 1977), Red-tailed Hawks in western North America (Mindell 1983), kestrels in Europe (Mead 1973), and Common Buzzards in Europe (Salomonsen 1955).

Leapfrog migrations are also manifested by differences in seasonal timing of migration. Individuals from northern populations tend to migrate earlier than individuals from southern populations. Geller and Temple (1983) hypothesized that Red-tailed Hawks from the northern extent of the species' range in interior North America initiated migration earlier than individuals from farther south. Their evidence was a significant decrease in the size of birds (wing chord) as the autumn migration season progressed. Larger individuals, presumably from more northern populations of Red-tailed Hawks, preceded smaller individuals from southern populations. A similar trend has been suggested by Mueller, Berger, and Allez (1981a) for Cooper's Hawks migrating in autumn through Wisconsin. The gradual change of wing size from larger to smaller birds during autumn migration as shown by Geller and Temple (1983) may indicate a clinal variation in the seasonal timing of migration that reflects latitudinal rather than strictly populational (subspecific) differences. Furthermore, size differences among age-sex classes (i.e., adult male, adult female, immature male, immature female) may have confounded these results for Red-tailed Hawks. If males and females could be distinguished by morphological characters (which is not always possible at this time for Red-tailed Hawks), a separate regression could be constructed for each age-sex class that would make the pattern easier to interpret.

Leapfrog migration patterns have been noted in Europe for raptors and other migrants. Pienkowski, Evans, and Townshend (1985) and earlier Salomonsen (1955) discussed the importance of body size in the evolution of leapfrog migration. It should be noted that for some species in Europe (Eurasian Kestrel, Common Buzzard, and Peregrine Falcon) and North America (Peregrine Falcon) the northern populations are smaller than southern populations. Thus, size may or may not be important in the evolution of leapfrog migration.

In addition to leapfrog patterns, there are other patterns that are often species specific and have not been well studied. In species such as the Tawny (or Steppe) Eagle, with subspecies that range over enormous geographic areas, variations of migration pattern include autumn migration out of Eurasia into Africa, intratropical migrations that are synchronized with rainy and dry seasons, and others. One of the most unusual variations of seasonal timing of migration is that of Bald Eagles in North America. Whereas some populations of eagles are nonmigratory, others are highly migratory. Eagles in northeastern North America make "normal" autumnal migrations south in September through November and back again in March and April (Bent 1937). Southern Bald Eagles (which are smaller than northern individuals), however, move north in late spring and early summer after breeding and then move south in late August to October to begin breeding in Florida during the winter (Broley 1947). Recent radiotelemetry studies (M. Fuller, personal communication) show that there is a continuous change in the distance and direction of migration tendencies moving from northern to southern populations. Eagles breeding between the extremes do not move as far as those at the most northerly and southerly extensions of the range. For example, some eagles breeding in South Carolina move north after breeding, like the Florida birds, but may not move as far north. Birds from New England move south after breeding, but not as far south as Ontario or Quebec birds. Northward movements following the breeding season are not uncommon, as is the case with Red-tailed Hawks in Wisconsin (Brinker and Erdman 1985) and Common Buzzards in Europe (Olsson 1958). Brinker and Erdman demonstrated that these movements were made by immature birds and that they occurred with prevailing southwest winds. This does not explain the migration pattern of southern Bald Eagles.

Differential Migration by Age and Sex Classes

Differential migration by age and sex classes of migrant hawks is manifested by seasonal differences of migration timing, geographic distribution during the non-breeding season, and habitat segregation where age-sex classes overlap during the non-breeding season. Of these, differential seasonal timing of migration has been documented for more species than geographic differences or habitat segregation because migration studies are easier to conduct.

Evidence for differential seasonal timing of migration by age-sex classes (appendix 2) has been acquired from hawk migration counts, banding studies during migration, and studies done on the breeding grounds. In spring migration back to the breeding range adults almost always precede immatures (appendix 2), although no studies of species in which sexual maturation occurs within the first year seem to have been done. In some species, such as the Osprey and possibly the Bald Eagle, this pattern is so pronounced that immatures do not migrate back to the breeding range until they are nearly two years old or more. For most species the difference between the arrival time of age classes is from a few days to more than two months.

The picture for differential timing by sex is not as clear. In most complete migrants or long-distance partial migrants, the sexes seem to arrive on the breeding grounds paired or so close in time that no difference can be discerned (appendix 2). Among partial migrants that do not migrate long distances, males often arrive on the breeding grounds before females, although in a few species females either do not migrate (Enderson 1964; Mearns 1982) or arrive first. The difference between the arrival dates of the sexes, however, is probably no more than two weeks.

In autumn there is no predominant trend in whether adults precede immatures or vice versa (appendix 2). Haugh (1972) stated that "immature birds of most species migrate earlier in the fall than adults do," and he seems to be correct, at least among North American hawk migrants. Newton (1979) made a distinction between long-distance (both partial and complete) migrants and shorter-distance (partial) migrants with regard to timing of autumn migration. He noted that among long-distance migrants it was the adults that preceded immatures and that immatures migrated earlier among shorter-distance migrants. The evidence in appendix 2 shows that Newton (1979) was correct, but quantitative evidence is available for only a few species. Until more species are studied the question should be considered open.

Evidence is less available for differential geographic distribution of age-sex classes during the non-breeding season than for differential seasonal timing of migration. The phenomenon has been termed latitudinal segregation by Myers (1981) and differential distance of migration by Gauthreaux (1978a, 1982a). Whatever terminology is used, the assumption is that age-sex classes must travel different distances to their non-breeding ranges after leaving the breeding range. Most of the data that suggest differential geographic distribution among age-sex classes of raptors come from winter population studies conducted at single locations. Unequal sex or age ratios noted in these studies have been interpreted as skewed sex ratios in the population (or species) as a whole and as differential migration. The species for which unequal sex or age ratios have been reported are listed in appendix 2. Because few studies have employed statistical analyses or interpreted the results as differential migration, I have had to interpret the data as presented. The data should be considered only suggestive at this time, to be used as a starting point for more rigorous studies. Studies of differential migration should employ the techniques and analyses used by Ketterson and Nolan (1976, 1983) for Northern Juncos, Moore (1976) for Herring Gulls (*Larus argentatus*), Myers (1981) for various shorebirds, Howell (1953) for Yellow-bellied Sapsuckers (*Sphyrapicus varius*), Rabenold and Rabenold (1985) for Carolina Juncos, and Kerlinger and Lein (1986) for Snowy Owls (*Nyctea scandiaca*). Moore's study is a classic and should be given particular attention because of the use of banding data and large sample sizes. Moore found a reduction of migration distance among gulls from the first through fourth year of age.

Only one conclusion can be made with confidence about differential distribution of hawks during the non-breeding season. When it does occur adults generally winter north of immatures. As with differential timing of migration, the trend seems to be more evident among partial migrants than among complete (or long distance) migrants. No consistent difference between the sexes is obvious.

As with differential timing and geographic distribution, few studies have directly investigated whether age-sex classes use or select different habitats during the non-breeding season. Differences have been reported between the sexes of the American Kestrel (Enderson 1960; Koplin 1973; Mills 1975, 1976) and the Sparrowhawk (Marquiss and Newton 1982). In both species females tend to spend the non-breeding season in more open habitats than males. Differences in habitat selection between the sexes are also suspected among Peregrine Falcons in south Texas and Central America (Hunt, Rogers, and Stowe 1975) and perhaps among Merlins (S. A. Temple, personal communication).

Differential migration is widespread among the falconiforms. The evidence is fragmentary and sometimes contradictory. For example, references (appendix 2) suggest that both male and female Prairie Falcons arrive first at Colorado breeding sites in spring. Because Enderson (1964) presented data that show females arrived first on breeding sites, I favor his results over Webster's (1944) claim that males arrive first. Appendix 2 shows similar discrepancies (e.g., Sharp-shinned Hawks). Also of interest is the reversal of trends among falcons. Male Gyrfalcons, Merlins, and American Kestrels seem to arrive at breeding sites before females in spring, whereas for Prairie and Peregrine falcons there is some evidence that females arrive first. Furthermore, Cade (1960) reports that male and female Peregrine Falcons arrive at tundra breeding sites as pairs. Why is there such variation among related taxa? The variation may be attributed to real differences within a species or population of that species as a result of population density or nest-site limitation, or it may be a result of sampling bias. So the evidence yields ambiguous or doubtful results, which may be due to small sample sizes or to samples of only a single population. Obviously, much more research is needed to resolve these discrepancies.

Changes in the Geographic Pattern of Migration

The geographic distributions of migratory hawks during the non-breeding season (usually winter) should not be considered fixed or unchanging. In recent years geographic changes in the pattern of migration of numerous hawks have become evident. Many of these changes can be linked to modifications of the environment by man. For instance, some Merlins that breed on the Canadian Great Plains (subspecies *Falco columbarius richardsoni*) have begun to winter in or near Calgary, Saskatoon, and other cities (personal observations). Formerly, these birds probably migrated farther south. Food sources such as granaries and planted (exotic) shrubs and trees now provide winter food for prey species such as House Sparrows (*Passer domesticus*), Starlings (*Sturnus vulgaris*), and Bohemian Waxwings (*Bombycilla garrula*). In an apparently adaptive fashion, Merlins have responded to the increased prey availability by changing their migratory behavior.

A similar northward shift of wintering range for the Red Kite in Europe (Juillard 1977) is due to an increase in garbage dumps. Bald Eagles in North America probably winter farther north or in greater local densities at higher latitudes than in earlier times, especially where dams and spillways keep water from freezing. At Calgary, Alberta, up to a dozen Bald Eagles winter along the Bow River (personal observations), where water is kept from freezing by treated wastewater that is dumped into the river. The eagles seem to prey on Mallards (*Anas platyrynchos*) and other ducks (and some fish) that congregate at open water. In addition, they forage on the prairie more than 75 km from Calgary, where they feed on road kills and probably white-tailed jackrabbits (*Lepus sp.*). These changes again demonstrate the variability and plasticity of migratory behavior among individuals and populations within a species.

There are at least two other species in which recent geographic shifts in winter range have been reported. Before the 1950s, few Broad-winged or Swainson's hawks wintered north of the Mexican border. Swainson's Hawks are now known to winter in small numbers in both south Texas and south Florida (Browning 1974), as are Broad-winged Hawks in south Florida (Brown and Amadon 1968). These changes presumably have occurred in the past three decades. Although some individuals of both species winter in Florida, it is not known if they are simply immature birds that have not found their normal wintering ranges south of the American border. There is evidence that a majority of these birds are immature (museum skins examined by Browning 1974). Three immature Broad-winged Hawks tagged by Tabb (1979) during winter were recovered in subsequent years in Central America where they were wintering or migrating. Of 156 Broad-winged Hawks Tabb captured in Florida, 133 (85.3%) were immature. These findings are consistent with the hypothesis that the individuals of these species that terminate their migration in Florida have possibly made "mistakes" during migration. Migrants entering the Florida peninsula might be reluctant to cross the Caribbean or the Gulf of Mexico and terminate their migration within the peninsula. Thus, Swainson's and Broad-winged hawks that winter in Florida may be extralimitals, occurring only by accident.

That some migrants vary the distances they fly is not surprising. In much of the earlier literature, migration was considered a fixed, instinctual behavior that was not subject to environmental modification (see Berthold 1975 and Gauthreaux 1982a for reviews of environmental modification of migration). Recent papers by Terrill and Ohmart (1984) and Rabenold and Rabenold (1985) demonstrate that migration in some passerines is an extremely "plastic" behavior. They have shown that the seasonal timing of migration is not fixed and that there are differences between years. In mild years Yellow-rumped Warblers (*Dendroica coronata;* Terrill and Ohmart 1984) and Carolina Juncos (Rabenold and Rabenold 1985) did not migrate as far from the breeding ground as in years with colder weather. For warblers, the readiness to migrate continued throughout the winter, and birds migrated again when climatic conditions became stressful. The juncos studied by Rabenold and Rabenold returned to their breeding sites earlier after mild winters than after "hard" winters. Unpublished radar studies (S. A. Gauthreaux, personal communication) show that massive migrations occur during the entire winter even in the Deep South. The species composition of these movements is unknown, although characteristics of radar echoes suggest passerines and waterfowl. Observers at Hawk Mountain have also found that "hard" weather seems to "stimulate" movements of Red-tailed Hawks, Rough-legged Hawks, and Bald and Golden eagles after the "normal" autumn migration period (L. Goodrich and S. Senner, personal communication). The same species as well as Sharp-shinned Hawks, Northern Harriers, and others have been noted migrating well into December at Cape May Point, New Jersey (C. Sutton, F. Nicoletti, and P. Dunne, personal communication). Hawk counts at Hawk Mountain and Cape May are now continued into December in some years. These movements may continue into January and February. Variability in timing of migration among individuals of the same species undoubtedly is present in many falconiform migrants, but it has yet to be studied adequately.

Explanations of Differential Migration

The short review above shows that differential migration is important in hawks and other birds. At least six hypotheses have been invoked to explain why males and females or adults and immatures migrate at different times and to different places. Implicit in all hypotheses is the assumption that there are costs (reproductive fitness) associated with migrating too far, not far enough, too soon, or too late. The hypotheses are as follows.

THE SOCIAL DOMINANCE HYPOTHESIS Gauthreaux (1978a, 1982a) and others (Mueller, Berger, and Allez 1977; Newton 1979) have proposed that differences in seasonal timing of migration and winter distribution (including habitat differences) among age-sex classes of partial migrants are related proximally to social status. Because size (sexual dimorphism) and experience (age) often reflect social status (subordinance/dominance), Gauthreaux predicted that the smaller sex (females in most passerines and males in most raptors) as well as younger individuals should migrate farther from the breeding grounds or migrate earlier in autumn. Thus, the resource-holding potential— the ability to maintain a territory or access to food while in a flock— determines which birds migrate, which migrate the earliest in autumn, and which migrate the farthest. The hypothesis predicts variance from year to year depending upon the density of the populations involved and the availability of resources.

THE BODY SIZE OR PHYSIOLOGICAL HYPOTHESIS The body size hypothesis predicts that the largest members of a population can spend the non-breeding season (winter in this case) at more northern, or colder, latitudes (or higher elevations). The rationale for this is that larger individuals are able to tolerate colder temperatures and can fast longer than smaller individuals. For species in which a large size dimorphism exists between the sexes or in which adults are larger than immatures, differences in timing of migration or winter distribution can result. For most raptors the hypothesis predicts that females should migrate later and occupy more northern latitudes during winter. For some passerines the difference between body size of males and females may not be large enough to confer a difference in ability to fast and tolerate cold temperatures (Stuebe and Ketterson 1982). It is not known if this is true for raptors, where greater size dimorphism exists.

THE ARRIVAL TIME HYPOTHESIS King, Farner, and Mewaldt (1965) proposed a hypothesis, later formalized by Myers (1981), that related unequal sex ratios of birds during winter to timing of arrival on the breeding grounds. They reasoned that individuals of the sex establishing the breeding territory in spring would benefit from remaining close to the breeding grounds by undertaking shorter migration in spring. Such individuals presumably would acquire the best-quality breeding sites. Thus, intrasexual competition (one component of sexual selection) for breeding sites shapes the distribution pattern of the sexes during the non-breeding season. Myers (1981) made no mention of how the hypothesis explained differences between adults and immatures or between immature males and females. It is probable that for species that breed in cavities, on cliffs, or in other limited situations, early arrival is crucial (Lundberg 1979).

THE CHARACTER DIVERGENCE HYPOTHESIS Habitat differences between wintering male and female American Kestrels have been explained as character displacement (Koplin 1973). Character divergence or displacement among raptors has been invoked to explain the large size dimorphism between the sexes (Storer 1966; reviewed by Newton 1979 and Cade 1982) and has been related to differences in prey size. The differences in habitat and prey-size selection in these cases are hypothesized to reduce intraspecific competition, in this case intersexual competition.

THE FEEDING EFFICIENCY HYPOTHESIS Rosenfield and Evans (1980) and Duncan (1982) proposed a hypothesis to explain the differential timing of migration between immature and adult Sharp-shinned Hawks. They suggested that because immature Sharp-shinned Hawks are less efficient at capturing avian prey than adults, immature birds should be under some pressure to migrate earlier, following the masses of migrating passerines.

THE MOLT HYPOTHESIS Until very recently, the molt hypothesis was not considered formally as an explanation of differential timing of migration for falconiforms. Smallwood (1988) hypothesized that the differential timing of migration seen in American Kestrels was a result of the need to complete molt before initiating autumn migration. This explanation may apply to differences in autumn migration timing related to age, sex, and population. For example, if breeding adults of a northern population delay molt until after breeding, they may not complete it until later than more southerly populations that were able to initiate breeding earlier in the season. Similarly, if adults delay molt of flight feathers until after breeding, they may have to initiate migration after immatures.

Many raptors that are captured during autumn migration show interrupted molt patterns. This itself may be an adaptation so that birds can migrate before hard weather or food scarcity. The timing of molt in relation to migratory movements has been examined by Alerstam (1985). Among Common Terns (*Sterna hirundo*) molt of flight feathers occurs "slowly and successively on the non-breeding grounds," promoting efficient flight during the long molt period. For Arctic Terns (*S. paradisaea*) molt occurs rapidly after a long migration. Arctic Terns migrate farther than Common Terns, which may explain the large difference between these closely related species. Comparisons like this one may prove fruitful among the falconiforms (T. Alerstam, personal communication).

THE MIGRATION THRESHOLD HYPOTHESIS Baker (1978) presented a complex, multifactorial approach to differential migration that includes aspects of most of the hypotheses described above. Baker's model states that individual migrants possess an inherited threshold for migration and that they will not migrate until that threshold is exceeded. The threshold is based on the suitability of the breeding or natal site in relation to the suitability of an alternative habitat and the cost of round-trip migration to this habitat. Expressed mathematically, migration will occur when $h_1 < h_2 M_R$, where h_1 is the suitability of the breeding or natal site, h_2 is the suitability of the "winter" or non-breeding season site, and M_R is a migration factor measured as potential reproductive success (fitness). Simply put, an individual should migrate when winter habitat suitability, including the cost of migration, outweighs the suitability of the natal site. Suitability is measured by habitat quality, experience, distance of the winter site from the natal site, population density, and resource-holding potential (dominance status). Because individuals of different age-sex classes vary with respect to these variables, differential migration can evolve. That is, suitability is dependent upon the age-sex class of the individual. Although Baker's hypothesis is "all-encompassing" and may be "untestable" (Ketterson and Nolan 1983), it can be used in an a posteriori fashion. Ketterson and Nolan (1983) conclude that data from studies of differential migration of some passerines fit Baker's model.

Of the six hypotheses, three merit further consideration. Baker's (1978) complex hypothesis is difficult to comprehend and has been criticized as lacking predictive power (Ketterson and Nolan 1983). The character divergence hypothesis is nearly impossible to test; as presented, it smacks of the "for the good of the species" type of thinking that has been criticized as group selectionist. Character divergence (usually invoked for differences in morphology *between* species) is difficult to document (Grant 1972). Although differences in winter distribution, prey selection, and habitat selection could result in reduced competition between the sexes, the selective pressures that shaped these differences can be explained more parsimoniously by other hypotheses. Finally, the feeding efficiency hypothesis is limited to a few species and is not general enough to be powerful.

Recent discussions of differential migration considered only the social dominance, body size, and arrival time hypotheses (reviewed by Ketterson and Nolan 1983). There is ample evidence consistent with the predictions of all the hypotheses. Migratory male falcons tend to winter north of females, supporting the arrival time hypothesis. This is also the case with some passerines, shorebirds, owls, and woodpeckers (Howell 1953; King, Farner, and Mewaldt 1965; Ketterson and Nolan 1976; Lundberg 1979; Myers 1981). Since males are larger than the females in some of these species, the distribution patterns also fit the body size and social dominance hypotheses. In shorebird species in which females are larger than males and males winter north of females, the arrival time hypothesis is clearly favored. Other species in which the larger sex winters farther north and the smaller sex presumably establishes the breeding territory include Snowy Owls (Kerlinger and Lein 1986), Rough-legged Hawks, Goshawks, and some others (see appendix 2). Social dominance may explain the differing winter distributions among the age-sex classes of these species.

Because evidence supports more than one hypothesis, it is probable that the selective pressures that are responsible for the differences between the age-sex classes vary among species as a function of the particular ecology of the species. In some falcons where cliff sites or nesting cavities are limited, selection for early arrival on the nesting grounds may be of primary importance. In other species where densities are high or there are ample nest sites, social dominance combined with physiological differences between the sexes may be important. It is obvious that focused research is necessary before a more complete answer emerges. Hawks and owls offer excellent opportunities to test these hypotheses for several reasons: females are larger than males; there is a large variance in size dimorphism ranging from nearly 2:1 to almost 1:1 between the sexes; there are numerous species that are partial migrants; males usually arrive first in spring; and there is a delayed maturation among many species of migrants. These factors allow a researcher to tease the hypotheses apart, especially when predictions about non-breeding distributions of age-sex classes are identical.

Summary and Conclusions

The migration of hawks is widespread and variable. Of the 285 species of Falconiformes, at least 133 (46.7%) undertake some form of migration. Partial migration is the commonest type (91 species), whereas complete migrations (18 species) and "local"/irruptive movements (24 species) are less common. Migration includes travel across latitudinal, longitudinal, and elevational gradients. These movements occur seasonally: usually during spring and autumn in temperate, arctic, and subtropical climes or during the rainy and dry seasons in tropical and subtropical climes. The tendency to migrate is related positively to the distance of a breeding population from the equator, especially in the Northern Hemisphere. Variability of seasonal timing of migration and geographic distribution of migrants during the non-breeding season (termed differential migration) was shown to occur at the population level and among age-sex classes of given species. Three explanations of differential migration were presented that relate the phenomenon to social dominance, intrasexual competition for nest sites, and physiological tolerance of harsh weather.

This chapter is meant to be an overview of the geographical ecology of migration within the order Falconiformes. The review is incomplete for two reasons. First, little is known about the migration of falconiforms outside Europe, North America, and to a lesser extent, Africa. Even within these continents we know little about migration as related to population or age-sex class. Second, the ecology and geography of migration is a vast topic about which entire volumes will be written.

2

Methods of Studying Migrating Hawks

Most scientific endeavors are limited by technology and by methodological (and philosophical) approach (Kuhn 1962). How data are collected, the type of data collected, and how they have been used to answer questions give an outsider insight into a field of research. The field of hawk migration is no exception.

Two methods have predominated in raptor migration. These include counting raptors as they pass along topographic leading lines (Dobben 1955; Malmberg 1955; Mueller and Berger 1967a) and banding of raptors, often at these same sites. Counting and banding have histories dating from the nineteenth century. Radar, radiotelemetry, cine-theodolite, and motor-glider methods, special marking projects, and various visual techniques have been used more recently to study raptor migration (table 2.1). Counting and banding still predominate, although other methods are used to answer specific questions about migratory flight behavior. In this chapter I describe and evaluate the methods used to study raptor migration and review the types of questions each method was used to address.

Hawk Migration Counts

For hundreds of years aggregations of migrating raptors at coastlines and peninsulas have attracted and fascinated observers (Brown and Amadon 1968; Heintzelman 1975; Newton 1979). It was not until the 1930s that systematic counts of migrating raptors were conducted at Hawk Mountain along the Kittatinny Ridge in eastern Pennsylvania (Broun 1935, 1949) and at Cape May Point, New Jersey (Allen and Peterson 1936). In Europe before 1950, counts were conducted at Falsterbo (Rudebeck 1950; Ulfstrand 1958, 1960) and Ottenby (Edelstam 1972) in southern Sweden, and in the Netherlands (Dobben 1953). Following these pioneering studies, systematic hawk migration counts were conducted only at a few sites in North America and Europe until the late 1960s and early 1970s (but see Broun 1949; Muel-

Table 2.1 Methods Used to Study Hawk Migration

Method/ Technique	Frequency of Use	Uses and Advantages	Disadvantages
Hawk migration count	Very common	Document occurrence of migration at a location Quantify seasonal timing of migration at a given location Document habitat usage Inexpensive and easy to use	Cannot be used to study behavior Biased to low flying migrants Observer fatigue
Trapping/ banding	Common	Determine where migrants come from and are going Determine migration pathways Measure migrant morphology, body condition, ectoparasites, blood chemistry, etc. Relatively nonintrusive and does not harm birds	Labor intensive Poor return/recovery rates yield small sample sizes No knowledge of where migrant originated or destination until data are acquired Potential age/sex bias
Marking projects	Uncommon	Document use of habitat Document movements of individuals Inexpensive	Poor rate of resighting Removal of markers by birds Labor intensive (capture)
Radiotelemetry	Uncommon/rare	Document flight behavior of individual migrants for many days or entire migration Document habitat usage, stopover time, and behavior during stopover	Expensive and labor intensive Time intensive for amount of data acquired Difficulties associated with following migrants
Motor glider/ aircraft	Rare	Document flight behavior for varying distances Examine geographic distribution of migrants	Labor intensive and expensive Modifies behavior of migrants Biased to high flying migrants
Direct visual	Uncommon/rare	Document flight behavior Inexpensive and adaptable	Inaccurate measurements Biased to low flying migrants Observer fatigue
Cine-theodolite	Rare	Document flight behavior	Expensive and labor intensive Inaccurate except for short distances
Radar	Uncommon	Document flight behavior Document geographic distribution	Expensive and labor intensive Not readily mobile Simultaneous visual observations necessary Biased to high flying migrants Measure behaviors for short periods

ler and Berger 1961) when environmentalists became aware of declining raptor populations (table 2.2; fig. 2.1).

By the mid-1970s hawk counting had gained popularity in North America, and an association of hawk watchers, the Hawk Migration Association of North America (hereafter HMANA), was formed. Since then hawk migration counts have been conducted by volunteers and a few professionals at dozens of locations in North America, Europe, Israel, Egypt, and elsewhere. The growth of hawk watching is attributable to greater environmental awareness and the work of the Cape May Bird Observatory (New Jersey Audubon Society), Hawk Mountain Sanctuary Association, HMANA, and other organizations. Financial support for these projects has come from private donations,

Figure 2.1 Map of Central and North America showing locations where large numbers of hawks can be seen and counted during migration. Numbers on map match those of lookouts listed in table 2.2 except for locations in Asia and Europe.

local Audubon and bird groups, the National Audubon Society, New Jersey Audubon Society, United States Fish and Wildlife Service, and various state, federal, and provincial agencies. By 1983 more than 800 HMANA members and others reported hawk counts from over 100 sites in North America (gleaned from HMANA newsletters). The number of Hawk watchers continues to grow, and visitors to lookouts such as Cape May Point number in the tens of thousands each migration season.

The traditional hawk migration study involves counting raptors as they fly past a lookout. Count sites have almost invariably been situated along "leading lines" (Mueller and Berger 1967a), topographic sites such as ridges, coastlines, and peninsulas that funnel or divert large numbers of migrants. Some of the well-known sites in North America, Europe, and the Middle East are shown in figure 2.1 and listed in table 2.2. These sites were chosen by observers because large numbers of raptors could be counted with ease. Sites where few migrants were observed have been abandoned, as can be seen in past issues of the HMANA newsletter. Procedures are nearly the same as those established in the 1930s by Broun (1935, 1949) and recently standardized by the HMANA. From early morning until late afternoon observers scan the sky with unaided eye and binoculars, noting the number and identity of migrants passing each hour or half-hour. General behavior of migrants is also noted (direction of flight and altitude), as are local weather conditions. Data are recorded on standardized data sheets (fig. 2.2) and kept in HMANA files.

In most hawk count studies the dependent measure is the number of hawks counted by hour or day. Hourly or daily counts are assumed to be a measure of the volume or number of migrants aloft, analogous to the "amount of migration" in studies of passerine migration (Able 1973; Richardson 1978). A large variability among hourly or daily counts of migrating raptors has been described and is often attributed to time of day (Broun 1949; Mueller and Berger 1973; Heintzelman 1975) or weather (Broun 1949; Mueller and Berger 1961, 1967b; Haugh 1972; Alerstam 1978a; Titus and Mosher 1982). Variability in migration counts may also be associated with interobserver reliability (Sattler and Bart 1985a), the number of counters present (Kochenberger and Dunne 1985), method used to detect migrants (binoculars vs. unaided eye), the presence of a banding station (Kerlinger and Gauthreaux 1984), and observer fatigue (Sattler and Bart 1985b). With the exceptions noted, sources of variance have not been studied quantitatively, although projects are now being conducted by the Cape May Bird Observatory, Hawk Mountain Sanctuary, Whitefish Point

Table 2.2 Selected List of Sites That Are Well Known for Large Aggregations of Migrating Hawks

Site	Approximate Number per Season	Dominant Species	Topographic Situation	Season	Reference
NORTH AMERICA					
1 Balboa, Panama	>500,000	BW, SW, TV	Isthmus	Fall	Smith 1973, 1980
2 Point Diablo, CA	<10,000	SS, RT	Coast/crossing	Fall	Binford 1977
3 Goshutes, NV	>10,000	SS, RT	Mountain/ridge	Fall	Hoffman 1985
4 Rio Grande Valley National Wildlife Refuge, TX	>50,000	BW, SW, TV	Isolated forest groves	Fall, spring	Kerlinger and Gauthreaux 1985a,b HMANA newsletter
5 Corpus Christi, TX	>50,000	BW	Near coast	Fall	HMANA newsletter
6 Hawk Ridge, Duluth, MN	>50,000	BW, SS, RT	Ridge near coast	Fall	Hofslund 1966
7 Whitefish Point, MI	>20,000	BW, SS	Coast/crossing	Spring	Grigg 1975
8 Cedar Grove, WI	>10,000	SS, BW, RT	Coast	Fall	Mueller and Berger 1961
9 Hawk Cliff, Ontario	>50,000	BW, SS, RT	Coast	Fall	Haugh 1972; Field 1970
10 Derby Hill, NY	>30,000	BW, SS, RT	Coast	Spring	Haugh and Cade 1966
11 Hawk Mountain, PA	>20,000	BW, SS, RT ++	Ridge	Fall	Broun 1935, 1949
12 Cape May Point, NJ	>70,000	SS, BW, AK ++	Coast/crossing	Fall	Allen and Peterson1936; Dunne and Clark 1977

13 Mount Wachussett, MA	>10,000	BW, SS, RT	Mountain	Fall	Hopkins et al. 1979
14 Mount Tom, MA	>10,000	BW, SS, RT	Mountain/ridge	Fall	Bates 1975
15 Hook Mountain, NY	>10,000	BW, SS, RT	Hilltop/ridge/river crossing	Fall	Thomas 1975
16 Raccoon Ridge, NJ	>15,000	BW, SS, RT	Ridge	Fall, spring	F. Wolfarth, personal communication; Dunne 1977, 1978
17 Bake Oven Knob, PA	>20,000	BW, SS, RT	Ridge	Fall	Heintzelman 1986
18 Veracruz, Mexico	>100,000	BW, SW, TV	Coast/Mountain	Spring	Thiollay 1980
EUROPE/AFRICA/ASIA					
19 Strait of Gibraltar	>125,000	HB, BK + +	Coast/hill/crossing	Fall, spring	Evans and Lathbury 1973; Bernis 1973
20 Bosporus—Europe	>35,000	HB, BK, SH	Coast/crossing	Fall	Porter and Willis 1968
21 Eilat, Israel	>500,000	HB, BK, LK	Coast/mountains	Spring	Safriel 1968; Christensen et al. 1981
22 Kfar Kassem, Israel	>200,000	HB, BK, LSE, LSH	Mountains/near coast	Fall	Leshem 1985
23 Falsterbo, Sweden	>25,000	HB, CB, SH	Coast/crossing	Fall	Ulfstrand et al. 1974; Alerstam 1978a
24 Strait of Malacca, Strait of Singapore, Cape Rachado	>100,000	JS, CH + +	Coast/crossing	Fall	Baker 1981

Key to abbreviations: TV = Turkey Vulture, BK = Black Kite, HB = Honey Buzzard, CH = Crested Honey Buzzard, SS = Sharp-shinned Hawk, LSH = Levant Sparrowhawk, JS = Japanese Sparrowhawk, SH = European Sparrowhawk, LSE = Lesser Spotted Eagle, BW = Broad-winged Hawk, CB = Common Buzzard, SW = Swainson's Hawk, RT = Red-tailed Hawk, AK = American Kestrel, LK = Lesser Kestrel, + + = many more species; CA = California, TX = Texas, MN = Minnesota, MI = Michigan, WI = Wisconsin, NV = Nevada, MA = Massachusetts, NJ = New Jersey, NY = New York, PA = Pennsylvania.

Bird Observatory, and New England Hawk Watch to investigate this question.

Counts have been used by researchers to study daily timing of migration, seasonal timing of migration, and amount or volume of migration in relation to weather. At least four studies have examined the

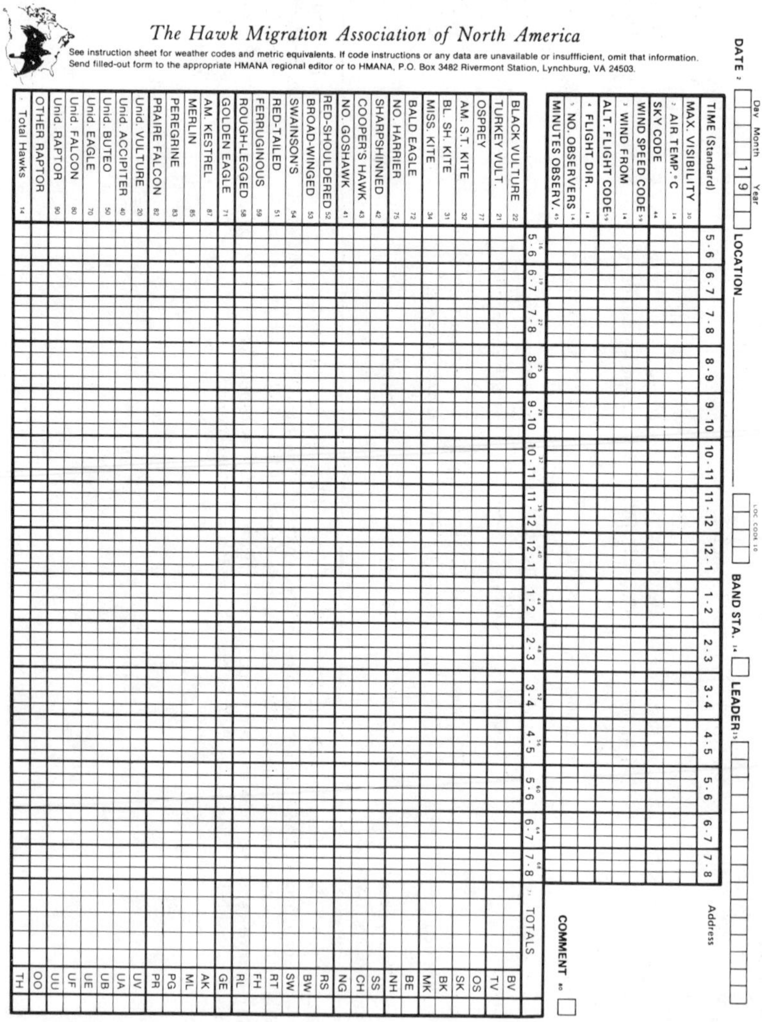
The Hawk Migration Association of North America

See instruction sheet for weather codes and metric equivalents. If code instructions or any data are unavailable or insufficient, omit that information.
Send filled-out form to the appropriate HMANA regional editor or to HMANA, P.O. Box 3482 Rivermont Station, Lynchburg, VA 24503.

DATE (Day Month Year 19) LOCATION LOC CODE BAND STA. LEADER Address

TIME (Standard)	5-6	6-7	7-8	8-9	9-10	10-11	11-12	12-1	1-2	2-3	3-4	4-5	5-6	6-7	7-8
MAX. VISIBILITY															
AIR TEMP. °C															
SKY CODE															
WIND SPEED CODE															
WIND FROM															
ALT. FLIGHT CODE															
FLIGHT DIR.															
NO. OBSERVERS															
MINUTES OBSERV.															

	5-6	6-7	7-8	8-9	9-10	10-11	11-12	12-1	1-2	2-3	3-4	4-5	5-6	6-7	7-8	TOTALS	
BLACK VULTURE																	BV
TURKEY VULT.																	TV
OSPREY																	OS
AM. S.T. KITE																	SK
BL. SH. KITE																	BK
MISS. KITE																	MK
BALD EAGLE																	BE
NO. HARRIER																	NH
SHARPSHINNED																	SS
COOPER'S HAWK																	CH
NO. GOSHAWK																	NG
RED-SHOULDERED																	RS
BROAD-WINGED																	BW
SWAINSON'S																	SW
RED-TAILED																	RT
FERRUGINOUS																	FH
ROUGH-LEGGED																	RL
GOLDEN EAGLE																	GE
AM. KESTREL																	AK
MERLIN																	ML
PEREGRINE																	PG
PRAIRE FALCON																	PR
Unid. VULTURE																	UV
Unid. ACCIPITER																	UA
Unid. BUTEO																	UB
Unid. EAGLE																	UE
Unid. FALCON																	UF
Unid. RAPTOR																	UU
OTHER RAPTOR																	OO
Total Hawks																	TH

COMMENT

Figure 2.2 Standard data collection form used by hawk counters. (Courtesy of the Hawk Migration Association of North America.)

diel distribution of migrant hawks and have come to slightly different conclusions. It is possible that differences reflect site-specific or count-method biases based on altitude of flight. These inconsistencies will be examined in chapter 8.

Hawk migration counts have been used frequently to examine seasonal timing of migration of a particular species. Seasonal timing is not defined in most studies but is assumed to be the time during spring or autumn when migrants pass a given count location. The seasonal timing of migration for a species is usually plotted as a histogram (fig. 2.3), with percentage or numbers of migrants on the *y*-axis (vertical) and date on the *x*-axis (horizontal; Haugh and Cade 1966; Mueller and Berger 1967b; Haugh 1972; Roos 1985; Heintzelman 1975; Alerstam 1978a; Geller and Temple 1983). The histograms show the duration of migration during given seasons, permitting comparison of different species or the same species among years. Although this approach is used by many researchers, it is misleading because it fails to provide basic information such as when migration is initiated and terminated and how long a bird takes to complete migration (including stopovers and flight time). For example, a Peregrine Falcon that passes

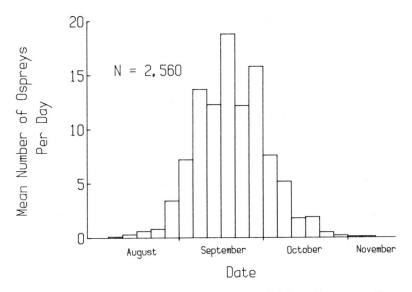

Figure 2.3 Seasonal timing of Osprey migration at Hawk Mountain Sanctuary, Pennsylvania. Bars represent the average numbers of hawks counted per day for five-day periods from 1980 to 1984. (Data courtesy of Hawk Mountain Sanctuary Association.)

Cape May Point, New Jersey, on 1 October could have initiated migration anywhere in an area from the north slope of Alaska to western Greenland (C. Schultz, personal communication) and possibly as far south as New York or New Jersey. Depending on its place of origin, this bird may have initiated migration in late September, one day before it was observed, or in August, a month before it was observed. As another example, Sharp-shinned Hawks passing Cedar Grove, Wisconsin, in autumn could have originated in British Columbia, Alberta, Saskatchewan, Manitoba, Minnesota, Wisconsin, Michigan, or Ontario (Mueller and Berger 1967b; H. C. Mueller, personal communication). These questions are ultimately related to the life history and fitness of the individuals involved.

Kerlinger and Gauthreaux (1985a) used counts of migrating Broad-winged Hawks to show that the peak of spring migration through south Texas occurred in late March and early April. They also showed that the adults moved through about two or more weeks before immature birds. What does it mean that birds move past a given location at a particular time? Because Texas is the approximate midpoint of the Broad-winged Hawk migration, the data reported by Kerlinger and Gauthreaux may have little meaning, but if they are considered along with data from Whitefish Point, Michigan, and Derby Hill, New York (fig. 2.1, table 2.2), they may become more meaningful. The peak of Broad-winged Hawk migration at Whitefish Point occurs during the first week of May, and at Derby Hill it occurs during the third and fourth weeks of April. The variation in timing of migration in south Texas suggests that differences there are a function of the time migration begins on the non-breeding range between populations that breed in central Canada and those from northeastern North America. A knowledge of the peak time of migration for a species is important for management purposes and may be used by birders to plan their field trips.

Examples like these demonstrate how histograms of seasonal timing of migration can be misleading. Counts at sites far from the initiation or termination of migration yield greater variance because birds passing such sites may come from many different populations and be going to many different destinations. Other methods, especially telemetry, are needed to determine timing of migration for specific populations of raptors.

Relating weather conditions to the amount of migration is probably the most-discussed topic in the hawk migration literature (Broun 1949; Mueller and Berger 1961, 1967b; Haugh 1972; Alerstam 1978a; Titus and Mosher 1982). Many studies have demonstrated correlations between the numbers of hawks passing a given location and selected weather variables. These studies differ from investigations relating the amount of passerine or waterfowl migration to weather in that hawk counts are taken at only one location, whereas passerine or waterfowl migration is measured with a radar over a large area (Richardson 1978). Furthermore, hawk migration counts are usually from leading line sites where hawks migrate in large numbers. The use of count data from one topographic site can yield a biased or incomplete picture of migration, since sites used for counts are very different from the topography hawks must pass over during migration. A topographic map of North America shows that long continuous ridges or coastlines oriented in a direction appropriate for migration occupy only a small portion of the continent. If hawks use leading lines during specific weather conditions, or if they are more visible with certain weather conditions, the results of some studies are questionable or of limited value.

Count studies from leading lines have also been used to make inferences about the behavior of migrants that were not flying along leading lines. The relations among numbers of hawks and particular weather variables have been interpreted in two ways. The first is that more migrants are aloft when larger numbers of migrants are counted than when fewer birds are counted. The second is that the larger number of hawks counted is a result of migrants' drifting with the wind until they reach a leading line, whereupon they reorient and follow the leading line (Mueller and Berger 1967a). Some studies have not interpreted the relation between numbers counted, weather, and topography (Titus and Mosher 1982).

When explaining the differences in counts associated with weather conditions, few authors (Murray 1964, 1969) attempt to invoke or test alternative hypotheses. Alternative explanations for the daily variability in numbers of hawks counted include real differences in the numbers of hawks aloft; drift (chap. 7); count bias based on differential altitude of flight that is correlated with wind or other meteorological variables (Murray 1964, chap. 8); pseudodrift (Evans 1970; Alerstam 1978b); and differential speed of migration (as with optimal drift, Alerstam 1978b; chap. 7). These alternatives are not mutually exclusive in that two or three may explain the variance in counts from day to day. Because these hypotheses have not been tested and probably cannot be tested with count data, the wind drift hypothesis has become widely accepted (Mueller and Berger 1967a; Haugh 1972; Heintzelman 1975). The wind drift hypothesis has only recently been tested for raptors by using directional data (Kerlinger 1984; Kerlinger and Gauthreaux 1984, 1985a; Kerlinger, Bingman, and Able 1985; chap. 7) as has been done for other birds (Gauthreaux and Able 1970; Richardson 1976; Bingman, Able, and Kerlinger 1982).

In the preceding paragraphs I questioned the use of hawk migration counts in scientific research. The criticisms may seem harsh, especially to those who use counts in their research, but the questions raised must be addressed before the utility of counts in understanding migration can be demonstrated. The criticisms put forward were logical arguments, although in chapters 7 and 8 I present empirical data to support the criticisms.

My harsh criticism of the use of hawk migration counts should not be viewed as a condemnation of the technique. Such counts have prepared the way for many important studies of migrating hawks. For example, Heintzelman's (1975) classic treatise on hawk migration is based on count data. When used judiciously, hawk migration counts will continue to be an important study technique. Readers who disagree with my negative comments should view them as a challenge to conduct better-quality studies of migrating hawks.

Nonscientific Uses of Hawk Migration Counts

Nonscientific uses of migration counts include conservation (management) and education programs. Broun (1949) and later Spofford (1969), Robbins (1975), Nagy (1977), and Roos (1978, 1985) postulated that counts of migrating raptors might be indicators of population trends. Using five-year averages of forty years of hawk counts from Hawk Mountain Sanctuary, Nagy (1977) demonstrated a long-term decline of Cooper's Hawks in eastern North America. The analysis of Dunne and Sutton (1986) of twelve years of count data from Cape May Point, New Jersey, shows a steady recovery of Peregrine Falcon, Merlin, Osprey, and Sharp-shinned Hawk populations following the reduction of organochlorine use in North America pesticides. The analysis of Roos (1978) is particularly noteworthy because he has examined long-term population trends of several species from northern Europe based on counts taken between 1942 and 1977 at Falsterbo, Sweden. Such analyses will prove indispensable to wildlife managers who wish to monitor long-term populations of raptors before making management decisions.

The large variability in count data obscures short-term trends, so counts from several different lookouts may be necessary to determine trends for shorter periods for some species (Robbins 1975). A detailed knowledge of the factors contributing to the variation in hawk migration counts is necessary before we interpret population trends from such counts. Hussell (1985) recently devised a multivariate approach to the variance problem. The model seems to correct for variance due to weather, but it needs empirical testing. Detailed analyses of the more than fifty years of data from Hawk Mountain are now being conducted (S. Senner, personal communication) and will provide a unique picture of the population trends of more than a dozen raptors.

Another example of how migration counts have increased our knowledge of raptor populations comes from Israel. In the 1970s the world population of Lesser Spotted Eagles was estimated to be about 10,000 pairs (Cramp and Simmons 1980). Since this estimate was made, counts of migrating Lesser Spotted Eagles from Israel show that the world population is five times greater than previously estimated (Y. Leshem, personal communication). Daily counts during autumn can exceed 45,000 individuals. It appears that the entire population of species like Lesser Spotted Eagle and Levant Sparrowhawk passes through the narrow land bridge between Eurasia and Africa. During this passage the Israeli researchers have attempted to census or sample populations with the objective of monitoring long-term trends (Y. Leshem, personal communication).

Migration counts have also been used to identify locations that are important as traditional hunting or roosting sites of migrants. The best example is from southern New Jersey at Cape May Point, where counting (with associated observations) and banding activities of the Cape May Bird Observatory (Dunne and Clark 1977) have resulted in the preservation of the Cape May Meadows, a traditional hunting and resting site of migrating Peregrine Falcons and other raptors. Similar work in Virginia, in Maryland (Ward and Berry 1972), and on Padre Island, Texas (Hunt, Rogers, and Stowe 1975), has identified important resting and hunting sites for Peregrine Falcons. Roosting by thousands of Broad-winged Hawks, Turkey Vultures, and Mississippi Kites at sites like Santa Ana National Wildlife Refuge (now called the Rio Grande Valley National Wildlife Refuge) in south Texas shows the importance of remnant woodlands. The area is nearly devoid of forest as a result of human activities. Without hawk migration counts, such sites would not be recognized as important to migrants. Counting hawks at a given site is an inexpensive and easy means of identifying important resting and hunting sites. I should mention that in the studies described above, behavioral observations were made in conjunction with counts.

Educating the public is an integral part of the programs of such traditional counting locations as Hawk Mountain Sanctuary and Cape May Bird Observatory. At many lookouts there are observers who, in addition to counting, answer questions about identification of birds, basic biology, migration, and population status of raptors. These observers, in many cases, are trained or amateur naturalists who devote considerable time to conducting and organizing hawk watches. While in the field they educate hikers, school groups, and other people who visit a lookout about wildlife, nature, and conservation. HMANA has also done much to inform the public by organizing hawk watches, publishing pamphlets and newsletters, and developing slide shows about raptor identification, migration, and conservation. Thousands of people are first introduced to raptors at count sites such as Hawk Mountain Sanctuary and Cape May Point. Recent educational programs initiated as an extension of raptor counting include internships for college students (sponsored by Hawk Mountain Sanctuary Association), a schoolyard hawk watch in New Hampshire (HMANA and the New England Hawk Watch), and a high-school hawk migration project in New Jersey (sponsored by Cape May Bird Observatory). Hawk watching programs are powerful tools for introducing people to raptors, wildlife, and environmental issues. (Although the impetus for the establishment of Hawk Mountain Sanctuary and Cape May Bird Observatory was the protection and study of migrating raptors, these institutions have broadened their purpose to include education, conservation, and research. Courses and workshops offered at these institutions are attended by a range of people from graduate students to laypeople.)

Capture and Banding of Hawks

As with hawk migration counts, hawk banding during migration has been conducted at leading lines where large numbers of hawks pass at low altitudes. Banding stations at many locations have been operating for more than ten years. Thousands of passing raptors are lured into various traps including mist nets, dho ghazas, bow nets, and bal chatris. Berger and Mueller (1959), Mueller and Berger (1970), Holt and Frock (1980), Evans and Rosenfield (1985), and Clark (1981, 1985a) describe trapping and banding procedures.

At Whitefish Point, Michigan, hundreds of raptors are banded during spring migration by dozens of banders (coordinated by the Michigan Audubon Society). Trapping at Whitefish is unusual because most birds are captured in unbaited mist nets. The largest hawk banding operation is at Cape May, where up to five trapping stations are used by dozens of banders (fig. 2.4A). More than 5,000 hawks of over twelve species are banded each autumn under the direction of W. S.

Clark. Ancillary to the massive banding operation, Clark and his colleagues give free public demonstrations of banding procedures during which he also talks about raptor ecology and conservation (fig. 2.4B). Over the years Clark has given hundreds of such demonstrations to thousands of people. Banding of more than 500 raptors per migration season occurs at only a few locations in the world: Hawk Cliff, Hawk Ridge, Cedar Grove, Whitefish Point, and perhaps Helgoland (near the northwestern coast of Germany). All these locations are near large bodies of water. Ridge sites along the Kittatinny Mountains in New Jersey and Pennsylvania and in the Goshute Mountains of Nevada also sometimes capture 500 to 1,000 migrating raptors per season.

The newest of the major banding sites is in Israel, where W. S. Clark and his colleagues have collaborated with the Society for the Protection of Nature in Israel (SPNI). With the benefit of Clark's expertise, the Israelis now capture more than 500 raptors per spring migration season. In the near future this number will undoubtedly increase. Recoveries of banded birds from Africa, Europe, and Asia will provide information on the origin and destination of the millions of migrants that pass through the Middle East land bridge.

In addition to capture at regular stations, banders work at other locations during migration and other seasons. Road trapping, dropping traps with live lures (bal chatris) from a moving vehicle to perched birds, is widely practiced among raptor banders (Berger and Mueller 1959). Road trapping is conducted in varying topographic settings at all times of the year. A modification of the method was reported by Ward and Berry (1972) and Hunt, Rogers, and Stowe (1975), who lure "passage" falcons along the barrier beaches of the Atlantic and Gulf coasts, respectively.

Finally, a large number of raptors are banded as nestlings. This method, as with road trapping, is conducted throughout much of North America and in some parts of Europe. Because of the wide geographic range of the banded nestlings and the known origin of the banded hawk, this type of banding offers some of the best opportunities for elucidating geographic patterns of migration.

After capture, the age and sex of each bird are determined (when possible), and birds are measured, banded, and then released. The procedure varies among banding stations and banders, although mass, wing chord, molt, and tail length are usually recorded (fig. 2.4C). Some stations (not those mentioned above) are guilty of "ringing and flinging": capturing, banding, and releasing without collecting additional information. Most banders record numerous morphological measurements (Mueller, Berger, and Allez 1977, 1979, 1981a, 1981b;

A

B

C

Figure 2.4 Banding of migrating hawks at Cape May Point, New Jersey. (A) Joanne Mason places a lock-on band on a first-year female Cooper's Hawk. (B) W. S. Clark holds a first-year male Goshawk. (C) Chris Schultz examines wing molt of an adult male Goshawk while Frank Nicoletti records data. (Photographs by Paul Kerlinger.)

Bildstein et al. 1984). There is still a need for more complete morphological measures from captured birds, including visible fat, wing span, tail area, wing area, and other measures.

It should be obvious that the capture and banding of raptors can yield useful data. Although most banders have few questions in mind, data collected during banding and the recovery or recapture of banded birds can answer questions related to many aspects of migration, such as the pathways, origin, and destination of migrants (Enderson 1964; Mueller and Berger 1967b; Gerrard et al. 1978; Melquist and Carrier 1978; Osterlof 1977; Griffin, Southern, and Frenzel 1980; Houston 1981; and several others), seasonal timing of migration by age and sex class (Mueller and Berger 1967b; Bildstein et al. 1984), and morphology. Other studies have addressed prey selection (Mueller and Berger 1970), blood chemistry (F. Sheldon personal communication), and contamination by toxic substances such as oil (Clark and Gorney 1986).

Banding as a research technique is not without problems. Many studies show that immature birds are easier to trap than adults, and males are sometimes more difficult to capture than females (Hunt, Rogers, and Stowe 1975). A trapping bias may also apply to individuals of the same age or sex class. For example, Weatherhead and Greenwood (1981) demonstrated that blackbirds trapped with bait (grain) were lighter than those trapped at random at a roost site. Because hawk banders usually use bait, banding data may be biased toward hungry or even starving birds. This may not be a problem with hawks, because many (particularly Sharp-shinned Hawks) are trapped with a full crops and fat reserves. However, this may not reflect the population of migrants that is passing. Further problems with banding data are small samples for most species, the reluctance of some banders to share data with scientists, and the large amount of time and effort necessary to acquire even a small sample. Whereas some migration studies can be conducted in a few banding seasons, questions related to the geographic origin and destination of migrants will require many years of labor before publication of data.

Special Marking Projects

Although banding is a means of marking individual birds, the identification number on the band cannot be read unless the bird is recovered or recaptured. Several studies have undertaken special marking projects as a part of regular banding activities. In these projects a marker was placed on the bird's wing, tail, or tarsus. Numbers on the markers, or the markers' shape, color, or placement on the bird's body, can be observed at distances of 100 m to 300 m. Placing markers on raptors is supervised in North America by fish and wildlife agencies, so that marking systems are unique.

The most commonly used marker is the patagial tag, which is riveted through the patagium (the skin between the wrist and shoulder). Each patagial tag has a unique color and number code, permitting recognition of individuals (see plate 3 in Newton 1979). Whereas patagial markers are ideal for observing birds within limited areas (i.e., territories), they are time consuming to place on birds, and the probability that someone will see and report them accurately during migration is remote.

A different system has been used at Cape May Point and along the Kittatinny Ridge in Pennsylvania. Migrant Sharp-shinned and other hawks were given unique markers: color, shape, and location of the marker on the tail identified the day and location of marking. At Cape May tail markers (fig. 2.5) were used to determine the behavior of migrating Sharp-shinned Hawks at or near the end of the Cape May peninsula, where banding and count studies have been conducted for several years. Researchers at Cape May were interested in determining how long migrants remained near the counting station. Researchers along the Kittatinny Ridge hoped that marked birds would show how fast migrants move along the ridge and how far they follow the ridge. Cooperation between banders, who placed markers on the birds, and hawk counters, who looked for those markers, is unique. Few cooperative ventures between these "factions" have met with success because of the disparate interests of the groups. The projects were possible through efforts by Cape May Bird Observatory and Hawk Mountain Sanctuary.

Radiotelemetry

Radiotelemetry may be the best means of studying the flight behavior of migrating birds. The technique entails placing a radio transmitter with a power supply on the back or tail (fig. 2.6A) of a raptor. Once in place, a radio receiver (fig. 2.6B) is used to follow the raptor during migration. Strong transmitters permit detection at 3–4 km or more over land (Cochran 1972; Holthuizjen and Oosterhuis 1985) and from 20 to more than 100 km over water (Cochran 1985; Holthuizjen and Oosterhuis 1985). The distance at which a transmitter can be detected depends on its strength, which is a function of the size of the battery or solar-battery pack a migrant can carry. Also of importance

Figure 2.5 Sharp-shinned Hawk at Cape May Point, New Jersey, with streamer made of plastic tape attached to a tail feather. Color and placement of the streamers allow researchers to determine when a migrant was captured and how long migrants remain in the Cape May region. (Photograph by William S. Clark.)

A   B

Figure 2.6 (A) Radio transmitter with wire antenna affixed to a central tail feather of
an immature female Sharp-shinned Hawk (photograph by Paul Kerlinger). (B) Researcher with a radio receiver (antenna) listening for signal from telemetry devices attached to migrating Sharp-shinned Hawks near Cape May Point, New Jersey (photograph by Clay Sutton).

is the height of the receiver antenna and the transmitter. Cochran
(1967, 1985) and Fuller and Tester (1973) reviewed the method.

Southern's (1964) report on the home ranges of Bald Eagles in winter was one of the first in which a raptor was studied with radiotelemetry. Craighead and Dunstan (1976) proposed that as technology improved, movements of migrants could be monitored by satellite.
Battery size (mass), longevity, power, and mode of attachment have
been the limiting factors. The transmitters in use today are not powerful or small enough to allow satellite tracking, except for some
larger birds like swans, pelicans, or eagles. Indeed, transmitters small
enough to be carried by birds less than about 500 g are incapable of
transmitting for more than thirty to sixty days and over distances
greater than 4–5 km with "normal" conditions. The migratory movements of a few raptors including Sharp-shinned Hawks (Cochran
1972; Holthuizjen and Oosterhuis 1981, 1985), Northern Harriers
(Beske 1982), and Peregrine Falcons (Cochran 1975, 1985), and Bald

Eagles (Harmata 1984; Harmata, Toepfer, and Gerrard 1985) have been monitored with radiotelemetry. At Cape May Point Holthuizjen and Oosterhuis (1981, 1985) monitored the movements of about forty female Sharp-shinned Hawks to quantify their movements within the Cape May peninsula and to determine how much time the migrants spent near the point before continuing migration. Their habitat requirements were also studied. The small number of Northern Harriers followed by Beske (1982) made short migratory flights before they were lost.

Radiotelemetry studies by Cochran (1972, 1975, 1985, 1987) are fascinating. After several failures (lost transmitters, lost migrants) Cochran followed a Peregrine Falcon from Green Bay, Wisconsin, to northern Mexico before ending his pursuit. He also followed a female Sharp-shinned Hawk from Cedar Grove, Wisconsin, to northern Alabama. Since then Cochran has radio tracked Peregrine Falcons along the Atlantic coast. It seems that males, especially adults, may spend a significant portion of their time migrating over the ocean, sometimes at night. This finding is something of a surprise because it was formerly believed that raptors did not migrate at night unless forced to do so. Most important, he was able to determine when birds initiated migration, when they landed, how high they flew, when they were soaring or flying in straight lines, the distance each bird traveled, whether or not a bird migrated on a particular day, flight directions, and flight speeds.

Although radiotelemetry offers a means of acquiring accurate data on flight behavior of individual birds, the method has several shortcomings. First, radiotelemetry systems are expensive. Each transmitter costs about $100 or more (U.S.), and receivers cost more than $1,000. Second, transmitters and receivers are subject to failure. Third, following birds with ground transportation is difficult. Cochran (1972, 1975) hired aircraft to fly aerial transects, sometimes at night, to locate his birds. Fourth, sample sizes are necessarily small, since few birds (Cochran 1985) can be followed at a time. Fifth, the researcher is never sure whether the telemetry unit has changed the flight behavior of the bird being studied. Before embarking on a radiotelemetry study, I advise that a researcher determine whether the merits of a project justify the cost and the difficulties.

Cine-Theodolite

The cine-theodolite is a combination 16 mm movie camera with a telephoto lens (1,000 mm) and a surveyor's transit (called a theodolite). As a flying bird is filmed, the azimuth and elevation are simultaneously recorded along with the time (in hundredths of a second). Robinson (in Hopkins et al. 1979), in an ingenious attempt to study flight behavior of migrating hawks, employed a theodolite at Mount Wachussett, Massachusetts. Shortcomings of the system were inaccurate measures of distance of the bird from the theodolite and the tediousness of analyzing the films. The theodolite could be useful for studying flight behavior of birds of known dimensions at distances of less than 200–300 m, especially if employed in a triangulation system with a second theodolite synchronized with the first.

Colin Pennycuick (1982a, 1982b, 1983) recently devised an ornithodolite that is similar to the cine-theodolite used by Leif Robinson. Using the device, Pennycuick has made detailed studies of various soaring birds in Panama.

Motor Glider

Only three research groups have reported the use of motor gliders or light aircraft to study hawk migration (Hopkins 1975; Welch 1975; Hopkins et al. 1979; Smith 1985a; Leshem 1987), although the method has been used to study the foraging flights and aerodynamic performance of African vultures (Pennycuick 1971a, 1972a). A motor glider is a sailplane equipped with a small engine and propeller. Hopkins et al. (1979) and Welch (1975) used a motor glider to find and follow flocks of Broad-winged Hawks and other migrating hawks in Connecticut and Massachusetts. They maintained radio contact with hawk counters on the ground to locate hawks more readily. The motor glider proved to be a difficult means of studying migration because it was expensive and was feasible only for short periods of time. The research group in Israel headed by Y. Leshem followed large flocks and made observations of them for up to 200 km. Data gathered with motor gliders include altitude, flock size and dynamics, climb rate in thermals, flight speed, sinking speed, and direction of flight. In addition, researchers can study the updraft sources used by migrants. The technique would be useful if time and money were available to allow an observer to follow migrants for long distances. The data would thus be comparable to data collected with radiotelemetry, except that a bird probably could not be followed for more than a few hours. Ideally, a motor glider could follow raptors equipped with radio transmitters. Readers interested in the technique should read papers by Pennycuick (1972a, 1972b), who followed vultures during long foraging flights over the Serengeti Plain in Africa.

Direct Visual Techniques

By direct visual techniques (DVTs), I mean methods in which visual observations (usually aided with binoculars, scopes, and equipment such as compass or stopwatch) are used for purposes other than counting. The goal of such studies has been to quantify and document the behavior of migrants. Frequently, DVTs are used in conjunction with counts of migrating raptors, radar, radiotelemetry, or banding studies.

The use of DVTs other than for counting can be divided into studies in which verbal descriptions of flight behavior are made and those in which quantitative data are collected. General descriptions of flight behavior have usually resulted from count-oriented studies at ridges such as at Hawk Mountain (Broun 1935, 1949; Haugh 1972) and coastlines such as at Cape May Point (Allen and Peterson 1936), Derby Hill (Haugh and Cade 1966; Haugh 1972), and Gibraltar (Evans and Lathbury 1973). Behavioral descriptions from these locations have been limited to soaring, gliding, and to a lesser extent, flocking and interspecific aggression.

More recently, quantitative description of behavior has been reported from leading line sites and other locations. Thake (1980) studied the flocking behavior of migrating Honey Buzzards on the island of Malta in the Mediterranean Sea by recording flock sizes. Kerlinger (1984, 1985a) examined the behavior of migrating hawks at the ends of two peninsulas in eastern North America to determine how and when they made water crossings. In other studies (Kerlinger 1982a; Kerlinger and Gauthreaux 1984, 1985a, 1985b; Kerlinger, Bingman, and Able 1985) DVTs were used (sometimes simultaneously) with radar to study flight behavior (see chaps. 6–12).

DVTs are the most economical methods available to the researcher. They require only pencil, paper, binoculars, and other small instruments. Significant questions can be answered without the aid of sophisticated equipment, although the accuracy of measuring some aspects of flight behavior is not as great as with radar or telemetry. Perhaps the most important function of these methods is to augment the more sophisticated techniques in migration research.

Radar

Radar was used to study bird migration beginning in the 1940s and 1950s (Lack and Varley 1945; Lack 1960; Eastwood 1967). It has been used to study passerines (Evans 1966; Gauthreaux and Able

1970; Gauthreaux 1971, 1972; Able 1970, 1972), shorebirds (Richardson 1979), and waterfowl (Bellrose and Graber 1963; Bellrose 1967; Kerlinger 1982b). It has been the most important tool for studying the flight behavior of migrants; without radar we would know little about migration. Radar, however, was virtually unused in hawk migration research before the mid-1970s. (For a review of how radar detects birds, insects, planes, and atmospheric phenomena, see Houghton 1964; Eastwood 1967; and Konrad 1968, 1970.)

"Early" radar studies of hawk migration include Houghton's (1971, 1974) research at Gibraltar and that of Richardson (1975) in southern Ontario, Canada. No visual confirmation of "targets" was made in these studies, so there is some doubt as to what types of birds were "observed" by the radars. At Gibraltar, visual observations showed that the migrants detected by Houghton's tracking and surveillance radars were a mixture of raptors, cranes, storks, and other soaring birds. Most of the radar echoes that Richardson dealt with were unquestionably Broad-winged Hawks, since there are few other migrants in southern Ontario that could have made such echoes (as will be shown in chap. 9). Furthermore, hawk counters at nearby Hawk Cliff on Lake Erie observed large numbers of hawks aloft on some of the days included in Richardson's data set. Richardson's paper is a milestone in hawk migration research because it was the first to employ radar to ask specific questions about flight behavior and because it documented how radar could be used.

After 1975 only a few research groups made use of radar. My studies with Able and Bingman at the State University of New York at Albany and with Gauthreaux at Clemson University involved various radars, and the data from these studies constitute most of the research in chapters 6 through 12 of this volume. Tracking radar studies from Schmid, Steuri, and Bruderer (1986) in Switzerland and Alerstam (1987) in Sweden have added significantly to our understanding of the aerodynamics and flight behavior of migrating raptors. Leshem's (1987) airport surveillance radar studies in Israel commenced in autumn 1986 and will provide a new perspective on migration in the Middle East. Comparison of the results reported by the American, European, and Middle Eastern researchers provides a means of investigating the general nature of the findings of the various research groups.

At least five types of radar can be used to study hawk migration: airport surveillance radar (ASR), weather surveillance radar (WSR), marine surveillance radar (MSR), vertical fixed-beam radar (VFBR), and tracking radar (TR). These radars can be used in different situations to answer a range of questions about migration.

Airport Surveillance Radar

ASRs (ASR-4, ASR-6, AASR, etc.) have been used numerous times to study bird migration (Gauthreaux 1970; Richardson 1972, 1978; and many others), but only once to study hawk migration (Richardson 1975). ASR can detect birds the size of hawks at distances of over 50 km and follow their movement for at least that distance. Beam configuration of this radar is fan shaped, usually about 1° to 2° in the horizontal field and from 20° to 30° in elevation (from the ground up to 20° to 30° above the horizon). For this reason the radar is suitable for determining a bird's position over the ground and then measuring flight direction, speed, and sometimes flock shape. These aspects of flight behavior can be read from the plan position indicator (PPI) or from pictures taken of the PPI (Polaroid, 35 mm, or 16 mm movie film—frame-by-frame time exposures). Unfortunately, the wide vertical dimension of the beam precludes determination of altitude. Airport radar can best be used to determine geographic patterns of migration and flight direction in relation to wind and topographic features. Simultaneous observations by hawk counters in the vicinity of the radar are necessary to confirm that the targets are hawks.

Weather Surveillance Radar

Although WSR (WSR-57, WSR-74) has never been used to study hawk migration, it offers a technology that can be used in the same manner as ASR. Students of passerine migration (Gauthreaux 1970, 1971, 1972; Able 1970, 1972) have long realized that WSR has certain advantages over ASR. The WSR beam differs from the ASR beam in that it is a narrow, conical beam about ½° to 1° in width. In addition to yielding information on direction of flight, shape of flocks, and flight speeds, the WSR, by virtue of its narrow beam, can measure the altitude of migrating birds. To determine altitude, the elevation of the radar beam can be changed so that altitude can be read directly from a range height indicator (RHI) in the same way meteorologists measure the height of thunderstorms (Gauthreaux 1970).

Access to both WSR and ASR is limited and often difficult to acquire. Understandably, air traffic controllers are reluctant to give ornithologists access to radar when planes are landing or taking off. Similarly, WSRs are needed to watch for violent thunderstorms and other life- or property-threatening atmospheric disturbances. Of the two, weather radar is probably the more accessible and versatile, although not as many units are in use.

Marine Surveillance Radar

The MSR is the most versatile and underused tool for studying bird migration. Williams et al. (1972) and Sielman, Sheriff, and Williams (1981) have used MSR to study the migration of passerines and other birds at several locations in eastern North America. It is also the least expensive and most portable radar available. The MSR is a smaller and less powerful version of the ASR. The beam configuration is similar to that of ASR (about 2° in the horizontal plane and somewhat greater than 20° in elevation). The MSR discussed below is mounted on the roof of the Avian Migration Mobile Research Laboratory (fig. 2.7), a 7 m motor home designed and constructed by Sidney A. Gauthreaux of Clemson University. The radar unit is a Decca 110 with a 3 cm wavelength beam and 10 kW power output (Gauthreaux 1985). The radar was field tested on grebes, ducks, and gulls flying near power transmission lines in California, Washington, Delaware, and Montana before being used on hawks.

Kerlinger and Gauthreaux used the Decca 100 (MSR) to study the spring migration of Broad-winged and other hawks through south Texas at Santa Ana National Wildlife Refuge (Kerlinger 1985a, 1985b). The radar was capable of detecting flocks of 500 to 3,000 Broad-winged Hawks at about 7 km (over 4 nautical miles). Although the identification of targets could not be confirmed at this distance, on many occasions the birds were cooperative and flew directly overhead. Echoes of hawks were distinct. Climbing in thermals, flocks formed echoes that appeared to be roughly circular on the PPI. They subsequently spread out in a linear fashion over 400 to 1,500 m during interthermal glides. Data gathered with the radar during our study in south Texas included flight direction, length of flocks when gliding between thermals, and altitude. Because altitude could not be determined with MSR directly, we constructed an inclinometer to measure the angle of the flock above the horizon. The inclinometer consisted of 20× binoculars, a protractor, a level, and a plumb bob mounted on a tripod. Altitude was calculated from this angle and the distance between flock and radar (Kerlinger and Gauthreaux 1985a). The radar detected many birds (even flocks of over 1,000 birds) that would not have been seen even if they were within range of binoculars or the naked eye. During all radar operations, simultaneous visual observation of the radar targets was made.

A

B

Figure 2.7 (A) Avian Migration Mobile Research Laboratory (property of Clemson University), with (B) marine surveillance (right) and vertical fixed-beam (left) radars on its roof. The "laboratory" was designed and constructed by S. A. Gauthreaux and was used to study hawk and passerine migration in New Jersey, Texas, South Carolina, Louisiana, and California. (Photographs by Sidney A. Gauthreaux.)

Vertical Fixed-Beam Radar

By substituting a 75 cm parabolic antenna for the normal T-bar antenna on a MSR (in this case a Marconi Seafarer [Raytheon], 3 cm wavelength, 10 kW) Gauthreaux constructed a VFBR (fig. 2.7). The "new" radar beam was a cone of about 4° that could be pointed in any direction. Aligning the radar vertically allowed precise measurements of the altitude of migrating birds. The VFBR is also mounted on the roof of Gauthreaux's mobile laboratory. The complete mobile laboratory consists of two radars (MSR and VFBR), a night-vision scope, video cameras, a video-beam splitter, a ceilometer, and video recorders. The Avian Migration Mobile Research Laboratory is one of the finest systems designed and used for studying bird movements.

The VFBR was excellent for measuring the altitude of migrating hawks (Kerlinger and Gauthreaux 1984). As with MSR, visual confirmation of targets is necessary. Thus, one person must watch the radar screen (a modified PPI) while a second watches the sky for radar targets. When the VFBR is used in the vertical mode the observer reclines so that targets can be identified as they pass through the radar beam.

Kerlinger and Gauthreaux (1985a, 1985b) used MSR and VFBR alternately during their spring migration study in south Texas. By combining radar and visual observations they gathered data on altitude and other aspects of flight behavior as well as determining the visibility of migrants to ground-based observers. The radars were used alternately because the altitude of migration became so high after 1000 to 1100 h that the MSR could no longer detect hawks. At that time the VFBR was used. Radars could not be used at the same time because of interference caused when two radar signals are used.

One drawback of the particular VFBR used by Kerlinger and Gauthreaux (1985b) was the narrow beam width (4°). A problem with using such a narrow beam for hawk migration studies is that sample sizes are small. For passerine migrants, which are more numerous, the 4° cone is adequate. Because of the versatility of the MSR/ VFBR system, beam width can be modified by changing the size of the parabolic antenna that is mounted on the MSR. There is an inverse relation between beam width and size of the parabolic antenna such that a smaller antenna enables the radar to "see" more of the sky (larger beam width). As beam width increases, however, the power of the beam decreases so that some birds may not be detected by the radar. Birds at the edge of the beam, small birds, and birds that are at great distances will be difficult to detect with increased beam width.

Thus, there is a trade-off between beam width and power. A conical beam of up to 10°–15° is useful for detecting hawk-sized birds.

Tracking Radar

Of the five types of radar discussed, tracking radar yields the most complete data on flight behavior. The TR "locks onto" moving targets and follows them through space, giving a precise location of the object being tracked. Although Houghton (1971, 1974) used this type of radar to track migrating raptors, Kerlinger (1982a) and Kerlinger, Bingman, and Able (1985) have done extensive radar tracking in which targets were identified to species. The tracking radar used for this work was an AN/MPQ 10 cm wavelength, 250 kW (Sperry gyroscope) military surplus radar (Able 1977). It was designed to track artillery shells and (fig. 2.8) is accurate to within 18 m in slant range and 0.08° in azimuth and elevation. It is owned and operated by K. P. Able and has been used since 1975 to track passerines migrating at night (Able et al. 1982) as well as loons (Kerlinger 1982b) and other birds migrating during the day. Other tracking radars have been used to study bird migration by Griffin (1973), Emlen (1974), Pennycuick, Alerstam, and Larsson (1979), Larkin (1982), Schmid, Steuri and Bruderer (1986), and Alerstam (1987).

As with MSR and VFBR, an observer using binoculars or a spotting scope makes simultaneous observations of the target being tracked. In the tracking radar studies described in this volume the observer noted species identity, number of individuals, and flight type (flapping, gliding, soaring in thermals, catching insects) and other behaviors including "doing something weird." At ten-second intervals a bird's position (slant range, elevation, and azimuth) was read from a control panel into a tape recorder. By tracking weather balloons with radar reflective material attached, wind direction and speed at the altitude of migration were determined. From TR data flight behavior such as altitude, flight direction, flight speed, and sinking speed were quantified. These were used to test various hypotheses about flight behavior and to quantitatively describe migratory flight (chaps. 6–12).

Radars of the type discussed above offer excellent means of studying hawk migration. Unfortunately, the cost of purchasing an ASR, WSR, and most types of TR are prohibitive without a large research budget. ASR and WSR use is limited to airports and weather stations where access is problematic. Tracking radars are more difficult to obtain, although military surplus units are occasionally available. Keeping surplus radars operating is not easy, since parts and service are generally not available. (Able obtained spare parts from four radars that arrived at his laboratory by mistake. It seems there was a computer error and the radars could not be returned. Perhaps more military material should be used for such peaceful endeavors.)

Figure 2.8 Tracking radar (property of State University of New York and Kenneth P. Able) used to study hawk, passerine, and loon migration in central New York State. (Photograph by P. Kerlinger.)

Another disadvantage with radar, at least the MSR, VFBR, and TR, is the power supply necessary for operation. Generators as small as 500 W suffice for most MSRs and VFBRs, although if video equipment is used to tape the PPI or computers for recording data, a different power source is necessary. A set of rechargable batteries can give the surgeless power required for recording equipment or computers and eliminate the noise and fumes produced by gasoline generators.

MSRs and VFBRs are substantially less expensive than other radars and range in price from about $4,000 U.S. to more than $10,000 (1984) depending upon the power of the radar, accessories, and the manufacturer (Raytheon, Faruno, Decca, etc.). To operate the smaller radars, a vehicle, trailer, and power supply (generator, batteries, or line power) are necessary. A second advantage of smaller radars is that they can be positioned exactly where the researcher needs to work.

When planning a research project involving radar, the researcher should think about what questions are to be addressed and which radar is best suited for answering them. It is conceivable that some projects may have to be deferred because of logistical (radar location and accessibility) and financial constraints.

Summary and Conclusions

Diverse methods have been used to study hawk migration. Hawk migration counts have been the predominant research tool, followed by trapping and banding. The count method has been used to examine seasonal and daily timing of migration and to relate the amount or volume of migration to weather. The reliability and validity of some hawk migration count studies have been questioned. Use of the migration count method should be considered carefully by a researcher before its use. Hawk migration counts may be most useful in monitoring populations and education rather than in furthering our understanding of migration.

Banding of hawks may answer questions about the geographical origins of migrants passing a specific location, their destinations, and the approximate pathways used. In addition, banders can acquire morphological and physiological data during the banding process (mass, fat, wing area, wing span, tail area, wing chord, etc.). The banding of hawks has not realized its potential, partly because low rates of band recovery ($< 1\%–5\%$) and recapture yield samples that are too small to answer many important questions, as well as owing to a lack of cooperation among banders.

Radiotelemetry, radar, and direct visual techniques have yielded the best data for answering questions about the flight behavior of migrating hawks, but logistical and financial constraints should be considered when planning radar or radiotelemetry studies. The advantage of these methods is the precision of their measurements. Direct visual techniques are of limited accuracy for some questions but are adequate for measuring flock sizes, flight direction (using hand-held compass or theodolite), water crossing tendency, and other flight behaviors.

Before considering a field technique, a researcher should determine the question to be answered. After a question has been posed, a technique or combination of techniques can then be selected. A paucity of clearly defined questions is one reason for the limited usefulness of some research based on hawk migration counts. The future of hawk migration research depends on researchers' defining their questions and collecting data with techniques that permit them to answer those questions.

3

Natural Selection and the Study of Migration

As shown in chapter 1, the migration of falconiforms is extremely diverse. Despite this diversity, all migrants share a common problem—how to move from location A to location B. Alerstam (1981) has drawn a useful analogy between migrating birds and pilots of small aircraft. Both migrants and pilots are limited by the performance of their "aircraft" (i.e., how fast and how high they can fly and the rate at which they consume fuel), the amount of fuel they can carry, and the weather at the time of flight. Given these constraints, both must adopt a strategy that will allow a speedy journey that can be completed with the fuel available.

Pilots must make numerous decisions when flying from point A to point B. First, they must decide when to take off. If the trip takes six hours in still air, pilots must consider wind speed and direction in their "simulation" of the ensuing flight. Second, they must decide how much fuel is needed. Filling the fuel tanks for a short trip is wasteful because surplus fuel requires additional fuel to transport. A safe strategy is to take on slightly more fuel than is needed. Third, pilots must consider the weather between A and B. If thunderstorms or other "weather" are evident along the route, they must change flight plans. Fourth, pilots must decide how high to fly. This decision is based on turbulence, wind direction and speed at various altitudes, the fuel required to climb, and how high an air traffic controller directs them to fly. Fifth, cruising air speed and heading must be considered. A slow air speed could delay arrival or allow a plane to be blown off course by crosswinds. Speeds that are too fast waste fuel. Finally, the overall flight plan must incorporate a means of dealing with weather changes, engine malfunction, or other unexpected contingencies. The set of decisions pilots use is complex but necessary to promote a safe and successful flight.

A six-hour flight was considered in the analogy presented above. For a small, single-engine aircraft, a flight from Boston to Detroit takes about this much time, depending on wind speed and direction and type of aircraft. If a pilot wants to fly from Montreal to Peru in the same plane, a different flight plan is needed in which more than a half-dozen six-hour flights would be required. In addition to plotting a complete course between Montreal and Peru, the pilot must determine whether to fly over the Gulf of Mexico or the Caribbean (a dangerous task in a single-engine aircraft), where to refuel, and how many times to refuel, as well as making all the decisions outlined above for a short flight. A knowledge of prevailing winds and weather at various altitudes along the flight route would also be handy, but these are usually available from airport weather stations. The longer flight becomes more complex and mistakes are more critical.

Flying from Montreal to Peru is analogous to the migration of many birds. The major difference between the aircraft's flight and that of a raptor is air speed. Because the cross-country speed at which raptors fly during migration is less than one-third that of a small aircraft, a raptor requires more time to complete the same flight. That is, more stops are necessary, and refueling takes hours or days instead of minutes. The migration of a Sharp-shinned Hawk from Montreal to Tennessee is probably analogous to the flight of a small airplane between Montreal and Peru.

A successful flight for both aircraft and bird relies on a sound flight plan or strategy. In the case of aircraft, a flight plan is decided beforehand using maps, meteorological information, information on the airplane's performance (mileage, speed, ability to float, ability to fly on instruments, whether it has a pressurized cabin, etc.), and often a computer. Migrating birds do not possess flight plans per se (at least we do not know if they do), but they do have strategies that increase their chances of completing a safe and rapid migration. I do not mean to imply that birds consciously plan a migration. Strategy is a term evolutionary biologists use to describe behavior, physiology, reproductive biology, and other biological attributes of organisms that have been shaped by natural selection. That is, they are products of evolution. Whether they use conscious processes is irrelevant. In this volume a migratory strategy is defined after Alerstam (1981) as "a harmonious mixture of rigid and flexible behavior adapted to a bewildering number of factors affecting the safety and economics of the migratory journey." Alerstam asserts that natural selection works "to maintain and refine the level of adaptation in the bird's journey." Thus, a migratory strategy is a set of behavioral decisions, as well as morphological and physiological adaptations, shaped by natural selection that allow birds to complete their migratory journeys safely and swiftly. Determining migration strategies will help us understand how natural selection has shaped the physiology, morphology, and behavior of migrants.

Natural Selection and Migration

That natural selection has shaped the migratory strategies of birds is a safe, yet relatively untested, assumption. There are strong logical arguments for selection of migratory behavior. In its crudest and most expeditious form, natural selection eliminates those individuals (and their genes) that make gross errors during migration. For instance, a bird with no fat reserves that attempts to fly across the Gulf of Mexico against a 30 kph wind will likely die. Its future reproductive success will be zero. A less clear-cut case is that of a hawk migrating from Texas to New York in spring. If it flies northwest it will probably never find its breeding grounds, nor will it leave any offspring (that year). Similarly, if it begins migration late it may not breed. In these cases natural selection operates either by eliminating individuals from the breeding population or by limiting the number of surviving offspring an individual produces.

Natural selection among migrants often has immediate and severe consequences. Reports of large-scale mortality events abound, although few reach the reviewed literature. Most of these reports deal with birds attempting to cross large bodies of water such as the Great Lakes (Saunders 1907) or the Gulf of Mexico (Gauthreaux 1972) during foul weather. Conversations with personnel from offshore oil-drilling platforms have revealed that "little yellow birds" land on the rigs and that many die. Not all of these birds are observed during storms or adverse weather. The fat condition of passerine migrants arriving at the northern Gulf coast in spring shows that many birds are severely fat depleted even when weather has been favorable for crossings (Moore and Kerlinger 1987). Unlike birds that arrive with large fat deposits, thin birds may have started their journey with only a small amount of fat. Thus, these thin birds might have been selected against if they encountered adverse weather.

Mortality of migrants over land is not as well documented as with water crossings. The few published cases of mortality of migrants over land are associated mostly with bizarre and brutal weather (Roberts 1907).

Discussions of raptor mortality during migration have been restricted mostly to human persecution (Broun 1935, 1949; Beaman and Galea 1974; Newton 1979), and there are few data pertaining to natural losses during migration. Most of the evidence available pertains to water crossings (chap. 10), although there are a few reports of mortality during migration over land. For example, Smith, Goldstein, and Bartholomew (1986) state that 27 (2 adults, 25 immatures) Broad-winged Hawks received in Panama "all lacked fat" and "appeared to be dehydrated." Because Smith did not report information on bodily injuries, season of the year, gut contents, or absence of toxic substances, we can only speculate that the poor body condition of these birds was a result of starvation during migration. Smith (1980) has also reported that some Swainson's Hawks arrive in Argentina in such poor condition that they can be captured by hand. Geller and Temple (1983) have reported that many immature Red-tailed Hawks from northern populations are thin during the autumn migration through Wisconsin. They hypothesize that these birds migrate shortly after fledging and are susceptible to starvation. Other thin or emaciated raptors have been reported by banders, especially those working with Northern Goshawks during irruptions (Mueller, Berger, and Allez 1977).

There is no doubt that migration is a "perilous" time in the life of birds (Mayr 1982). Mass mortality during migration is a dramatic example of selection. Individuals need not die for selection to operate. A question that is important to the student of bird migration is: Is there a difference in net (lifetime) reproductive success between a bird that migrates quickly and efficiently and another that is less efficient? For example, will a female Broad-winged Hawk that arrives in the Adirondack Mountains of New York on 22 April with 60 g of stored body fat realize a reproductive advantage over one that arrives on 5 May with only 20 g of fat or one that arrives on 22 April with no body fat?

Although empirical studies have not demonstrated a link between reproductive success and migratory flight behavior, it is easy to see how a relation can exist. Variance in speed and energy efficiency of migration means that some individuals arrive on the breeding or wintering grounds before others and in better physical condition. There are advantages to arriving earlier (although not too early) and fatter than conspecifics following migration in both spring and autumn. Migrants that arrive early have priority of access to nesting sites, winter territories, mates, and other resources. Those that arrive later have more competitors for limited resources. This was discussed in chapter 1 as it relates to differential migration and the arrival time hypothesis. An obvious advantage of migrating quickly is completing migration before the termination of migratory restlessness (*Zugenruhe*) and the end of a breeding season. For individuals of some species early arrival may mean the opportunity for a second or replacement clutch. All of the reasons for early arrival could be important at one time or another for most raptors.

At some northern breeding sites it may be disadvantageous to arrive too early. Snow, ice, and cold may select against early arrival, especially among species dependent upon the emergence of insects and ectothermic vertebrates from torpor. Migrants that depend upon avian prey that also migrate, mammals that hibernate, or fish that become unavailable because of ice may also face starvation if they arrive too early. Thus, there is a trade-off between arriving too early and too late as shown in figure 3.1. Migrants that arrive within the range of optimal arrival time presumably realize greater fitness than those that arrive too early or too late. Quick and efficient migration is paramount to arriving at the correct time.

Variance in energy efficiency among individual migrants will also result in a variance of stored body fat, both during migration and shortly following the completion of migration. Fat birds are generally more successful than conspecifics with less fat. Fat reserves are particularly important for migrants that must travel through unpredictable or inhospitable habitats such as deserts, oceans, some islands, and far northern or southern temperate latitudes. Inhospitable habitats for some species include prairies, forests, or mountain ranges. There are several documented cases of birds' being stranded in inhospitable habitats because they lack the body fat that enables them to leave (Spendelow 1985).

The efficient migrant will also not make as many stops during mi-

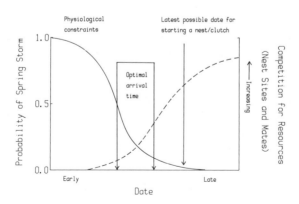

Figure 3.1 Theoretical model of the costs (risks) and benefits of arriving early at the breeding grounds in temperate regions.

gration to replenish fat stores metabolized during flight. These stop-overs require 3–4 days among small passerines (Cherry 1982; Moore and Kerlinger 1987) or weeks in the case of larger birds. Fat reserves are especially important for birds like geese (Ankney and MacInnes 1978; McLandress and Raveling 1981) and possibly Sandhill Cranes (*Grus canadensis;* Krapu et al. 1985) that depend on it for reproduction. Fat for these and other migrants may mean a difference in the number of eggs individuals can yolk and in the energy available for incubation as well as for attracting a mate (song and display), defending a territory, building a nest, or even guarding a mate from potential cuckolding. In autumn and winter, migrants like the Snowy Owl may use accumulated fat stores to fast for long periods of harsh weather (Kerlinger and Lein 1988). Similarly, stored fat enables "hard weather" movements following the completion of "normal" autumn migration (Terrill and Ohmart 1984). Thus, stored fat is used as fuel for migration, as a hedge against a harsh or unpredictable environment, and as an energy source for migrants after they arrive at breeding and non-breeding sites.

Among the falconiforms one species has been studied enough to allow generalizations regarding fat condition at the onset of reproduction and reproductive success. Female European Sparrowhawks increase body weight by some 15% (40–50 g) before egglaying and incubation (Newton 1986). This surplus consists mostly of fat and is used for fasting when food is unavailable during incubation. Although the population Newton (1986) examined is not highly migratory, it does inhabit an area in which unfavorable weather makes food unavailable for short periods. Study of migratory populations of this species would be enlightening.

In sum, there is empirical evidence that the flight behavior of migrants has been and continues to be shaped by natural selection, and there are logical reasons why this is so. Unfortunately, it is difficult to test empirically the relation between migratory flight behavior and reproductive success. The result of selection will be a strategy that allows a migrant to travel the maximum distance in the minimum amount of time, using the least amount of fuel. By virtue of quick and efficient migration, a migrant will realize several advantages over conspecifics that will ultimately result in a greater number of offspring.

A Reductionist Approach to the Study of Migration

If a migration strategy is the product of natural selection, on what aspects or components of an animal does selection operate, and how can a biologist determine if and how selection has operated? To answer these questions, as well as to elucidate migration strategies, a reductionist perspective is desirable. A reductionist approach is "achieved by dissecting a complex problem and partitioning it into its components" (Mayr 1982). The problem we are interested in here is migration strategy. Because a strategy cannot be measured directly, components that together make up the phonemenon must be determined so that they can be measured. The components of migration must be measurable so that a researcher can describe the variance within and among populations of migrants.

It is common knowledge that animals possess morphological (Savile 1957; Dingle 1980), physiological (Berthold 1975; Blem 1980), and behavioral (Able 1977, 1980; Alerstam 1979a; Gauthreaux 1972) adaptations for migration. If we assume that natural selection operates on these three aspects of a migrant's biology, at least as they pertain to migration, we have reduced migration to three components or aspects researchers can focus on. Still, a researcher cannot measure these three components of migration directly but must take a further reductionist step by examining specific components or questions within each of these categories. The procedure of focusing on aspects of a problem that can be measured in a research context is what Nagel (1961) and Mayr (1982) call explanatory reductionism as opposed to true reductionism, which focuses on the molecular or submolecular level of a problem. With the explanatory reductionist approach we still operate on the organismic level, although we examine questions, problems, and hypotheses pertaining to physiology, morphology, and behavior of migrants. To answer questions or test hypotheses, we must define measurable units that when collected are called data.

The path diagram in figure 3.2 illustrates a reductionist strategy and identifies three levels in the reduction hierarchy. Physiology and morphology have been placed at the top of the hierarchy because they are the primary constraints on migratory behavior, especially flight. Aspects of physiology such as flight energetics, fat deposition, and hormonal regulation limit the type of flight a bird can use (gliding vs. flapping), its maximum speed, the altitude at which it can fly, how long it can fly, when it flies, and other aspects of migratory flight. At the same time the flight of migrants is also constrained by morphology. Wing length, wing shape, wing area, tail area, tail length, mass, and musculature determine a bird's aerodynamic performance—that is, its flight capabilities. The physiology and morphology compartments in the path model are connected by a two-headed arrow to show that the two compartments are not independent. For example,

flight energetics are determined in part by body size and wing morphology. The biologist who is interested in migration can study one or
more of the aspects of morphology or physiology listed in the boxes
in figure 3.2.

At the second level of the hierarchy is behavior. Within this box is
a list of behaviors that have been studied by migration biologists. During migratory flight a migrant must make decisions about each of the
behaviors listed. By decision, I do not mean a conscious process (sensu
cognitive animal psychologists; Gallup 1970; Griffin 1984). Instead, I
refer to the process described by McFarland (1977). A decision according to McFarland is "some process or mechanism that determines
which activity is to have priority at any particular time." Thus, the
animal has a choice of several alternative means of doing something
(behaving). For example, migrating hawks have the option of gliding
and soaring, intermittent gliding, and continuous flapping or powered
flight. By using one of these three types of flight (or combinations of
them) to the exclusion of the others a migrant, by definition, makes a
decision. In a similar manner the migrant makes a decision when it
changes from one type of flight to another.

Once a migrant has chosen the type of flight it will use, it must then

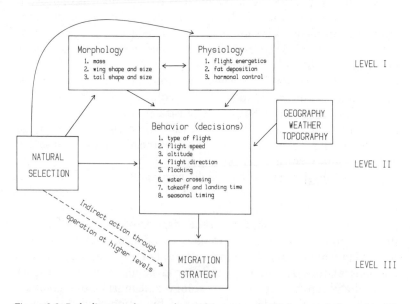

Figure 3.2 Path diagram showing the explanatory reductionist approach used in this
volume to study migratory flight behavior.

make several other decisions (fig. 3.2). These include how fast to fly, how high to fly, what direction to fly, whether (and when) to fly with other migrants (flock), whether (as well as when and where) to cross large bodies of water, when to take off and when and where to land, when to feed or if it should feed, and probably others. Usually, migrants are faced with several of these decisions simultaneously. If a migrant makes decisions in an appropriate manner it will complete migration quickly and efficiently. To understand flight behavior is to understand how migration occurs. Considering each aspect of flight behavior separately is a reductionist approach to studying bird migration.

In the case of the specific migration behaviors listed in figure 3.2 all are subject to changes (decision making) based on weather, topography, and geography at a particular location and time in migration. Many migratory flight decisions have been shown to be determined by these variables. That is, migrants modify their behavior during migration as changes of weather, topography, and geography dictate. Thus, the entire suite of migratory flight behaviors must continually change. The set of decisions a migrant makes determines its ultimate fitness, a result of natural selection on the behavior of individuals.

Whether the hierarchical levels presented in figure 3.2 reflect the way natural selection operates on the various components of migration is a question for future research. The model needs scrutiny and consequently will be enlarged, changed, or even "trashed." Most important, the model is a heuristic tool to be used to determine a research strategy. In other words, potential researchers can use it to formulate questions and organize their thoughts on how natural selection operates on migrants.

Students of hawk migration have seldom examined flight behavior. A few qualitative decriptions of some of the components of flight behavior listed in figure 3.2 have appeared in the literature (Allen and Peterson 1936; Broun 1949; Mueller and Berger 1961, 1967a; Haugh and Cade 1966; Haugh 1972; Evans and Lathbury 1973; Heintzelman 1975). Before the late 1970s only two studies focused on quantitative aspects of flight behavior. Broun and Goodwin (1943) studied flight speed of hawks migrating along the Kittatinny Ridge in Pennsylvania, and Richardson (1975) used radar to study flight direction of flocks of hawks in southern Ontario. For this reason, little was known about the behavior of migrating hawks until recently, at least compared with that of passerines and other migrant groups.

The reductionist approach has been criticized by organismic biologists as sterile and insensitive to the emergent properties of biological systems or processes (reviewed by Lorenz 1981; Mayr 1982). Emergent properties of systems are greater than the sum of the parts or components. The criticism is legitimate, although the explanatory reductionist approach is a compromise between true reductionism (the molecular level) and a holistic approach. Migration strategies should be viewed as emergent properties because they are sets of behaviors (fig. 3.2) that are used during migration along with morphological and physiological adaptations. The only means a researcher has for elucidating migratory strategies is to study the various behavioral, morphological, and physiological components that together constitute migration and then examine combinations and interactions of these components in order to determine emergent properties or strategies. Relations and interactions among the components cannot be studied directly with a reductionist approach. It is these relations and interactions that are the emergent properties or strategies of migration.

Students of migration have relied upon an explanatory reductionist approach, although to my knowledge they have never referred to it as such. For example, researchers focus on the orientation abilities of migrants. To do so they attempt to control for other aspects of migratory behavior by experimental or statistical means. Researchers such as Kramer (1957), Sauer (1961), Emlen (1967), Gauthreaux and Able (1970), Able (1977), Richardson (1976, 1979), and others have explored the orientation abilities of migrants by ingenious use of orientation cages (called Emlen funnels) and radar. In their studies they controlled for other aspects of migratory behavior so that they could ask questions solely about orientation abilities. Some researchers go as far as incorporating one or two other behavioral or physiological aspects of migration in their studies. Able (1977), for example, has examined and analyzed numerous aspects of flight behavior simultaneously by tracking individual birds with radar under "known" wind conditions. He examined air speed, orientation in relation to wind, and altitude simultaneously to determine how behavioral components of migration were related. Richardson (1976) also considered the simultaneous operation of air speed, altitude, and orientation. Pennycuick (1978) and Alerstam (1979a) have developed theoretical (mathematical) models that incorporate many variables. Empirical data can be compared with the predictions of the models, which incorporate many different behavioral components of migration, such that specific hypotheses regarding flight behavior and perhaps emergent strategies can be tested.

The most comprehensive study of the flight strategy of migrants is Alerstam's (1985) comparison of Arctic Terns (*Sterna paradisaea*) and

Common Terns (*S. hirundo*). Using tracking radar with simultaneous visual observations, Alerstam quantified the air speed, energetic expenditure, flocking behavior, altitude of flight, and flight direction of terns migrating in southern Sweden. His study is one of the first to consider several aspects of flight behavior simultaneously and integrate them into a strategy that he then compared with predictions made from aerodynamic models. In addition to elucidating the flight strategy of the species, Alerstam's approach permits a comparison of congeners with different migratory and breeding biology.

In the remainder of this book I use an explanatory reductionist approach. The goals are to describe quantitatively the flight behavior and flight morphology of several species of hawk migrants, compare flight behavior and morphology among species, and test hypotheses regarding specific components of migratory flight behavior. Chapters 5–12 address specific aspects of migratory flight. The final chapter will be devoted to elucidating the migratory strategies of raptors, posing a hypothesis for future migration research, and discussing unanswered questions.

The approach used in this volume and the many references made to natural selection may seem simplistic or naive to some readers. To the scientist who studies life-history strategies, mating systems, or other aspects of modern ecology, the evolutionary approach is second nature. Migration researchers, with some exceptions (e.g., Alerstam 1981, 1985; Gautheraux 1982a), have not always benefited from this approach. It is only by dissecting flight behavior into its basic components and studying each separately that we will fully comprehend the behavioral strategies of migrants and gain some insight into the way natural selection has shaped these strategies.

Description of Species to Be Examined

In the chapters that follow I examine and compare numerous aspects of the flight behavior of migrating hawks. Most of the data are restricted to several species of North American raptors that I have studied: Broad-winged Hawk, Red-tailed Hawk, Osprey, and Sharp-shinned Hawk, and to a lesser extent American Kestrel and Northern Harrier. I chose these species because they are common and large samples can be acquired. In addition, they vary in size (Snyder and Wiley 1976) and morphology, distance of migration, prey utilization, taxonomy, and habitat requirements, allowing comparison of a wide array of migrants. Because of these differences the comparisons of migratory flight behavior among species will be more meaningful than a comparison of ecologically or morphologically similar species. Presumably, natural selection has shaped each species to deal with differing ecological settings as well as the contingencies of migration. Below are brief descriptions of the species, including breeding and non-breeding range, migratory tendency, prey type, habitat, and body size. The following paragraphs are not meant to be a substitute for field guides or natural history accounts of raptors but constitute an introduction to the characters that will be the focus of the remainder of this volume. For identification of raptors see Brown and Amadon (1968), Cramp and Simmons (1980), Porter et al. (1981), Clark and Wheeler (1987), Dunne, Sibley, and Sutton (1988), and standard field guides.

Broad-winged Hawk

More Broad-winged Hawks are counted during migration than any other raptor species in the New World. These small, chunky birds breed in forested habitats from Alberta and Saskatchewan (uncommonly) east to the Canadian Maritime Provinces and southward into the southeastern United States. Most migrants move into Central America and as far south as Brazil and Peru. Thus, this hawk is a complete, long-distance migrant that may fast for extended periods during migration (Smith 1980; Smith, Goldstein, and Bartholomew 1986). This small (males average 450 g, females 490 g) generalist (eating mammals, birds, herps, and insects) buteo occupies secondary forest and edge habitats in the tropics during the non-breeding season.

Red-tailed Hawk

This large buteo ranges in mass from about 700 g for males to about 1,300 g for females. Its breeding range extends from central Alaska in the west to Nova Scotia in the east and south through North America, where it prefers edge, lightly forested, and open habitats. During the non-breeding season some birds occur in southern Canada in the east and far west, although not in large numbers, and into the central-northern Great Plains of the United States. Whereas northern breeders make long-distance migrations into the southern United States and Mexico, individuals of some southern populations remain on nesting territories year round. Thus, the Red-tailed Hawk is a partial migrant. Although little is known about differences between the winter ranges of males and females (differential migration), immature birds move farther south than adults (Brinker and Erdman 1985; Kerlinger, unpublished data). The species is primarily a mammal eater, although it does take birds and herps regularly.

Sharp-shinned Hawk

This species is the smallest woodland hawk in North America, ranging from about 95 g for small males to over 180 g for large females. Sharp-shinned Hawks are bird eaters that inhabit woodland and edge habitats. Their range extends over much of North America from Alaska to the southern United States and into Central America. The species is partially migratory, with northern populations leaving areas north of the United States, except British Columbia and perhaps Newfoundland. Some northern breeders migrate into Mexico and Central America, especially those from western North America. In eastern North America migrants seldom leave the United States (Clark 1985a). Birds that breed south of about 40° N may not be migratory or may be partially migratory with a differential migration by age or sex. Hawk watchers at sites like Cape May Point, Derby Hill, Hawk Ridge, Whitefish Point, and other locations are rewarded by seeing thousands of these small raptors during migration. Along with Red-tailed and Broad-winged hawks, they are the most often viewed migrating raptor in North America.

American Kestrel

Breeding in North America from northern Alaska to northern Ontario and southern Quebec, the kestrel is a bird of open and edge habitats. It preys on mammals, insects, and birds (Collopy and Koplin 1983). Male kestrels (about 110 g) are somewhat smaller than females (about 120 g) so that a moderate sexual size dimorphism exists. During the non-breeding season kestrels occur northward into southern Canada in the east and west, but not in between. They are partially migratory, with males wintering north of females (appendix 2). Some individuals migrate long distances to winter south of the United States, whereas others do not move at all.

Northern Harrier

The Northern Harrier is a partial migrant that breeds south of the tundra from northwestern Alaska to central Quebec and southward into Mexico. It nests and forages in prairies, marshes, and edges of farm fields. After autumn migration, a few harriers are found in the northern Great Plains, but most occupy the central and southern United States, Mexico, and the islands between North and South America. Mammals and birds are their dominant food. Sexual dimorphism exists, with males averaging about 350 g and females about 530 g.

Osprey

The largest of the species I examine in detail, the Osprey is sexually dimorphic (males from a low of about 1,300 g to large females up to 1,800 g). The northern extent of the breeding range extends from northwestern Alaska across North America to Churchill, Manitoba, and across northern Ontario and southern Labrador. A cosmopolitan species, it breeds along lakes, bays, rivers, and other bodies of water. Because the Osprey eats almost no prey other than fish, it leaves most of the United States and all of Canada during the non-breeding season. Florida Ospreys may not migrate. The winter range extends north to central California and the far southeastern United States. Most birds move to Central and South America, or to some Caribbean islands. Birds banded in the United States have been recovered in Brazil and Peru, but most winter near northern South America (Poole and Agler 1987). Most Ospreys are complete migrants, flying farther than all but a few North American raptors.

In this book I refer to several other raptors, but none was studied as intensively as the species listed above. Readers are encouraged to look elsewhere for more detailed natural histories of these and other species (Bent 1937; Friedman 1950; Dementiev 1951; Craighead and Craighead 1956; Grossman and Hamlet 1964; Brown and Amadon 1968; Brown 1970, 1976; Ali and Ripley 1978; Newton 1979; Cramp and Simmons 1980; Cade 1982; American Ornithologists' Union 1983).

From the descriptions given above it should be obvious that the raptors I focus on throughout the book are diverse. The species range in size from slightly less than 100 g to over 1,800 g; some are generalized carnivores, whereas others are food specialists; and some prefer open habitats, whereas others prefer edges or forests. With the exception of Broad-winged Hawks and Ospreys, all of the other species are partial migrants (chap. 1). Variation of migration tendency and distance as well as ecological and morphological attributes makes comparisons among species and among populations of the same species interesting but difficult (control, in the scientific sense).

Despite differences in ecology and morphology, all of these organisms must fly from a breeding or natal site to a non-breeding location(s) during migration. During this flight they encounter similar problems and must make decisions based on similar constraints. Just as natural selection has shaped the ecology and morphology of these birds, it has also shaped their migratory flight behavior. Thus, it is necessary to determine how these species (populations of individuals) behave during migratory flight and then to compare the behavioral strategies used by each species in light of constraints such as weather, physiology, and morphology.

Summary and Conclusions

The migration of birds is presented as the problem of flying from point A to point B. As such, migrants must have a flight plan or strategy like that of an airplane pilot, during which decisions continually must be made. A strategy is a set of behavioral, morphological, and physiological attributes or adaptations that are the product of natural selection. Conscious thought processes need not be invoked to study decision making or migration strategies. Although natural selection has presumably shaped the migratory strategies of birds, there is little documentation of selection. To study migration and determine migratory strategies, I adopt an explanatory reductionist approach in which specific behavioral, morphological, and physiological components or aspects of migration are examined and related. I examine one or more components of migration in each of the remaining chapters. Many of the data presented are from studies of a diverse group of North American raptor migrants that vary with regard to geographic distribution, prey utilization, habitat preferences, size, taxonomy, and distance of migration. By describing quantitatively the migratory flight of each species, we can compare different species and test hypotheses regarding decision making. Ultimately, information from these sources will promote the elucidation of emergent properties or flight strategies of migrants.

4

Structure of the Atmosphere

No study of bird migration is complete without consideration of weather. More specifically, to study flight behavior of soaring migrants a researcher must have a rudimentary knowledge of atmospheric structure. Several reviews have dealt with the influence of mesoscale weather phenomena on the migration of birds (Lack 1960; Haugh 1972; Heintzelman 1975; Alerstam 1978a; Richardson 1978), but few have examined quantitatively or comprehensively how smaller microscale phenomena influence flight behavior (Cone 1962a; Haugh 1972, 1974; Heintzelman 1975; Larkin 1982). By mesoscale (and sometimes macroscale), I mean frontal systems or synoptic patterns with associated precipitation, relative humidity, barometric pressure, temperature, and wind. Frontal systems with high and low pressure cells influence migration over hundreds or thousands of square kilometers. For example, an autumn cold front can dominate the weather encompassing a large portion of eastern North America for several days. Microscale phenomena (Oke 1978; Stull 1988) are orders of magnitude smaller in size (<5 km) and shorter in duration (<30 min) than meso- and macroscale phenomena and are embedded within high or low pressure cells that dominate them. Wind shears, wind gradients, thermals, and ridge updrafts are examples of microscale phenomena. Haugh (1974) in his studies of migrating hawks referred to them as "local ephemeral weather conditions."

The portion of the atmosphere that is of interest to students of hawk migration includes the lower troposphere and most of the turbulent surface layer (fig. 4.1). This portion ranges up to 2 km or more above the earth and is called the planetary or atmospheric boundary layer. Atmospheric scientists refer to most phenomena that occur within this region as microscale phenomena (Oke 1978), and until the past thirty years the boundary layer remained relatively unstudied. Much recent interest in the boundary layer has resulted in a proliferation of information including an entire journal, *Boundary-Layer Meteorology*. Readers interested in air movements within the boundary layer are referred to this journal as well as to Oke (1978) and Stull (1987) for detailed introductions.

This chapter is devoted to atmospheric structure as it relates to migration by soaring birds. In particular, I focus on updrafts and horizontal wind. Readers are assumed to have some knowledge of meso-/macroscale meteorology.

Structure of the Atmosphere

The boundary layer is a vital, pulsating fluid driven by solar energy and modified by the earth's surface as well as large atmospheric cells (high and low pressure). Although continual structural changes occur on the order of minutes to hours and meters to kilometers, a diel pattern of activity characterizes the boundary layer (fig. 4.1). Rapid and constant changes make the boundary layer difficult to describe and study, especially during daytime. Migrating birds must adjust their behavior in accordance with these changes.

Two flow patterns in the boundary layer are of concern to migrating birds: vertical currents or updrafts and horizontal winds. Vertical currents are especially favorable for soaring birds because they can use them to gain altitude and avoid powered flight. Strong updrafts are ideal for soaring migration because they minimize the time a bird must spend climbing and allow more time for gliding in the appropriate direction for migration. Downdrafts, often associated with up-

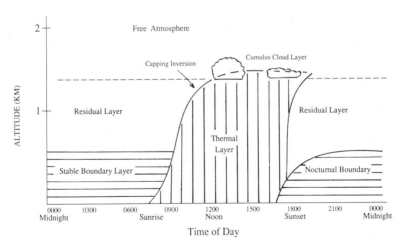

Figure 4.1 Diel pattern of boundary layer development in fair weather over land.

drafts, must be avoided. On days when updrafts are weak or scarce, few raptors migrate except those few that use flapping flight.

Horizontal winds are important because they determine the ground speed of migrants and influence their headings and flight paths over the ground. Wind has also been hypothesized to be an orientation cue for some migrants (Able 1980). Horizontal winds are of additional importance to raptors because they create updrafts when deflected off hillsides, ridges, tree rows, and other structures. Strong horizontal winds that flow in the direction of migration are obviously favorable, whereas opposing winds are unfavorable. Winds perpendicular to the desired flight path may also be unfavorable. Thus, migrants must contend with both vertical and horizontal components of the wind that vary from place to place and moment to moment.

Horizontal Wind

The speed and direction of horizontal wind are of importance to migrants. Not only does horizontal wind determine the track and ground speed of a migrant, but it may be used as an orientation cue similar to the sun, stars, or magnetism (Able 1980). By setting a flight path at a constant angle to the wind a migrant may be able to maintain a constant path over the ground (if wind does not change direction or strength). Horizontal wind, however, is seldom smooth or laminar.

Wind at a given location is influenced by topography, mesoscale weather cells, and turbulence from solar heating. Gauthreaux (1978b) has emphasized the importance of prevailing winds in determining directional strategies of migrants and has published maps of prevailing spring and autumn winds in North America (fig. 4.2). From Gauthreaux's maps it is evident that migrants must pass through changing wind fields during migration. Although the maps yield an overview of wind patterns, a researcher studying migration should learn more about the details of wind patterns near a study site.

The term prevailing wind rarely is defined, but it seems to indicate that wind direction at some time and place is predictable. I interpret this as meaning statistically predictable as opposed to random. Prevailing winds presumably are those that predominate at a given time of year such that the probability of the wind's being in the prevailing direction is high. Some meteorologists use the term "constancy" to describe the predictability of wind at a given location.

To show what I mean by predictability of prevailing wind, I have analyzed wind directions from three locations in North America (tables 4.1 and 4.2). Circular or Rayleigh statistics (Batschelet 1981; Zar 1984) can be used to calculate a mean direction and confidence

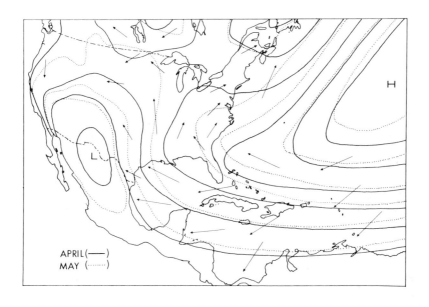

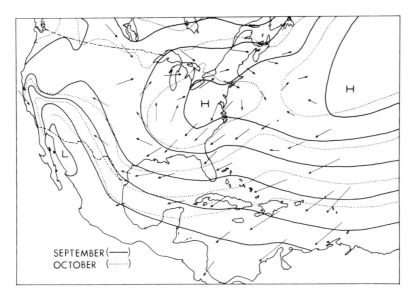

Figure 4.2 Prevailing wind patterns in North America during spring and autumn migration periods. (From Gauthreaux 1978b; courtesy of International Ornithological Congress.)

Table 4.1 Summary of Wind Direction by Month during Spring and Autumn Migration, 1984, at Albany, New York, and Lake Charles, Lousiana

	Wind Direction			
	Surface		900 mb (1,000 m)	
Location and Month	Mean ± 95% Confidence Interval	Length of Mean Vector (r)	Mean ± 95% Confidence Interval	Length of Mean Vector (r)
Albany, NY				
April	311° —[a]	0.22[b]	67° ± —[a]	0.07[b]
May	259° ± 37°	0.41**	282° ± 78°	0.39*
September	248° ± 44°	0.35*	296° ± 21°	0.66**
October	326° ± 75°	0.27[b]	308° ± 32°	0.47**
November	244° ± 37°	0.40**	285° ± 28°	0.53**
Lake Charles, LA				
April	187° ± 40°	0.38*	238° ± 45°	0.37*
May	145° ± 27°	0.54**	175° ± 35°	0.45**
September	67° ± 34°	0.43**	127° ± 29°	0.50**
October	122° ± 18°	0.73**	180° ± 16°	0.79**
November	73° ± 40°	0.38*	258° ± —[a]	0.19[b]

Source: Taken from National Climatic Center data.

Note: Sample size ranged from 28 to 31.

[a]Data are not oriented significantly, so confidence intervals cannot be computed. Mean direction is given even for data that are not oriented significantly.

[b]Not significant ($p > 0.05$).

*$p < 0.05$; **$p < 0.01$ (Rayleigh test).

intervals and to test whether the data show a "meaningful" mean direction or are distributed randomly around a circle (no true mean direction). Monthly mean wind directions from surface and 900 mb levels (about 1,000 m above the ground) are presented with 95% confidence intervals and the lengths of the mean vectors for the three locations. Both 95% confidence intervals and length of mean vectors are indicators of variability of wind direction during that month. If there was no variation in wind direction (wind on all days being from the same direction), the confidence interval would be 0° and the mean vector length would be 1.00. If confidence intervals (e.g., >60°) are large and mean vector lengths (e.g., $r < 0.30$) are small, variation of wind direction within a month is great. Mean vector length ranges from 0.00 to 1.00 and can be interpreted like a correlation coefficient. Large mean vectors indicate that wind direction is predictable or relatively constant.

Winds for most months from the three sites were oriented significantly, and the mean directions were not due to chance (tables 4.1 and

Table 4.2 Summary of Wind Direction during Spring Migration at Brownsville, Texas, 1972–1981 and during the 1982 Migration Season (28 March–16 April)

Month and Altitude	Mean Direction ± 95% Confidence Interval (Wind 1972–1981)	Length of Mean Vector (*r*)	Sample Size	Percentage of Days with Unfavorable Winds
February				
Surface	358° ± —[a]	0.28[b]	10 yr	50
900 mb[c]	358° ± 9°	0.97**	10 yr	21
March				
Surface	323° ± 5°	0.99**	10 yr	32
900 mb	352° ± 5°	0.99**	10 yr	19
April				
Surface	325° ± 5°	0.99**	10 yr	30
900 mb	349° ± 5°	0.99**	10 yr	13
28 March – 16 April 1982				
Surface	309° ± 33°	0.55**	20 days	30
900 mb	333° ± 25°	0.68**	20 days	15

Source: Kerlinger and Gauthreaux 1985a.

[a]Mean direction was not significantly oriented, and 95% confidence interval was not computed. Mean direction is given even for data that were not oriented significantly.

[b]Not significant ($p > 0.05$).

[c]Millibars of mercury, equivalent to a height of 1,000 m.

**$p < 0.01$ (Rayleigh test).

4.2). Exceptions include surface and 900 mb winds in April and surface winds in October at Albany, New York, and 900 mb winds in November at Lake Charles, Louisiana. Mean vector lengths and 95% confidence intervals show that there is a fair amount of variability in the directions of daily winds. West winds predominated at Albany during most months, whereas at Lake Charles winds were often from the southeast (table 4.1). Late March–early April (spring Broad-winged Hawk migration season) winds from Brownsville, Texas (table 4.2), show mean directions from the south-southeast with a variability similar to those at Lake Charles and Albany.

Kerlinger and Gauthreaux (1985a) presented a detailed analysis of wind direction for the spring migration period of Broad-winged Hawks in south Texas. Using surface and upper air winds (from 0600 radiosonde), they showed that winds changed between February and April, becoming more predictable and favorable for migrants after February. They also showed that winds aloft were more predictable than surface winds (table 4.2). Migrants flying before mid to early

March experienced more days with north wind than migrants flying later in the year. This pattern is a result of the establishment of the so-called Bermuda High pressure cell that is extremely influential during spring in the southeastern United States and Central America (Buskirk 1980).

The predictability of wind direction also changes depending on time scale. If monthly mean directions from several years are considered instead of daily directions for one month, much of the variance disappears. When ten years of means from the months of February, March, and April from Brownsville are examined (table 4.2), the length of the mean vector that results is incredibly high ($r = 0.99$). This second-order analysis (calculating means and statistics from means) shows that wind direction is highly predictable from year to year but obscures the daily variance of wind direction. Wind maps published by Gauthreaux (1978b; fig. 4.2) also mask the true variability of wind direction. Individual migrants must deal with daily and moment to moment changes during migration. Variations of wind direction are important for making decisions about flight behavior such as direction, a migration route, and when to fly (day and hour). The important thing to note in all of these analyses is that though wind direction is usually predictable in a statistical sense, there is always variance. Using meteorological data in this manner allows researchers to understand the air birds fly through.

The average motion of horizontal wind is often disguised by turbulence, especially in the daytime. At night vertical air currents are reduced and horizontal wind is more laminar. Near the surface of the earth the flow of horizontal wind is impeded by friction. Obstacles that slow the flow of air create drag, which in turn produces horizontal eddies and turbulent mixing. Vegetation, ridges, and hills create more friction than do prairies or water. Thus, wind speed increases with altitude as friction decreases (fig. 4.3). Because of hills and vegetation, the wind gradient over land is different from that over water (fig. 4.3). Maximum wind speed is reached near the top of the tropopause. Differences of wind speed with altitude are called wind gradients. In regions where air currents of differing speeds or directions come together turbulence is also created, sometimes in the form of wind shears. Both gradients and shears are dangerous for powered aircraft and sailplanes.

Sample wind gradients from the surface to 700 mb are shown in figure 4.3 from two locations in North America. In all locations the wind gradient is greatest between the surface and 950 mb. At Albany wind speed continues to increase to the 900 mb level, whereas at

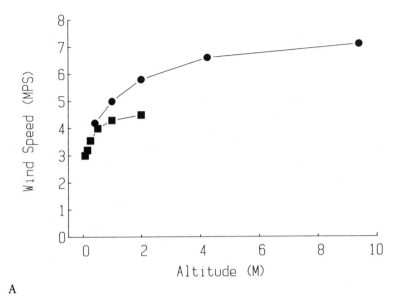

A

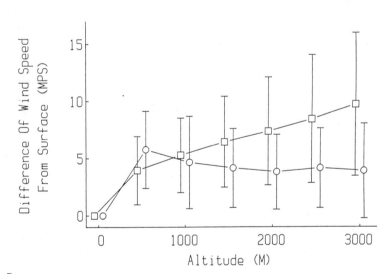

B

Figure 4.3 (A) Sample wind gradients in the first 10 m above the water (squares) and above land (circles). (Drawn from data in Hendriks 1972.) (B) Evening wind gradient (mean ± SD) from the surface to 3,000 m during spring migration at Albany, New York (squares), and Lake Charles, Louisiana (circles). (From Kerlinger and Moore 1989; courtesy of Plenum Publishing Corporation.)

Brownsville and Lake Charles wind strength decreases between the 950 mb and 900 mb levels. There are two explanations for the decline. First, wind direction changes with altitude. Surface winds at Lake Charles have a strong easterly component. At levels higher than 700 mb (approximately 3,000 m) the influence of geostrophic winds from the west is evident and wind becomes stronger. Second, nocturnal jets (rapidly moving sheets of air) often occur within the first 500 m above the surface, and winds above them are actually slower. Thus, real gradients are seldom as smooth as depicted in meteorology texts but have slight nuances that depend on local weather, topography, and synoptic conditions.

Wind speed and direction also differ between day and night. For example, surface wind speed is almost always less at night than in the daytime. The variance in direction of horizontal wind is less at night due to an absence of thermal cells that entrain massive volumes of air. These thermals create ephemeral, small-scale changes in wind direction and speed. The result is a more uniform or laminar flow of air during the night. Surface winds in the morning and at dusk are also not as strong or as turbulent as winds at midday. There is a similar variation in wind speed that is correlated with season of the year. During summer horizontal winds are not as strong as during late autumn and winter, at least at temperate latitudes. Changes of wind direction caused by turbulence and updrafts are a problem for some migrants because they must constantly adjust heading and air speed to maintain a straight course over the ground. For soaring birds this is a problem, but it must be tolerated because they need updrafts to gain altitude. Kerlinger and Moore (1989) have hypothesized that diel timing of bird migration evolved in reponse to vertical and horizontal wind patterns and that many migrants that use powered flight migrate at night to avoid turbulence and warmer temperatures that characterize the daytime boundary layer.

Vertical Wind

Turbulence is something jet passengers do not like to hear about and hate to experience. "Air pockets" and "bumps" are disconcerting and dangerous. One of the reasons jets fly so high is to be in the smooth, laminar air that is characteristic of high altitudes. Turbulence includes a wide variety of effects including updrafts generated by horizontal wind and by thermal sources. To a sailplane pilot or a soaring bird, however, some turbulence is a gift from the gods, or at least, an uplifting experience. Without it the sailplane must land and the soaring bird must land or resort to flapping flight. Therefore a stable atmosphere is desirable for jet travel but is a disaster for soaring flight.

Updrafts are synonymous with an unstable boundary layer that results mostly from thermal convection and mechanical (deflected wind) sources (Oke 1978; Stull 1988). Thermal convection results from differential heating of the earth's surface, whereas mechanical updrafts are a product of the deflection of horizontal wind by discontinuities on the earth's surface such as hills, ridges, trees, and sometimes clouds. This type of updraft is variously referred to as wind lift, ridge lift, orographic lift, or declivity currents. Although soaring birds use both thermal and orographic updrafts during migration, thermal convection is used more often.

Thermal convection has been described numerous times in the literature on hawk migration and soaring. A thermal forms at the earth's surface after solar energy has been absorbed. A result of this heating is a nonuniform distribution of surface temperatures. As the air above the earth becomes heated, temperature (ΔT), and pressure (ΔP) gradients are formed such that the warm air rises and is replaced by cooler air (fig. 4.4). The result is a warm plume or column that grows to hundreds or even thousands of meters in height. In earlier descriptions of the atmosphere, ornithologists have depicted thermals as a vortex (like a donut) instead of a column (e.g., Heintzelman 1975).

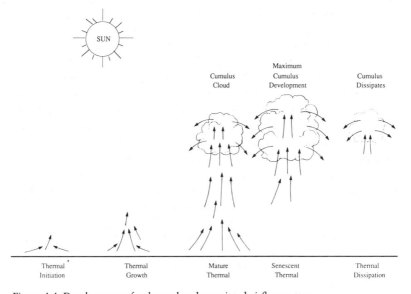

Figure 4.4 Development of a thermal and associated airflow pattern.

While thermals are still "attached" to the ground the columns lean in the direction of the wind, eventually breaking from the surface and moving with the ambient wind. The thermal dissipates when temperature and pressure gradients are zero in a layer called the "capping inversion" (Stull 1987; fig. 4.1). This type of thermal, called a type I thermal by Hardy and Ottersten (1969), is thought to be the basic convective unit of the boundary layer (Oke 1978).

The development and strength of thermals in the convective mixed layer (fig. 4.1) are determined by several factors, the most important being input of solar energy. The amount of solar energy that is transformed into a thermal is related to aspect (i.e., north vs. south sides of a hill), horizontal wind (mixing), cloud cover, geographic location, and type of surface that is absorbing (albedo) and reflecting (emissivity) solar radiation. Albedo and emissivity values of different vegetation types, snow, ice, water, sand, and other materials vary greatly (Oke 1978). Briefly, flat surfaces are less conducive to thermal formation than hilly surfaces; water is less conducive than land; and ice and snow are less conducive than bare earth. Oke (1978) and Stull (1988) present thorough reviews of the principles involved in thermal formation.

Thermal convection within the boundary layer usually follows a predictable diel pattern (figs. 4.1 and 4.5). In the early morning near sunrise, surface temperatures are at their minimum after emission of long-wave radiation to the atmosphere (radiative cooling) during the night. If the air above the ground is warmer than the ground, as often happens at night in the first 100 m, a radiative inversion is produced, forming a stable or nocturnal boundary layer (Stull 1988). The downward flux of heat also results in the morning inversion. After sunrise, air near the surface remains warmer than ground temperatures and remains stable. In other words, thermals do not form. Depending on cloud cover, aspect, surface type, surface wetness, wind, and some other factors, surface temperature rises during the morning. When surface temperatures exceed air temperatures, the boundary layer becomes unstable and thermals form. If ground temperatures do not exceed air temperatures or if air temperatures do not exceed the adiabatic lapse rate with altitude, an inversion will maintain a stable boundary layer or limit the altitude to which thermals rise. The adiabatic cooling rate (adiabatic lapse rate) of air is the rate at which air cools with increasing altitude without input of heat. Stull (1987) and Oke (1978) present comprehensive explanations of atmospheric stability, thermal formation, turbulence, and adiabatic cooling.

On most days ground temperatures rise from early morning

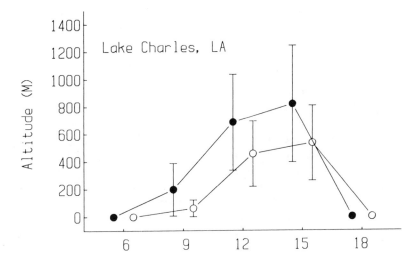

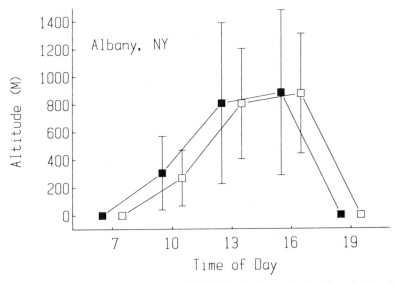

Figure 4.5 Daily pattern of convective field height during spring (solid symbols) and autumn (open symbols) migration seasons at two sites in North America as determined from morning soundings (see Kerlinger and Moore 1989; courtesy of Plenum Publishing Corporation). Hourly mean altitudes ± SD are plotted for all days during spring and autumn migration 1984. Convective field height was determined as outlined in text.

through midday. As ground temperatures rise thermal convection increases, as does the strength of vertical updrafts within the thermals. In temperate latitudes convective fields are well developed by midmorning (ca. 3–5 h after sunrise) and continue until the sun ceases to heat the surface layer.

The patterns of convective depth shown in figure 4.5 were not determined by measurement of the atmosphere. Instead, the height of convective activity was estimated using a method employed by atmospheric scientists and soaring pilots (World Meteorological Organization 1978, herafter WMO; Cipriano and Kerlinger 1985). The method uses the vertical temperature profile of the boundary layer and ground temperatures. The temperature profile was measured by radiosonde at about 0600–0700 h, from the surface to more than 3,000 m above the ground. Surface temperatures were measured every three hours at the same site. Using these two parameters, the height to which a convective element rises is determined as follows (refer to fig. 4.6):

1. Acquire Skew-t, log-P graph paper.

2. Plot the vertical temperature profile at different heights above the ground using pressure and temperature data as reported by radiosonde (from WBAN-33—Summary of Constant Pressure Data, U.S. Department of Commerce, Environmental Data Service, National Climatological Center, Asheville, North Carolina).

3. Plot hourly surface temperatures along the x-axis (horizontal or t-axis.

4. From each hourly surface temperature trace up a dry adiabatic line (the lesser of the two slopes that run from lower right to upper left on the graph paper) to the sounding temperature profile line.

5. From the point where the two lines intersect, move to the y-axis (vertical axis is altitude and pressure) on the right side of the page.

6. The value on the y-axis is the altitude to which thermals can rise, or the convective depth at that time of day.

This method is presented in detail in the WMO soaring handbook (1978), Cipriano and Kerlinger (1985), and meteorology texts.

Two profiles are plotted in figure 4.6 representing a "bad" day (profile B) and a good day (profile G) for soaring. On profile B an inversion was present and the atmosphere was stable; thermals were scarce and weak and did not ascend to great heights. Profile G, on the other hand, represents a good day for soaring because the morning inversion is shallow to nonexistent, with temperature decreasing rap-

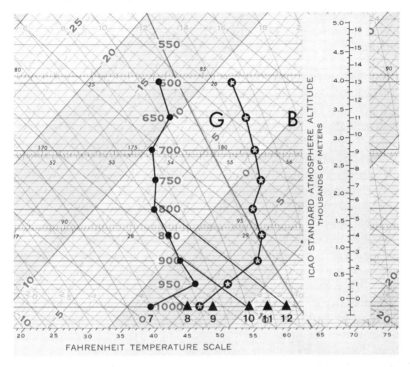

Figure 4.6 Skew-*t*, log-*P* plots showing temperature profile at dawn from the surface up to 600 mb (about 4,000 m). Plot G is for a day with good thermal development, whereas plot B is for a day with poor thermal development (an inversion below 900 mb, about 1,000 m; redrawn after Cipriano and Kerlinger 1985, courtesy of the Hawk Migration Association of North America).

idly with altitude. Thermals on this day would be abundant, strong, and large, making soaring conditions excellent.

A less complex and less precise method of determining the quality of soaring conditions involves "eyeballing" weather maps. The method described below is most useful during autumn and spring in the eastern part of the United States and Canada. Because prime soaring conditions depend on cool air above warm surface air, soaring will be best at the north, east, and center of a high pressure cell. These conditions prevail after the passage of a cold front. Contrary to popular belief, brisk winds that follow a cold front do not always inhibit thermal formation. These winds can promote convective activity. Soaring conditions are poorest in low pressure areas because cloud cover and precipitation inhibit heating of the surface layer. Along the western and southern edges of high pressure cells thermals do occur, but they are less powerful because of the input of warm air from the south and east. These warm winds create more stable conditions, including inversions.

Classification of Convective Elements

Although the principles of thermal formation as presented above apply for most thermal convection, it is impossible to characterize a "typical" thermal. The statement that "the isolated thermal is the ultimate form of all convective elements in the atmosphere" (Woodward 1949) is somewhat misleading. Until this time I have used a typological approach to describe thermals, but thermal elements vary in size, shape, strength, and duration. They also vary with geography, topography, time of day, altitude, and meteorological conditions.

Table 4.3 lists several types of convective elements that have been described in the atmospheric literature. In addition, I have attempted to rate them in terms of how likely they are to be used by hawks (and other soaring birds) during migration. My assessments were based on how each type rises, the strength of updrafts, and the frequency and dependability of occurrence. It is difficult to determine which types are used by hawks and how often because thermals cannot be seen, nor can they be measured with ease. These types of thermals have been described by soaring pilots who "feel" their way through the boundary layer, powered plane pilots, atmospheric scientists using various instruments, and meteorologists who have watched clouds, dust devils, thunderstorms, and other visible phenomena. In addition, theoretical models provide considerable insight into the structure of thermal convection.

All of the types of thermal convection listed in table 4.3 are probably used by soaring migrants at one time or another. I suspect that type I thermals are used more by soaring migrants than any other updrafts. Type I thermals include those capped by cumulus clouds and those that are not (see table 4.3 for references). Henceforth, they will both be referred to as type I thermals. There are several reasons they are used by migrants more than other types of thermal convection. Type I thermals are common and occur almost daily. They are also large and provide the strong lift necessary for gaining altitude quickly. These are the types of lift most often used by sailplane pilots as well.

The other types of thermal convection listed in table 4.3 are not really suited for soaring migration. Thunderstorms with associated cumulonimbus clouds are too violent to be used. Updrafts within thunderstorms may preclude controlled flight by birds. Birds that enter thunderstorms may be carried to incredible heights (ca. 5–10 km),

Table 4.3 Summary of the Various Types of Convective Elements and Their Frequency of Use by Migrating Hawks

Types of Convective Elements	Frequency of Use by Migrating Hawks	Reference[a]
Type 1 thermals (with or without cumulus clouds)	Most frequent	Warner and Telford 1967[b]; Hardy and Ottersten 1969[c]; Hardy and Katz 1969[c]; Konrad 1970[c]
Mesoscale cellular convection (linear thermal arrays, type II thermals, organized convection, etc.)	Frequent (?)	Hardy and Ottersten 1969[c]; Konrad 1968[c]
Dust devils	Infrequent	Kaimal and Businger 1970[d]
Sea-breeze updrafts	Locally common	World Meteorological Organization 1978
Thunderstorms (cumulonimbus clouds)	Rare/never	Oke 1978
Thermals over water	Infrequent/rare	Konrad and Kropfli 1968[c]; Woodcock 1975[e]
Lee waves and mountain waves (lenticular or wave clouds)	Uncommon (?)	Scorer 1978

[a]References are to the meteorological literature; the assessments of use are my own.
[b]Measurements made with aircraft.
[c]Measurements made with radar.
[d]Measurements made with wind instruments from a tower.
[e]Measurements made by watching gulls soar over the ocean.

experiencing below-freezing temperatures, precipitation, and low oxygen partial pressure. Downdrafts associated with strong convective activity have been responsible for several jet crashes in recent years, attesting to their power.

Smith (1980) has reported that some buteos use thunderstorms during migration in Central America. It is not clear if this behavior was intentional or if the birds were using updrafts associated with the edge of the thunderstorm. Smith also suggests that Broad-winged Hawks could use convective currents associated with thunderstorms to make long distance water crossings. By soaring to great heights within a thunderstorm, Smith suggests Broad-winged Hawks may be able to glide much of the way from Florida to Cuba and from Cuba to Yucatán (and back). It is doubtful that selection favors this behavior. Recently Smith (1985b) recanted his ideas about these migrants using thunderstorms.

Finally, thunderstorms are too rare to be relied upon, as is the case with dust devils. Besides being rare, dust devils are small and short-lived and do not rise to great heights. Thus, they are unsuitable for migrants.

The remaining types of thermal convection are suitable for migration when and if they occur. Special conditions are necessary for usable thermals to form over water (Woodcock 1975), although in the trade zones weak thermals with updrafts of 0.5–2 mps (meters per second) form regularly (Malkus 1953, 1956). These weak trade zone thermals are capped by small cumulus clouds at heights of about 600 m. At temperate latitudes conditions promoting the formation of thermals are relatively rare. For thermals to form over water, wind speed, air temperature, and water temperature must all be within a narrow range (Woodcock 1975). Because water has a high heat capacity compared with air, it is difficult to heat a body of water enough to produce thermals. To raise the temperature of water a certain amount takes three times as much energy as for the same volume of soil. (This also explains why it takes so long for thermals to form over ground that is damp from rain or dew.) The maximum efficient production of thermals over water occurs when winds are about 5 mps and when the water temperature is at least 1–2°C greater than the air temperature. With winds 2–5 mps, thermals can occur if the minimum difference between water and air temperature is inversely related to wind speed. This relation is not entirely linear, however. With wind speed less than 1 mps, a temperature difference of 3–6°C was needed to "induce" thermals (Woodcock 1975; but see Agee and Sheu 1978). These findings were a result of studies of the soaring behavior of gulls over the ocean done since the 1940s (Woodcock 1940; Woodcock and Wyman 1947). Several other factors influence thermal formation over water: evaporative cooling, mixing of the water by wind and waves, diffusion of energy through a great volume of water, and great penetration and absorption of solar radiation. The formation of thermals over water is often seasonal. Thermals form over water most often in late autumn and early winter in the North Atlantic Ocean.

Sea-breeze updrafts and convergence lines are common but occur only along coastlines when temperature, wind, and sky cover are correct (Pielke 1974, 1984; Lyons 1972; WMO 1978). These thermal elements are linear and parallel to a coastline, unlike type I thermals, which are either circular or elliptical in horizontal section. Conditions necessary for the production of a sea-breeze updraft are: higher temperatures over land than over the adjacent water, a light wind flowing from land toward the water (not always necessary), and an unstable air mass over land (thermal conditions). The distance from the shoreline at which sea-breeze updrafts occur varies with weather and topography. These types of updrafts may be quite common along the shores of the Great Lakes (Haugh 1974; Lyons 1972) and the Atlantic coast (Pielke 1984). During autumn light northwest-to-west winds favor the formation of strong sea-breeze updrafts along the east coast of the United States at locations such as Cape May Point, New Jersey. The sea-breeze updrafts at Cape May are enhanced by airflow from Delaware Bay and the Atlantic Ocean forming a linear convergence line in the peninsula. At such times they pass relatively unnoticed by counters. Thus, at peninsular locations like Cape May, sea-breeze updrafts can form with onshore or offshore breezes or no breeze at all.

The formation of linear thermal arrays (sometimes called thermal streets, cloud streets, and mesoscale cellular convection) are also dependent on special atmospheric conditions (Scorer 1978). Linear thermals are simply type I thermals aligned with the wind in parallel rows. These may or may not be capped with cumulus cloud, depending on the amount of water in the air and the condensation point. Satellite photographs show that organized convective systems extend up to 500–800 km in length and cover hundreds of square kilometers (Kuettner and Soules 1966). Cloud streets can be seen from the ground, but are more easily observed from an aircraft flying at 10,000–13,000 m. For linear arrays of thermal to form, wind direction must be constant with height, at least up to the limit of convection; wind speed should increase with height above the ground but should not exceed about 10–11 mps at the top of the convective layer; and the convective layer should be limited to 1,500–2,000 m above the ground (Konrad 1968; WMO 1978; Scorer 1978).

Convective streets are spaced by a factor of 2.0 to 2.8 times the height of the convective depth (Konrad 1968). For example, if thermals rise to 1,200 m, they will be spaced by about 2,400–3,400 m. In some arrays thermals are no longer distinct but are continuous roll vortices. Soaring migrants could use convective streets to make long glides without resorting to soaring or powered flight if thermals were aligned in the appropriate direction for migration. In the United States and Canada thermal streets are most often aligned from northwest to southeast or from west to east—the direction of prevailing winds. In North America these arrays form most often in the northern portion of high pressure cells behind cold fronts. Because thermal streets are not often aligned in a direction appropriate for migration, it is doubtful that hawks engage in long distance glides in these convective units (but see Haugh 1974). Instead, it is likely that they are used similar to the way that type I convective cells are used. In this way, raptors will not leave their preferred migratory flight paths.

Development and Properties of the Convective Field

As I describe quantitatively the development and structural properties of type I thermal convection, readers will become aware of the complexity of thermals and appreciate how difficult it is for a migrant or a sailplane pilot to use or study them. The properties of thermals that are of interest to students of hawk migration are time of onset, diameter, maximum vertical extent, maximum altitudes, vertical velocities of updrafts, ascent rate of entire thermals, rate of growth and longevity, and general shape. Also of interest are variation of convective fields at different times of year and geographic locations.

Konrad (1970) proposed three stages in the "lifetime" or life history of a thermal: initial formation, growth, and dissipation. Although these are not discrete categories, they help us study and understand the entire process. During the initial formation stage, thermals are generally small, do not extend very high, and are continuous from the ground upward. In addition, they are not strictly vertical but are tilted in the direction of the wind in which they are formed. Several studies have tracked the growth of the convective field. Rowland (1973), working with radar and aircraft in Virginia, found that between 0635 and 1020 h the altitude of convective activity increased at a linear rate of 3.8 mpm (meters per minute). Thermals had reached 1,000 m above the ground by the end of this period. Between 0700 and 0800 h, Konrad (1970), also using radar, found a vertical growth rate of the convective layer of about 2.0 mpm up to about 500 m. Later in the morning (1000–1130 h) rates increased to 6.1 mpm (Rowland 1973) and 4.5–6.0 mpm (Konrad 1970; Konrad and Robison 1973), with thermals extending to more than 1,600 m. Lidar (a laser technology that yields radarlike images) also reveals similar rates of development of the convective mixed layer (Stull and Eloranta 1984; Wilde, Stull, and Eloranta 1985). Measurements by Hardy and Ottersten (1969) for this same time period were greater, ranging from 0.5 to 1.5 mps. No explanation was given for this large discrepancy.

The earliest thermals (before 0900 h) detected with radar measured 100–200 m in diameter (Konrad 1970; Konrad and Robison 1973). Smaller thermals are present as early as 0700 h, but they are difficult for the radar to resolve. Konrad and Kropfli (1968) discuss the "inherent difficulty in measuring thermal size, shape, and spacing" from aircraft and towers but do not discuss the possibility that small thermals are not detectable with radar. Because detection of thermals with radar depends on differences in the diffractive index of air within and around the thermal, it is possible that small scale differences in diffractive index that occur on the order of a few meters may go undetected by the radars used by atmospheric scientists.

From morning through midday the diameter of thermals increases as the convective mixed layer becomes deeper. These thermals yield distinct radar and lidar echoes (fig. 4.7). By noon the diameter of type I convective cells can exceed 1,000 m (Hardy and Ottersten 1969; Konrad 1970; Konrad and Brennan 1971) and extend more than 1,500 m above the ground (Konrad and Robison 1973; Rowland 1973). Warner and Telford (1967) report thermals 200–400 m in diameter to about 900 m above the ground, but they never observed thermals more than 1,000 m in diameter. All of these studies were conducted at temperate latitudes. Thermals do not always grow to be this large or ascend to these heights at midday. As was shown earlier in this chapter, inversions inhibit the formation and growth of thermals. Hall, Edinger, and Neff (1975), working in Colorado grassland, found that during an August inversion the convective layer reached only 600 m at 1300 h. In Australia the mean diameter of thermals was found, in one case, to be only 16 ± 27 m (SD), but that was only for thermals within a few hundred meters of the ground (Manton 1977). Again, smaller and less powerful thermals are difficult to detect with radar.

The depth of convective activity also varies with time of year. Because thermal activity is dependent upon an accumulation of surplus

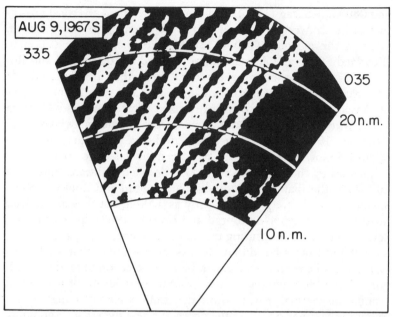

AUG 9,1967 S

335

035

20 n.m.

10 n.m.

A

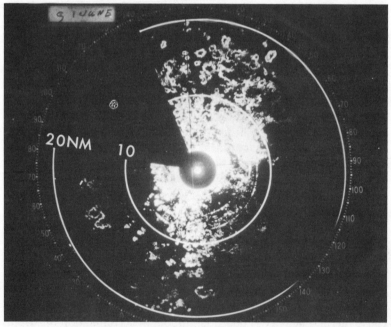

B

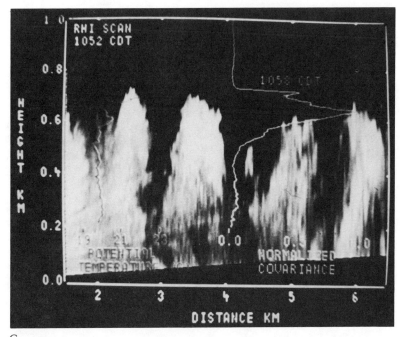

C

Figure 4.7 (A) Drawing of a high resolution radar screen (plan position indicator) showing aligned, individual convective elements (white, roughly circular structures; from Hardy and Ottersten 1969, courtesy of American Meteorological Society). (B) Photograph of radar screen showing individual convective units (from Hardy and Katz 1969; © 1969 IEEE). (C) Photographs of a lidar screen (RHI [range height indicator] on left) showing vertical thermal structure (courtesy of E. Eloranta).

energy (fig. 4.4; input exceeding output), factors associated with seasonality and the sun are important in thermal formation. These include the length of the daylight period, the angle of the sun above the horizon, and the material covering the surface of the earth (snow, grass, bare dirt, etc.). At temperate latitudes winter tends to be poor for thermals; days are short, the sun does not rise very high, humidity is lower, winds are stronger, and the surface is frequently covered with snow, ice, or water. Figure 4.8 shows that thermals are not as prevalent in winter as in summer and probably not as powerful. The method for determining convective depths in figure 4.8 is not sensitive to several factors (such as surface type) that influence convection, so convective depth during winter may be overestimated. Thermal soaring at this time of the year is poor. From November through early

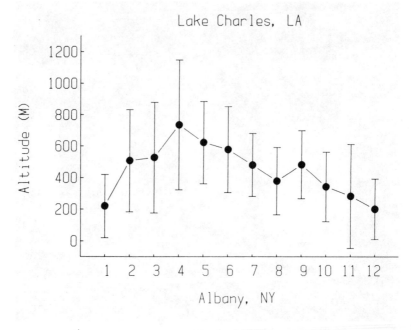

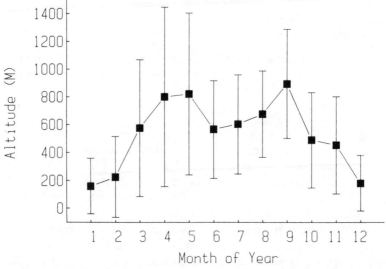

Figure 4.8 Pattern of convective field height throughout 1984 (monthly means ± SD; from Kerlinger and Moore 1989, courtesy of Plenum Publishing Corporation). Convective field height was calculated according to methods given in the text. The large variance is a result of the inclusion of all days (monthly Ns = 28–31).

March cumulus clouds formed by strong, moist thermals are absent from most northern locales, indicating that strong convective activity is lacking. Thus, during winter the atmosphere is more stable than during the rest of the year.

As days get longer in late winter and spring at northerly latitudes, thermals form more often and are more powerful (fig. 4.8). By late April into June thermals are strong and abundant. At temperate latitudes this is the best time of year for soaring flight. Thermals remain strong and abundant through the summer, although there is some reduction in activity. By late October and into November thermal activity abates as days become shorter and winds stronger. Strong thermal activity is not uncommon through October but is restricted to fewer days. As we will see in chapter 8, raptors migrating at some temperate latitudes can soar in thermals into November without problems.

Some thermals during autumn and spring are extremely powerful because masses of cool air pass over warm ground. For example, after the passage of a cold front in September or May the ground can be very warm as a result of clear skies. The large difference between surface and upper air temperatures is ideal for the formation of strong thermals. These are often clear air thermals (not capped by cumulus clouds) because they form in dry air. These conditions promote high altitude soaring flight by migrating hawks, making them difficult to see.

In and near the tropics thermals occur year-round because the daylight period does not fluctuate as greatly as at temperate latitudes. Furthermore, the sun remains high in the sky and the ground is relatively free from ice and snow. Differences in convective activity result from differences in humidity and cloud cover and between wet and dry seasons. These variations have not been studied intensively, and raptors would be ideal subjects for studying the convective process in tropical climates.

The shape of convective units has been debated for years. Two forms of convective elements have been proposed: a plume that is continuous from the ground upward, and a free-floating or isolated vortex. Both are partially correct, although the plume-column hypothesis is now in favor. Cone (1962b) presented a schematic figure of a free vortex ring (like a donut) that was shallow in vertical profile. This diagram has been cited in the hawk migration literature by several authors (Heintzelman 1975; Haugh 1972). Most, if not all, thermal convection starts as a plume or column near the surface of the earth. As shown in radar and lidar photographs (fig. 4.7), the plume can extend upward more than 1,000 m. Woodward (1949) states that the plume often leaves the ground as a "column or intermittent plume" and after leaving the ground ceases rotation around its vertical axis. Alternatively, the plume may, at times, break free from the earth before it has developed vertically. At this time the thermal would resemble Cone's classic figure. Intermediate forms are possible, but it is likely that the thermals used by migrating hawks most often are plumes (fig. 4.4). One should remember that the shape of thermals changes constantly from early development until senescence.

The pattern of airflow within thermals has not been investigated intensively, although some data and theoretical findings exist. The pattern is for air to rise in the center of a thermal and descend around the perimeter (fig. 4.4). Air drawn downward from the capping layer or free atmosphere (fig. 4.1) is called entrainment. Woodward (1949) presents evidence from laboratory studies that the isolated thermal is a vortex ring with air at the center rising at about twice the rate of the thermal cap. Radar and lidar photographs show that thermals are domed or an inverted U-shape in vertical section (Hardy and Ottersten 1969; Konrad 1970; Konrad and Robison 1973; Rowland 1973; fig. 4.7). This model of thermal structure is referred to as the "top hat" model.

The absolute height of thermals (distance from top to bottom of the plume) varies as the convective depth develops. Early in the morning thermals are less than 100 m, whereas by midday they can be upward of 1,000 m. Measuring the height of a plume is not easy. Cipriano (1975) investigated the problem by having two sailplanes fly in the same thermals during the midday period. Radio contact between the sailplanes enabled Cipriano to show that the vertical extent to which many thermals were soarable was nearly 600 m ($\bar{x} = 586 \pm 197$ m, SD, $N = 9$). These thermals were ascending in a convective field that was 1,600 m in depth. The vertical extent of the thermals exceeded 600 m, though they did not provide adequate lift for sailplane use.

Kerlinger, Bingman, and Able (1985) tracked hawks in thermals and also found that thermals extend vertically for several hundred meters. Dozens of hawks were tracked while they climbed more than 700 m, and a few climbed more than 1,000 m. This indicates that the thermals were regularly more than 400 m in vertical profile.

The rate of airflow in the center varies with size of the thermal. The best information regarding vertical ascent rate of air within thermals comes from different sources. Cipriano (1975) used a sailplane equipped with precise variometer and thermometer to study the airflow and termperature profiles within soarable thermals in summer in central New York. (A variometer measures changes in altitude.) Kerlinger, Bingman, and Able (1985 and unpublished data) used a tracking radar to measure the vertical ascent rate of hawks soaring in thermals during spring and autumn in central New York. Subtraction of the sink rate of the sailplane or hawk from the ascent rate yields the rate of vertical airflow within the thermal. By combining the data from Cipriano and from Kerlinger and his colleagues we are afforded a unique picture of the vertical flow of air within a thermal.

In figure 4.9 the vertical ascent rate of air within thermals is presented along with the turning envelope (diameter of circling) of the sailplane and hawks. Two things should be evident. First, the turning envelope of sailplanes is more than two times that of a hawk, and second, the air in which hawks soar has a greater vertical velocity. The difference in turning envelopes is correlated with the difference in size and weight between the sailplane and hawk. Large aircraft cannot turn in as small an area as smaller aircraft. Because of the disparity between turning envelopes and vertical velocities of air within thermals, I conclude that there is a gradient of vertical velocities that extends from the center or core of the thermal to the periphery. Although

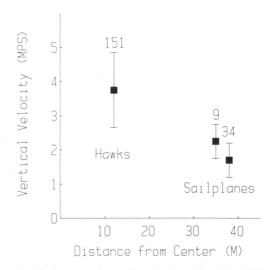

Figure 4.9 Vertical ascent rate (mean ± SD, Ns given above bars) of air near the center of soarable thermals. (Sailplane measurements calculated from Cipriano 1975; hawks from radar tracks of hawks from Kerlinger 1982a.) One meter per second was added to climb rates of sailplane and hawks to approximate their sink rate near minimum sink with a 30° bank angle.

thermals may be dome shaped (the top hat model) in vertical structure the distribution of the strengths of vertical air currents horizontally across a thermal resembles a normal curve or a distribution with a peak at the center. This is not an earthshaking conclusion; sailplane pilots have been painfully aware of it for years. I have soared in thermals and watched vultures climb "through" the wide circles in which the motor glider soared. This experience is frustrating to sailplane pilots because it makes them realize that stronger lift is available, but they cannot use it.

The vertical velocities shown in figure 4.9 overlap with published reports (Warner and Telford 1967; Hall, Edinger, and Neff 1975) but are somewhat greater than some others (Manton 1977). It is likely that the measurements made from planes may underestimate the vertical velocities at the core of thermals because it is difficult to locate the small cores. The data in figure 4.9 should be interpreted with care because they may not be representative of thermals in general. The following points should be considered: the sailplane and hawk data came from different thermals and times of year; the data were collected only from thermals that were soarable—raptors and sailplane pilots may ignore weak or small thermals; the data came mostly from midday; and the data are mostly from thermals that have reached altitudes greater than 300–400 m and do not include smaller thermals that never reach these altitudes. It is safe to say that most thermals used by raptors migrating in the midday period have vertical velocities ranging between 1 and 4 mps.

The life expectancy of thermals also varies greatly. Hardy and Ottersten (1969) report that large type I thermals (>1,000 m in diameter) have a 20–30 min life as "seen" by radar. Larger convective elements (mesoscale cellular convection or type II thermals) may last 4 h. Cumulus clouds are known to last for 5–20 min or more (Woodward 1949), so the convective cells that spawned them were undoubtedly longer-lived. Smaller type I cells may last only 5–10 min or less. It is likely that most type I convective cells exist somewhat longer than these examples because cumulus clouds and type I thermals represent only a portion of the life of a convective element.

Wind-Generated Updrafts: Orographic or Ridge Lift

Although thermal convection is used more than other updrafts, migrating raptors also use wind-generated updrafts, sometimes referred to as ridge or orographic lift. These currents are created by the upward deflection of wind (fig. 4.10) off topographic irregularities such as mountain ridges, hills, trees, bridges, buildings, ocean waves, boats, or any other discontinuity in the earth's surface. Migrating raptors use ridge lift in two ways: they soar in circles in these currents gaining altitude and then glide in the appropriate migratory direction, or they glide continuously through them gaining little or no altitude. When declitivy currents are isolated, such as those created by hills and tree rows, raptors cannot glide in them for long distances. Instead they soar upward until they reach a point where updrafts are no longer profitable, at which time they glide on.

Updrafts that form along mountain ridges are a better source of lift for migrating hawks than those at isolated hills or tree rows. Ridges such as the Kittatinny in eastern North America provide long, unbroken sources of lift in which hawks glide for several hundred kilometers. The Kittatinny Ridge has been called a "highway" for hawks (and sailplanes) because there are almost always updrafts present and because it is oriented in a direction (250°–260°) that is not inappropriate for autumn migration (Broun 1935, 1949). Prevailing west to north winds in autumn are deflected upward by this ridge, providing

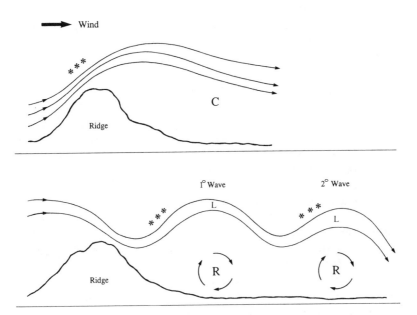

Figure 4.10 Schematic representation of ridge lift and wave lift. Declivity currents are generated by deflection of horizontal wind by obstacles (C = cavity, R = rotor, areas of turbulence to be avoided when ridge gliding; L = lenticular or lens clouds that form downwind from ridges where waves originate; asterisks indicate region that provides best lift for gliding or soaring.

a linear path of updrafts for hundreds of kilometers. In autumn a hawk can begin gliding on the Kittatinny Ridge in New York, just north of New Paltz (where the ridge is called the Shawangunk Mountains) and continue for several hundred kilometers until it has passed Hawk Mountain in southern Pennsylvania. The ridge becomes discontinuous southwest of Hawk Mountain, where fewer hawks use it. In this length of ridge more hawks are counted than on most other ridges in the world. The Kittatinny Ridge is unusual, however, in that there are very few ridges that are oriented appropriately for migration and few that extend for such long distances.

Other ridges that are known to be used by migrating raptors include several that are parallel to the Kittatinny Ridge, several in the western United States and Canada that are part of the Rocky Mountains or associated mountain ranges (Dekker 1970; Hoffman 1985; Hoffman and Potts 1985), the mountains associated with the Rift Valley in Africa (Brown 1970), and mountains in the Middle East (Christensen et al. 1981).

Ridge lift can be strong, as is evident from watching raptors glide and soar along ridges. Vertical air currents are frequently in excess of 3–4 mps, an estimate based on known sink rates of gliding raptors (chap. 6). Empirical studies of the strength of updrafts along various ridges would be helpful for evaluating their usefulness to migrating raptors. I have watched raptors glide for long distances (>1–2 km) without flapping or changing altitude.

Updrafts along ridges are often turbulent and difficult to use. In addition, downdrafts sometimes form downwind from ridges and mountains and vary in placement depending upon topography and wind. When sailplane pilots experience downdrafts at such low altitudes, they have little time to recover and may crash. Birds may be able to recover after flying into a downdraft because they have the ability to resume powered flight into a forest if necessary. They can also steer around trees or land if necessary, two luxuries that sailplane pilots sometimes wish they had!

Updrafts along ridges probably do not rise to the same height as thermals except at rare times and places. Early in the morning, ridge lift can extend to greater heights than thermals, which may not be present. The reason ridge lift does not exceed the altitude of thermals is that winds aloft are stronger than those at the surface (wind gradient) and the air that is deflected upward near the top of a ridge is not as strong as the horizontal wind at 100 m or more above the ridge. The faster air above tends to truncate ridge updrafts. Changes in wind direction (shear) with altitude tend to distort and disrupt these updrafts (Malkus 1953). Sailplane pilots have informed me that ridge lift seldom exceeds 300–400 m above a ridge. Most sailplanes that I have watched gliding in ridge lift have been within 200 m of the ridge, and some were below the top of the ridge. Thus, migrants flying in updrafts generated by wind along ridges and hills are normally constrained to fly at lower altitudes than migrants using thermals at midday.

In some topographic situations a rare updraft is formed that allows sailplanes to attain much higher altitudes than is possible with ridge lift. These updrafts, called lee waves, mountain waves, or standing waves (fig. 4.10), sometimes extend above 10,000 m, although they reach 2,000–3,000 m more frequently. Lee waves can be recognized by the formation of wave or lenticular clouds that form downwind from a ridge or mountain. The formation of a lee wave is dictated by a precise set of wind and temperature factors (Scorer 1978; WMO 1978). If these conditions are not met precisely, a wave will not form.

To reach a lee wave, glider pilots can either be towed (by powered plane) or use strong thermals (Harrison 1971; Kuettner 1972; Santilli 1972). I have sailed to nearly 3,000 m in a wave near Mount Greylock in the Berkshire Mountains of western Massachusetts. We reached the wave after climbing to about 1,500 m in a very strong thermal at 3–4 mps. After entering the wave we continued to climb at more than 3 mps—excellent conditions for soaring. Lift in lee waves ranges from 1–2 mps for smaller mountains to nearly 10 mps (up into the stratosphere) for larger mountains such as the Rockies. The lift in lee waves is the most powerful lift that can be used safely by a sailplane pilot or a bird. So great is the lure of waves that sailplane clubs locate their activities according to where waves occur. It is likely that raptors seldom use lee waves because they are difficult to reach, form infrequently, and are difficult to use. Nevertheless, Evans and Lathbury (1973) report that lee waves may help raptors migrate across the Strait of Gibraltar. Smith (1985b) reports that Broad-winged Hawks migrating through Panama glide extensively in wave lift.

Summary and Conclusions

The portion of the atmosphere in which raptors migrate is dynamic and constantly changing. Diel changes of temperature, wind, and updrafts are characteristic of the boundary layer. Horizontal winds are important to migrants because they determine ground speed and influence the direction and altitude of flight. Horizontal winds are more variable in direction during daytime and increase in strength with height (wind gradient). Updrafts are created by thermal and mechanical sources. Thermal updrafts such as type I thermals, dust devils, mesoscale convective cells (organized thermals, i.e., linear thermal streets and type II convection), thunderstorms, sea-breeze updrafts, and thermals over water are produced by differential heating of the air near the earth's surface. Migrating hawks use type I thermals more than other forms of updrafts because they are the most abundant and easiest to use. Type I thermals vary in size and shape, although the "typical" thermal used by migrants is an isolated plume column (and later a free thermal) that is from less than 100 m to nearly 1,000 m in diameter; extends upward more than 1,000 m; is characterized by vertical air currents ranging from 1 to 4 mps; rises more than 1,500 m above the earth; and varies in size and shape with time of day and season of the year. Ridges provide updrafts (wind or thermal sources) for raptors but are not frequently aligned in a direction that is appropriate for migration. The extreme variability of horizontal and vertical winds gives the researcher a clue to the complexity of decision making by raptors during migration. Successful migrants must adjust their behavior constantly in accordance with changing atmospheric conditions. Studies of flight behavior must focus on how migrants deal with these constantly varying conditions.

5

Flight Mechanics: Theory

Before we examine the flight behavior of migrating raptors, readers should understand the mechanics of avian flight—the aerodynamics and energetics of flight as they relate to morphology. This chapter presents an overview of these topics; for a comprehensive treatment of flight mechanics, see Pennycuick (1969, 1972b, 1975)[1] and consult Rayner's (1985a) excellent bibliography on vertebrate flight. There has been a proliferation of papers regarding bird flight in the past few years. Much of the recent work is theoretical (Rayner 1979a, 1979b, 1979c, 1985b; Ward-Smith 1984), although Torre-Bueno (1976, 1978), Withers (1979, 1981), and Marsh and Storer (1981), have made significant empirical contributions to our understanding of avian flight.

The chapter begins with a summary of the various types of flight birds utilize during migration and then examines flight morphology. After defining and describing the morphological attributes that are important for flight, I review the theory of flight energetics and gliding performance.

Types of Flight

Gliding and powered flight are the basic "types" of avian flight (fig. 5.1). Gliding flight is the simpler and less energetically costly of the two. Wings are held outstretched from the body in a fixed position during gliding, and no flapping occurs. The wings can be moved so that more or less surface area is exposed during a glide. The flight

---

1. Students interested in avian flight or migration must read the work of Colin J. Pennycuick. His theoretical and empirical research has focused on flight energetics of powered fliers, aerodynamic performance of soaring birds, flight morphology, and the ecological implications of aerodynamic performance. Pennycuick has dealt with such divergent groups as pigeons, tubenoses, raptors, and bats in both laboratory and natural settings on four continents.

path of a gliding bird is always downward with respect to air, although it can be upward with respect to the ground. Thus, during gliding flight a bird always loses altitude with respect to the air (fig. 5.1). Soaring flight is gliding in circles. The only way a gliding or soaring bird can gain altitude is to find updrafts that exceed the rate at which it is sinking in relation to the air. Raptors, pelicans, cranes, albatrosses, gulls, and some others use gliding flight during migration.

During powered flight the wings oscillate up and down to provide lift to keep the bird aloft and thrust to move it forward through the air. Powered flight can be subdivided into continuous flapping, bounding, and undulating flight based on kinematics (fig. 5.1). Shorebirds, ducks, loons, grebes, rails, and hummingbirds flap continuously during migration. During bounding flight flapping is interspersed with bouts of nonflapping when the wings are tucked in close to the body (fig. 5.1). Small birds such as warblers, finches, thrushes, tanagers, and most woodpeckers use bounding flight, climbing when flapping

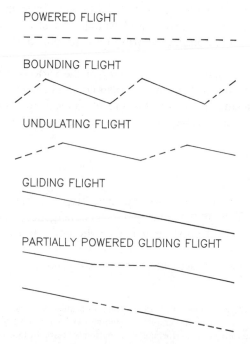

Figure 5.1 Schematic of various types of avian flight (solid line, no wingbeats, as in gliding or wings tucked into the body; dashed lines represent wingbeats). The angle of a line to the horizontal represents the flight path relative to the air.

and descending when their wings are held close to the body. During undulating flight flapping is interspersed with gliding, in which the wings are held outstretched from the body. Undulating flight is sometimes called intermittent-gliding (Videler, Weihs, and Daan 1983) or flapping and gliding flight (fig. 5.1). Larger birds such as hawks, gulls, many tubenoses, crows, and even large woodpeckers use undulating flight, as do martins, swifts, and swallows, which are small.

Although gliding and powered flight are discrete, there are times when some birds use a hybrid of the two. Larger birds like raptors and cranes sometimes mix gliding with bouts of flapping flight. Pennycuick, Alerstam, and Larsson (1979) have called this a partially powered glide. Flapping lengthens a glide or increases air speed during a glide. Hereafter, I will refer to this form of flight as a powered glide or power glide.

A species is capable of more than one type of flight; birds are plastic and change from one type to another when circumstances dictate. For instance, a Black Vulture must sometimes use continuous flapping to take off. It switches to undulating flight while climbing and then resorts to power gliding and gliding (and soaring) when it finds updrafts. If wind and updrafts are favorable, the same bird can commence gliding as soon as it jumps off the ground. Or it might flap continuously to cross a lake or river where updrafts are absent. In this example one species (individual) uses three types of flight in succession. Most raptors use all three (or four) of these types of flight depending upon the species and the circumstances. Raptors do not use bounding flight.

Despite the importance of correctly naming and describing avian flight types, classification schemes are not common in the ornithological literature. Avian flight has not been studied extensively, and there is considerable confusion as to what terminology is correct. The schema presented above will have to be amended as more is learned about the kinematics and energetics of avian flight.

Flight Morphology

With the rise of ecomorphology as a subdiscipline of ecology, more research is focusing on the relation between morphology and flight. The work of Pennycuick (1972a, 1982a, 1983) and Warham (1977) with seabirds and other species provides the best examples of morphological studies that give the detailed information necessary for studying aerodynamic performance and flight energetics. In addition, the work of Greenewalt (1975), Tucker and Parrott (1970), McGahan

(1973a, 1973b), Mueller, Berger, and Allez (1981b), Newton (1986), and a few others should be used as models by scientists who work with birds and wish to quantify morphology. Most researchers who measure birds are banders or taxonomists who have little interest in flight. For this text I have chosen to use the symbols and units of measurement as outlined by avian aerodynamicists (e.g., Pennycuick 1969, 1975, 1983).

A comparison of the structures of a sailplane and a bird reveals that they are similar, although birds are more complex. Figure 5.2 presents ventral views of a sailplane and a gliding bird showing several important aerodynamic features. Wings are airfoils that generate lift by deflecting air. Lift must be generated constantly to fight gravity, the constant adversary of flying machines. Tails are also airfoils that generate lift, provide stability, and assist in maneuvers.

The feature that makes avian flight different from that of aircraft is the plasticity of birds. Unlike ordinary aircraft, birds can switch from

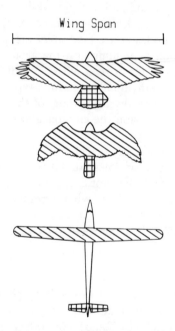

Figure 5.2 Outlines of a sailplane and a gliding bird (a Common Buzzard) showing the basic aerodynamic features. Diagonal lines indicate the measured wing area, whereas hatching indicates the measured tail area. The gliding bird is shown in a slow glide with wings fully outstretched (top) and a fast glide (flex gliding) with wings swept back (middle).

gliding to powered flight and vice versa. A second feature that distinguishes birds from planes is that birds can change wing and tail planform during flight. To increase speed a gliding raptor need only decrease wing span and wing area as well as tail area (fig. 5.2). By pulling the wings close to the body, raptors like Prairie Falcons become missilelike, enabling them to accelerate to great speeds when pursuing prey. This dive is called a "stoop." Seconds before impact with prey, the falcon opens its wings and tail to slow down (creating both lift and drag) and control its flight path.

Three additional features not shared with aircraft are: the ability to fly through forests, flight at very slow speeds, and the ability to land in trees or on the edge of a cliff. If these seem trivial, try to design an aircraft (even an unmanned drone) that can do these things. Even with space-age materials and powerful computers, engineers have not yet constructed flying machines that rival those shaped by natural selection.

Morphological features that are important for describing the aerodynamics and energetics of avian flight are listed below. With each I provide a definition, the symbol used in the aerodynamic literature, the units of measurement in which each is reported, and a brief summary of the availability of works on each feature in the literature.

Wing span (b) is the distance in meters between the tips of the outstretched wings (fig. 5.2). This feature is overlooked by most workers who measure birds, although some information can be gleaned from labels of museum specimens.

Mass (M) is frequently confused with weight (W), which is the product of mass and gravity (g = 9.81 mps^2). Whereas mass is reported in kilograms, weight is reported in newtons (N), a unit of force. Data on the mass of birds are available in the literature, although care should be taken because of variation between the sexes and between birds in migratory and nonmigratory condition.

Wing area (S) is the projected surface area of the wings in square meters (m^2), including the portion of the body intersected by the wings (fig. 5.2). Data on wing area are available for some species, but samples are small and often it is not clear if any portion of the body is included (but this may not matter; Greenewalt 1975).

Tail area is the projected or surface area of the tail measured in m^2 (hatched area in fig. 5.2). Tail area is a function of tail length and the degree to which the tail feathers are spread. No data on this are available in the literature.

Of these four features only mass does not vary during flight, although it decreases as migrants metabolize stored fat or become dehydrated. During gliding and soaring, birds constantly change wing and tail planform. During the slowest glides, as in soaring (fig. 5.2), the wings are held out as far from the body as possible to increase surface area. In many species the tips of the distal primaries are also opened, revealing spaces or slots (fig. 5.2). Greater wing area increases lift, which is needed at slow speeds. At faster speeds the wings and tail are closed to some extent, resulting in a functionally smaller wing span, wing area, and tail area. This was termed "flex gliding" by Hankin (1913).[2] During the fastest cross-country glides the tail is closed and the wings of a migrant are sometimes drawn in to less than 60% of maximum span. This decreases the drag that results from breaking the air with the leading edge of the wings and tail (profile drag) and from friction caused by air flowing across the wings and tail (parasite drag). It also reduces the amount of lift, thereby decreasing induced (see below) drag and increasing sinking speed.

The morphological characteristics examined above are those used commonly by aerodynamicists. Values for wing area, wing span, and mass are substituted into aerodynamic and energetic equations with little regard for details such as the type of surface (feathers, fur, fiberglass, metal, cloth, carbon polymers, etc.), shape of the wing (although area and span are related indirectly to shape), or elasticity of the materials the wing is composed of. (Eager students are referred to Mises' [1959] classic treatise on aerodynamics.)

Wing loading (Q) and aspect ratio (A) are additional morphological features. These are different from the measures given above in that they are composites. Wing loading, the more commonly encountered, is weight (mass times gravity, G) divided by wing area:

$$Q = mG/S. \tag{5.1}$$

Aerodynamicists usually use units of force (N, Newtons) per unit of wing area to report wing loading:

$$Q = N/m^2. \tag{5.2}$$

---

2. E. A. Hankin was a pioneer researcher in avian soaring flight. Using primitive equipment, he made measurements of the soaring dynamics of raptors and other birds in India. Readers who can find his book will be rewarded.

In the ornithological literature wing loading is most often given as mass per unit of wing area (in centimeters squared) or,

$$Q = g/cm^2. \tag{5.3}$$

Clark (1971) has made a plea for consistency of the units used to report wing loading, although he feels that g/cm^2 is the appropriate unit. Before Clark's plea, wing loading was variously reported as pounds per square foot (Pennycuick 1960), grams per centimeter squared, and centimeters squared per unit of mass (Poole 1938).

Aspect ratio is variably reported as the ratio of wing span to wing width, wing span to wing area, wing length to wing width, and wing span squared to wing area:

$$A = b^2/S. \tag{5.4}$$

It is a dimensionless term used to compare birds. For instance, seabirds have higher aspect ratios than land birds of similar size (Savile 1957). That is, their wings are disproportionately longer and narrower. In keeping with protocols established by avian aerodynamicists, I use the measures given in table 5.1.

Just as wing span, wing area, and tail area vary during flight, wing loading, and aspect ratio also change. Measurements given in the literature for wing loading and aspect ratio are for birds with wings fully outstretched. For a migrating raptor the wings and tail are held fully outstretched only during soaring flight and the slowest straight glides.

Table 5.1 Summary of Morphological Features Commonly Used in the Aerodynamic Literature

Morphological or Derived feature	Abbreviation[a]	Unit of Measurement	Availability in Literature
Mass	M	Kilograms	Many species
Wing area	S	Square meters	Some species
Wing span	b	Meters	Very few species
Tail area	S_t	Square meters	Very few species
Wing shape	None	No units	Few—written descriptions
Feathers	None	No units	Few—written descriptions
Descriptions of slotting	Wing formulas	No units	Nonquantitative wing formulas
Wing loading	Q	Newtons per meter squared	Some species
Aspect ratio	A	Dimensionless	Very few species

[a]Abbreviations follow Pennycuick (1969, 1975) where possible.

A raptor gliding between thermals or in ridge lift reduces its wing span, wing area, and tail area by sweeping its wings backward and closer to the body and closing the tail feathers (flex gliding). As it does, wing loading increases and aspect ratio decreases in a manner similar to a delta-wing jet fighter. The "new" configuration changes the bird's entire aerodynamic performance. For these new measurements the term "functional" can be added to wing loading and aspect ratio. Because of this change in flight morphology, it is possible for different species to have similar functional aspect ratio and wing loading even though they may not perform the same.

Readers, especially those who band birds, will note that wing chord was not mentioned in this discussion of flight morphology. When aerodynamicists refer to chord they mean the distance between the leading and trailing edges of the wing, a measurement that is not used by most ornithologists. Wing chord as used by banders and ornithologists is the distance (in millimeters) from the leading edge of the folded wrist to the tip of the longest primary feather. It is used by avian biologists to describe the size of birds, especially in taxonomic work and for determining sex. As such it is probably all right, but it is of limited use in aerodynamic studies.

Both aspect ratio and wing loading are related to wing shape. The high aspect ratio wing is long, narrow, and pointed as in seabirds (Savile 1957; Warham 1977) and birds of deserts and prairies (Janes 1985). The wing loading of these birds can be heavy, as with some falcons, or light as in frigatebirds. With wider wings the tips are more rounded. Forest dwelling birds have rounded wingtips, as do many soaring raptors. At least two morphological features influence wing shape: the relative length of the distal primaries (numbers four to ten) and the degree of slotting of the distal primaries. Slotted primaries are variously referred to as wingtip slots (Cone 1962b; Withers 1981), pinioned wings (Kokshaysky 1973), emarginate wingtips (Kokshaysky 1973), and separated primaries (Oehme 1977) (fig. 5.3). Distal primaries such as number nine tend to be longer in the high aspect ratio wing than primaries such as numbers seven and eight. Low aspect ratio wings, on the other hand, have primaries six through nine of similar length. When primary nine is long relative to eight, there is little slotting. When primaries six through nine are similar in length, wingtips are slotted. Thus, emarginate primaries are absent or reduced on the high aspect ratio wing.

The function of slotted wingtips has been debated (Savile 1957; Cone 1962b; Kokshaysky 1973; Oehme 1977; Withers 1981) but infrequently studied (Kokshaysky 1973; Withers 1981). Explanations

A   B

Figure 5.3 Tips of two types of bird wings: (A) rounded wingtip of a Sharp-shinned Hawk with deep slotting and (B) pointed wingtip of an American Kestrel with shallow slotting. (Photographs by Paul Kerlinger.)

focus on the reduction of induced drag, reduction of profile drag, maximization of lift coefficient, and increased aeroelastic properties of low aspect ratio wings (reviewed by Withers 1981). The explanation given most often is that slots reduce induced drag near the wingtips. As air flows over and around the tip of a bird's wing, large vortices are formed just behind the wingtip and create drag as they are pulled through the air. Slotted primaries may reduce this drag by acting as miniairfoils, creating small vortices that can be shed more easily than a large one. Newman (1958), studying the performance of Black Vultures, did not find evidence consistent with this hypothesis. Other researchers have found support for it (Withers 1981), although slotting has been thought to serve more than one function (Kokshaysky 1973; Withers 1981). The question is not resolved.

Energetics of Migratory Flight

The energy cost of powered (flapping) flight is a major factor influencing the behavior of migrants. Cost of powered flight is expressed most often as a power curve (fig. 5.4). As the air speed of a bird increases the energy cost of flight increases in a nonlinear fashion. At very slow and very fast speeds the energy required for flight is greater than at intermediate speeds. There are four attributes of a power curve that are used to describe the cost of flight for a particular bird: maximum range air speed (V_{mr}), air speed at minimum power required for flight (V_{mp}), minimum power required for flight (P_m), and maximum power available for flight (lines A and B in fig. 5.4). Lines A and B are included in figure 5.4 to compare two species with different maximum power available for flight. Species A is capable of a wider range of speeds than species B because it can generate more power.

The power curve is relevant only for birds that use powered (flapping) flight. Studies of numerous birds flying in wind tunnels (Tucker 1966, 1972; Pennycuick 1968a; Bernstein, Thomas, and Schmidt-Nielsen 1973) have validated the power curve concept and the calculations developed to describe the power curve of given species. Although undulating and bounding flight should require the same energy as continuous flapping flight, some studies suggest that these types of powered flight may be slightly less expensive (Videler, Weihs, and Daan 1983). Other authors maintain that there is no difference in energy cost between the two types of flight (Rayner 1985b).

The cost of hovering flight was included in figure 5.4 even though raptors are not capable of true hovering (defined as flight at zero air speed). It was included to show how expensive both hovering and

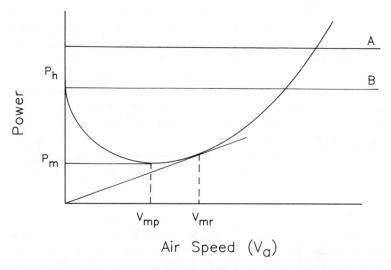

Figure 5.4 Power curve of a hypothetical bird (V_{mr} is the air speed at maximum range, V_{mp} is the air speed at minimum power required for flight, P_m is the minimum power required for flight, P_h is for true hovering flight). Lines A and B represent the maximum power available for flight of two different birds (see text).

slow flight are compared with normal powered flight and the maximum power available for flight. A second reason for including hovering is that it is misused in the raptor literature. According to Rayner (1979a) and other aerodynamicists, few birds are capable of true hovering flight. Weis-Fogh (1975) cites 100 g as the maximum weight for hovering flight, thereby eliminating nearly all raptor species.

Hovering is so energetically expensive that even small birds use it for only brief periods. Although most small birds normally use bounding flight, some are capable of short bouts of continuous flapping as in hovering. Small raptors like kestrels do not hover, although they can fly very slowly. Instead of flying at zero air speed, they actually fly at zero ground speed while maintaining a positive air speed. Minimum air speeds for kestrels have been measured at 4 to 6 mps during wind hovering (Videler, Weihs, and Daan 1983). At very slow air speeds they flap continuously, whereas at faster air speeds (in stronger wind) they incorporate short bouts of gliding flight. The latter also occurs when these birds are in updrafts. In faster wind the kestrel must again resort to continuous flapping to generate the thrust necessary for fast air speed. To confuse true hovering with wind hovering has serious implications for researchers making assumptions about the energy cost of flight.

In figure 5.5 hypothetical curves representing the cost of powered, gliding, and power gliding flight are presented for a hypothetical bird along with its standard metabolic rate (SMR). Of particular importance is where the curves are located on the y-axis (power) in relation to the other curves and to standard metabolism. These curves are only approximations of the energy required for flight at various speeds and are presented for heuristic purposes. Such comparisons are valid only within a given species, and perhaps within a given age-sex class. Comparing different species is difficult because of differences in size that require scaling.

The minimum cost of powered flight for the hypothetical bird shown in figure 5.5 is more than 11 times SMR, whereas gliding flight is only 2.5 times SMR. Thus, powered flight (at minimum power) for the bird in question is more than 4 times as expensive as gliding flight. Between the two is power gliding, for which the cost has been assumed to be 6 times SMR and more than 2 times gliding flight. Because there are no empirical studies of the metabolic cost of power gliding flight, the location and shape of the curve are suggestive.

Only a Laughing Gull (*Larus atricilla*) has been flown in a wind tunnel to measure the cost of gliding flight. Baudinette and Schmidt-Nielsen (1974) found that the cost of gliding is about 2–3 times stan-

dard metabolism, whereas the cost of flapping of a similar species was
at least 7–8 times standard metabolism. Pennycuick (1972a) suggested
a value of about 1.5–2 times standard metabolic rate for the energy
cost of gliding for vultures. Unlike other types of flight, the cost of
gliding flight presumably does not vary with air speed because no
wing movement is necessary.

Several researchers (Kuroda 1961a; Pennycuick 1972a, 1982a)
have found that soaring birds have anatomical adaptations that allow
them to "lock" their wings in place, freeing them from using muscles
for this. Among the procellariiforms (albatrosses and other tubenoses)
a "tendon sheet is associated with the pectoralis muscle," preventing
the wing from being elevated above the horizontal (Pennycuick
1982a). Vultures and some other soaring birds such as storks have a
subdivided pectoralis muscle in which the smaller, deep part is a slow
tonic muscle that may hold the wings in place when gliding (Kuroda
1961a; Pennycuick 1972a).

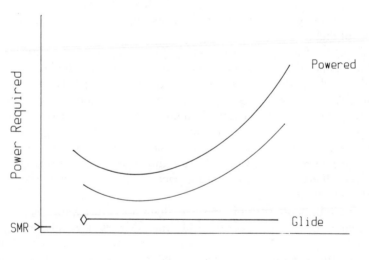

Air Speed

Figure 5.5 A comparison of the cost of powered flight (flapping) and gliding flight to
standard metabolic rate (SMR). Curve between these lines is the estimated cost of a
powered glide. Gliding was assumed to cost 2.5 times SMR, whereas powered flight
was assumed to cost 11 or more times SMR (approximate values derived from Pennycuick 1972a; Baudinette and Schmidt-Nielsen 1974; Gessaman 1980). Comparison of
flight cost in SMR multiples is for heuristic purposes only because the multiple varies
with size and shape among species.

So far the relation between size or body shape and the cost of powered flight has not been considered. To do so the power required for flight is partitioned into its various components. Pennycuick (1975) and others (Tucker 1974; Greenewalt 1975) have identified three components: the power needed to support the bird's weight, or induced power; the power to overcome parasite drag due to skin friction; and the power to overcome profile drag, the component of drag contributed by the leading edge of the bird as it moves through the air. Consult Pennycuick (1975), Greenewalt (1975), and Tucker (1974) for derivations of the equations used for determining cost of flight.

According to the theory of powered flight, weight and wing area are most important for determining power required for flight. Weight is the numerator in most powered-flight equations and is a power function. The power required for flight therefore changes with weight according to the proportion $P = W^{1.17}$. This means that the "ratio of metabolic rate in flight to metabolic rate is larger in large birds than in smaller birds" (Pennycuick 1975). (A recent paper by Masman and Klaassen [1987] reports a coefficient of only 1.04, considerably smaller than that calculated by Pennycuick.) Flight is disproportionately more costly for larger birds than for smaller birds. Herein lies the reason so many large birds utilize gliding flight. It is simply too costly to flap their wings. Similarly, it is relatively inexpensive for small birds to use powered flight. Nonlinear changes with size are called scaling effects. (Schmidt-Nielsen [1984] has written a readable volume explaining scaling effects as they relate to physiology and anatomy.)

Scaling of flight energetics and standard metabolism as a function of body size explains the admonition above about using multiples of standard metabolism to describe cost of flight. Although it is done often in the literature, researchers must be careful. If it is done among birds of similar size and shape errors will be minimal, but extrapolating among birds of disparate size is incorrect and should not be done.

Andersson and Norberg (1981) analyzed scaling of flight performance mathematically for raptors. Their particular interest was performance difference between males and females that result from reverse sexual size dimorphism (females being larger than males). They show that the cost of flight is considerably less for males than for females and that males are considerably more maneuverable, though they cannot fly as fast as females.

Wing shape and size also influence the energy cost of flight. Birds with long, pointed wings (high aspect ratio) generate more power at a lower cost than birds with short, rounded wings (low aspect ratio).

The reason for this is that longer, pointed wings are more able to shed tip vortices that create drag because they must be pulled along by the bird. If they can be reduced in size, they create less induced drag and are less costly to pull. Slotted wings of birds with low aspect ratio may reduce tip vortices by acting as independent airfoils creating many small vortices rather than one large one.

The flight energetics of one falconiform has been examined. Gessaman (1980)[3] studied the flight energetics of an American Kestrel in powered flight. He found such flight to be 12.5 times more expensive than standard metabolic rate and about 4 times as expensive as gliding flight, assuming that gliding is 2–3 times as costly as standard metabolism (estimate from Pennycuick 1972a).

Migratory Fat Deposition

Until now I have assumed that birds have limitless energy available for flight. Obviously this is not true. The distance and duration of a single flight for powered migrants is constrained by either fat reserves or evaporative water loss (Yapp 1962; Pennycuick 1969; Blem 1980). Fat is the fuel used for long migratory flights. Most warblers and shorebirds deposit fat rapidly and make long flights using stored fat. Small warblers (10–15 g) use only 2–4 g of fat when they cross the Gulf of Mexico, a distance of more than 1,000 km, during spring migration (with favorable winds; Moore and Kerlinger 1987). This is remarkable. Imagine traveling 200–300 km on 1 g of fat that was deposited during only 2–4 days of foraging!

The abundant literature on deposition and use of fat by migrants is incomplete. Little is known about fat deposition by falconiforms during or before migration. Do they gain weight for migration? It is likely that they do, but few empirical studies have addressed the question. Gessaman (1979) demonstrated that American Kestrels deposited fat during early autumn. Using standard fat extraction techniques ($N = 15$ males, 8 females), Gessaman found that females were slightly fatter than males (7.0% vs. 5.3% of body weight). These are 2%–4% greater than midsummer values.

Studies by Geller and Temple (1983) and Clark (1985b) documented mass and fat condition of hawks during migration. Using an ordinal scale to rate fat condition, similar to the scales used for pas-

3. Readers interested in raptor physiology (metabolism and energetics) should see papers by James Gessaman (and his students) of Utah State University. He has conducted fascinating research with Bald Eagles, Snowy Owls, and other species.

serine migrants, Geller and Temple found a large variation of fat condition among Red-tailed Hawks during autumn at Cedar Grove, Wisconsin, and Clark found the same among Merlins at Cape May Point, New Jersey. Clark stated that there was no relation between visible fat condition and mass of migrating Merlins, but crop and stomach contents were not determined.

Authors from both New World and Old World maintain that migrating raptors take on fat deposits during migration. Amur Falcons are known to be extremely fat and good to eat before their long flight over the Indian Ocean from India to East Africa (Ali and Ripley 1978). Similarly, Swainson's and Broad-winged hawks are thought to acquire fat deposits that permit fasting for one to two months during migration between North and South America (Smith 1980; Smith, Goldstein, and Bartholomew 1986). Their conclusions were based upon the absence of food in the guts of birds found in Panama and the absence of fecal material in large roosts of migrants. Finally, Glutz von Blotzheim, Bauer, and Bezzel (1971) reported a large variation of weight among migrating Honey Buzzards that they attribute to massive fat deposition for migration.

There is also anecdotal evidence that suggests raptors deposit migratory fat. W. S. Clark has shown me small fat deposits on Sharp-shinned Hawks and Merlins (Clark 1985b) migrating at Cape May Point, New Jersey. They occur as subcutaneous strips along the sides of the abdomen, extending up the torso under the wings. These deposits were barely visible through the feathers and skin, unlike the deposits of small passerines, which can be seen through the skin with ease. Clark asserts that Red-tailed Hawks, American Kestrels, and other species deposit fat in the same anatomical location.

Because raptors use gliding flight, most species may not need to accumulate as much fat as powered fliers like warblers and shorebirds. Whereas some warblers and shorebirds deposit more than 50% of their body weight in fat (Blem 1980), many raptors may need a maximum of 10%–20%. Long flights over water barriers and deserts by some raptors are not uncommon (chap. 10). These flights do not offer the opportunity to forage and require powered flight. If a researcher were to investigate fat deposition by a falconiform, the species to examine would be the Amur Falcon, which makes long water crossings; Swainson's Hawk, which is the longest-distance soaring migrant (land birds) in the New World; or perhaps raptors that move through the Middle East land bridge. Forests of the New World are poor feeding habitats for Swainson's Hawks in the same way that deserts in the Middle East and Africa are for many raptors. These types of barriers and long distance flights should be strong selective forces on migratory physiology and behavior.

Smith, Goldstein, and Bartholomew (1986) have simulated the physiology of soaring migration of Swainson's and Broad-winged hawks from the United States and Canada into Central and South America. These migrants fly one-way distances of about 9,000 and 5,000 km, respectively. Using estimations of gliding energetics from Baudinette and Schmidt-Nielsen (1974) and Pennycuick (1972a), and their own estimations of migrants' daily flight distance, Smith and his colleagues predicted the quantity of fat these birds must deposit to complete their journey without feeding. For a Broad-winged Hawk flying from New Brunswick, Canada, to Ancón, Panama (6,700 km), a minimum of 155 g of fat is necessary. The trip was estimated to take 29 days at a flight speed of about 240 km per day. This amounts to an increase of 25%–40% of lean body weight. One of their models, called a nomogram, is presented in figure 5.6. Although their estimates may be reasonable, they need to be tested. The models of Smith and his colleagues will be useful to future researchers. Kirkwood (1985) has also discussed the food requirements for deposition of energy reserves by migrating raptors.

I cannot understand how selection could favor individuals who fast during migration. When weather is adverse for migration most migrants do not fly, at which time they can forage. Because they are usually not efficient foragers or adept at migration, many immature birds must forage during their migratory journey. If birds fast during their entire migration they risk arriving without energy reserves at unfamiliar or unpredictable sites (breeding and non-breeding seasons). Fat is a hedge against starvation in unpredictable and harsh situations. Another argument in favor of depositing fat is that low body weight makes for slower glides between thermals. As migrants lose weight their migration becomes slower. To fast for more than a month is an incredible feat. There is no reason why falconiform species should not deposit fat for migration. Predatory birds such as Snowy Owls (*Nyctea scandiaca*) deposit massive fat pads on their abdomens (Keith 1964; Kerlinger and Lein 1988) and still capture prey successfully.

I have seen many species of raptors, including Swainson's Hawks, feeding during migration. During spring migration in south Texas these animals feed on vertebrates and insects. Observations of Ospreys carrying fish or Sharp-shinned Hawks capturing songbirds during migration are not uncommon. Foraging by Broad-winged Hawks seems to be common, at least during part of their southward migration. Shelley and Benz (1985) showed that during autumn migration in Pennsylvania, this species forages frequently. They found that during the first half of September about 8% of the Broad-winged Hawks observed had full crops, whereas during the second half of the month the total was 17%. They suggested that fewer invertebrate prey were

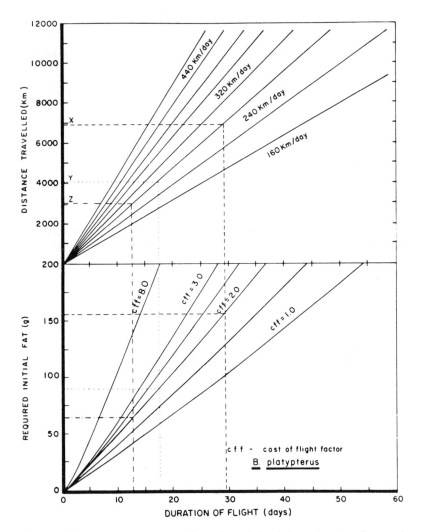

Figure 5.6 A nomogram, or energy simulation, of the gliding migration of a Broad-winged Hawk. (From Smith, Goldstein, and Bartholomew 1986; courtesy of the American Ornithologists' Union.)

available as the season progressed and that hawks relied more on vertebrates.

Aerodynamic Performance: Gliding Flight

The aerodynamic performance of a gliding bird or sailplane is usually presented as a glide polar diagram in which (fig. 5.7) air speed (V_a) is plotted against sink rate (V_z). All values are with respect to the air through which a bird flies, not to the ground. A glide polar is expressed algebraically by fitting a curve to real data (Pennycuick 1972a, 1975). There are three important points on a glide polar (fig. 5.7): the minimum air speed at which a bird can fly (V_{min}), or stall speed; the lowest sink rate (V_{zm}) at the corresponding air speed (V_{ms}); and air speed at the best glide ratio (V_{bg}), with the corresponding best glide ratio. Glide ratio (glide angle) at any point on the aerodynamic performance curve is equal to a ratio of V_a to V_z. The best glide ratio on the glide polar (the maximum ratio of V_a to V_z) is determined by drawing a tangent to the curve from the origin. The corresponding air speed is the best glide air speed, and the ratio of V_a to V_z at this point is the maximum lift to drag ratio (L/D_{max}).

After construction of a glide polar and fitting of a curve, lift (L), drag (D), and their coefficients (C_L, C_D) are calculated. For these cal-

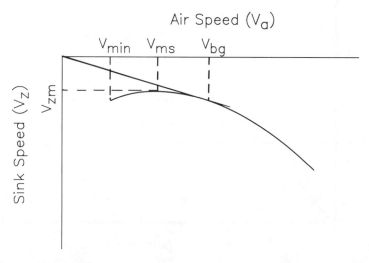

Figure 5.7 Polar (performance) diagram of a gliding bird or sailplane showing minimum sink speed (V_{zm}), minimum air speed (V_{min}), air speed at best glide (V_{bg}), and air speed at minimum sink (V_{ms}).

culations weight and wing area at the corresponding air speeds are needed. To determine L and D the following equations are used:

$$L = W \cos \theta, \tag{5.5}$$

$$D = W \sin \theta, \tag{5.6}$$

where θ is the angle between the bird's flight path and the horizontal. This angle is given by the relation:

$$\theta = \arctan V_z / V_a \tag{5.7}$$

For glide angles less than about 8°, values of lift deviate slightly from W. Coefficients of lift and drag are then calculated using the following equations:

$$C_L = 2L/(\rho SV^2), \tag{5.8}$$

$$C_D = 2D/(\rho SV^2), \tag{5.9}$$

where ρ is equal to the density of air. These coefficients are nondimensional, showing how much lift or drag a bird generates at a given air speed. They are used to make comparisons of performance of individuals at different speeds and to compare performance among species.

The ability to glide at both fast and slow air speeds (part of performance) must be considered when examining the aerodynamic performance of migrating soaring birds. Performance is also important during prey pursuit or soaring while looking for prey. Wing loading is important for determining how fast and slow a bird can fly, along with coefficients of lift and drag at these air speeds. Minimum air speed (V_{min}), determined from the wing area, maximum lift coefficient (C_{Lmax}), and weight, is given by:

$$V_{min} = (2W/\rho C_{Lmax}S)^{1/2}. \tag{5.10}$$

This equation shows that heavy birds are not capable of gliding as slowly as lighter birds with similar lift coefficients and wing areas. By flying slowly, a soaring bird can use isolated updrafts that a heavier bird might not be able to use.

For determining the terminal or maximum air speed that a gliding body can achieve, Tucker and Parrott (1970) give the equation:

$$V_{term} = (2W/\rho S_w C_{Dp})^{1/2}, \tag{5.11}$$

where S_w is the wetted wing area (approximately equal to 2.04 times wing area, S) and C_{Dp} is the coefficient of profile drag (which is not easy to determine). From equation 5.11 we see why birds reduce their wing area during stoops and interthermal glides. Reducing wing area (in the denominator of equation 5.11) increases terminal or inter-thermal gliding speed.

Aerodynamic theory was developed for fixed-wing aircraft, which cannot change wing span and wing area like a bird. These changes influence functional wing loading and aspect ratio. Readers now should understand how important it is for a soaring bird to change its wing loading and aspect ratio during flight. The theory and equations given above can be used to calculate values of lift and lift coefficients if wing area at a given speed is known. At the slowest speeds a gliding bird's wings are fully open (100% span). As air speed increases to the air speed at best glide, wing span is reduced to about 85%–90% (Tucker and Parrott 1970, for a Lanner Falcon; McGahan 1973a, for an Andean Condor). During the fastest interthermal glides wing span is reduced by more than 50% (Tucker and Parrott 1970; personal observations of raptors). Pigeons flown in a wind tunnel reached speeds of over 20 mps with wing span reductions of more than 50%. Migrants usually do not reduce wing span more than 35%. The ability to vary wing span with concordant changes in aspect ratio and wing loading promotes a flatter glide polar. The result is a wider range of aerodynamic abilities than would be possible if the wings were held outstretched to 100% span at all times.

One other topic must be considered in a treatment of aerodynamic performance: the ability to climb while soaring in a thermal. The rate of climb in a thermal depends on several factors. Most important is the strength and size of the thermal. If thermal strength is constant, the sink rate and turning radius (r) of a bird determines ascent rate in the thermal. The way sink rate influences climb rate is clear, but the effect of turning radius is not obvious until we consider the structure of a thermal. Because the strength of updrafts decreases with distance from the center of a thermal (Pennycuick 1975; fig. 4.9), the ability to soar in tight circles is a distinct advantage. This ability is determined by morphology as follows:

$$r = (W/S) \, (2/\rho g C_L \sin\phi), \tag{5.12}$$

where g is gravity and ϕ is the angle of bank while soaring in circles. The turning radius is proportional to wing loading (W/S) and lift coefficient such that relative turning abilities of different species can be predicted from morphological determinations. Thus, turning envelopes at a given angle of bank can be calculated from glide polar and morphological measurements.

Bank angles of soaring birds range from 20° to 35°. To climb in small thermals a bird (or sailplane) must use a large angle of bank or run the risk of "losing" the thermal core. By increasing bank angle the bird sacrifices lift by increasing air speed and sink rate. If it does not increase air speed it may stall and lose altitude at an enormous rate. Thus, there is a trade-off between the strength of a thermal and the rate of sink at a given bank angle. If thermals are small but powerful, there is no problem with a steep bank angle. However, if they are small and weak there is no advantage to using a steep angle. For soaring pilots and migrating hawks large, powerful thermals preclude the use of steep bank angles.

Why have all of this theory and these equations been presented? Wouldn't it be enough to present some of the actual data regarding aerodynamic performance of a few species of migrant raptors? Possibly, but the goal of science is to explain how and why a phenomenon occurs. In this case the phenomenon is bird migration, and we are interested in generalizing our findings to other raptors and nonraptors. The powerful theory developed by Pennycuick, Tucker, and their colleagues allows us to make these generalizations. By presenting theory along with empirical data that validates theory, the researcher is armed with a powerful theoretical tool for examining and comparing the performance of many species of birds. In addition, the same tools can be extended to ecological and behavioral research other than migration. These include foraging, habitat selection, and applications we have not yet considered. The misuse of aerodynamic terminology and theory in the literature might be corrected by a more rigorous approach. Finally, a knowledge of flight energetics and aerodynamics is prerequisite to the study of many aspects of avian behavior and ecology.

Ecology, Flight Morphology, and Natural Selection

It should be apparent that a bird's morphology determines how much energy it needs for flight, how it performs, and how behavioral responses can modify these relations. Ultimately, a bird is constrained by morphology and physiology. If we assume that morphology is the product of natural selection, the question of importance to evolutionary biologists is, What selective pressures have shaped the morphology of these birds? I have posed this question to introduce the idea of multiple, competing selective pressures and the importance of migration as a selective force on a bird's morphology and physiology.

Flight morphology, including wing shape and size, is a product of combinations of selective pressures. Selective pressures that probably account for most of the variance in a raptor's morphology are habitat, prey, migration (fig. 5.8), and perhaps social displays. Individuals that can maneuver in their habitats better than others, those that can catch more prey or catch prey more economically, and those that migrate more quickly and use less energy will realize greater fitness than less adept individuals.

The relative importance of the selective pressures shown in figure 5.8 is unknown. Evolutionary biologists and ecomorphologists seek approximations of the path coefficients r_1, r_2, and r_3. Ideally, their squared values should add to 1.00. If they do, it means that 100% of the variance in morphology would be explained by habitat, prey, and migration. Obviously, we will never know the exact values of path coefficients. If we could construct an experiment to determine these values, we would find that habitat, prey, and migration are intimately linked (double-headed arrows in fig. 5.8). To the student of natural history this should be obvious. Sharp-shinned Hawks live in the forest and eat forest birds, just as Peregrine Falcons live in the open and eat birds of open country. Although we could fit a predictive model, an explanation of the sources of variation would be tenuous.

Qualitative predictions about flight morphology (mass, wing span, wing shape, tail length) can be made if we know an organism's habitat, prey, and migration. For example, the greater the distance an individual migrates, the greater r_3 will be in figure 5.8. Thus, the strength of selection as a result of migration increases with migration

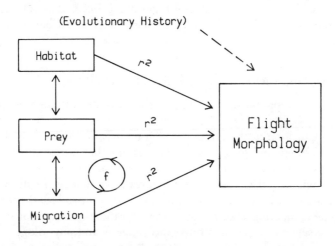

Figure 5.8 Path diagram showing probable selective forces that have shaped the flight morphology of raptors (r^1, r^2, r^3 represent path coefficients). Circle with arrows represents a feedback loop between different compartments.

distance. There are other factors that may act indirectly to shape the morphology of raptors, including atmospheric structure, topography (hills, plains, water), and the interaction of weather and topography. Incorporating these factors in the model (fig. 5.8) would be laborious and tentative. Topography and atmospheric phenomena presumably are related to selection of morphology through the primary variables habitat and migration (fig. 5.8).

There are limits to the course and speed of natural selection. The primary constraint on the effectiveness of selection is the amount and type of variance within a population. Although selection can change the morphology of a species, its evolutionary past (genetic makeup) precludes unbridled change. This is analogous to "genetic inertia," especially if coadapted genes are involved.

The take-home lesson from this is that the morphology of migrating raptors is dependent upon the ecology and evolution of the species in question. Whereas morphology may be limited by a species' ecology and evolutionary past, behavioral adaptations may be nearly as important as morphological and physiological adaptations for flight.

Summary and Conclusions

Five aspects of avian flight were reviewed: types of flight, flight morphology, energetics, gliding performance, and ecomorphology. Raptors utilize four types of flight: continuous flapping; undulating (sometimes called intermittent gliding), during which bouts of flapping alternate with gliding; power gliding; and gliding (and soaring) on outstretched wings. The importance of wing span, wing area, tail area, mass, wing loading, aspect ratio, and shape of wingtips was discussed in relation to energetic cost of flight and aerodynamic performance.

The large size of many raptors makes powered flight disproportionately more costly than for smaller birds, so they frequently resort to gliding during migration. The energy cost of flapping flight was presented as a power curve with power regressed on air speed. At very slow and very fast air speeds the energy required for flapping flight is greater than at intermediate air speeds. The energy costs of powered and gliding flight are 7–14 and 1.5–3.0 times standard metabolic rate, respectively. The power required for gliding flight is independent of speed, whereas the cost of powered flight increases as a power function of air speed. Empirical determinations of the cost of gliding flight are not available for raptors. Power required for flapping flight is known only for the American Kestrel and Eurasian Kestrel.

Gliding performance of a bird or aircraft was presented as a glide polar diagram in which sinking speed is regressed on forward air speed. Sink rate is greatest at very slow and very fast air speeds. The energetic and gliding performance of a bird depends primarily on morphology. The flight morphology of birds is the product of competing selective pressures such as prey type, habitat, and migration. The study of the relations between morphology and ecology is called ecomorphology. Prey capture, habitat, and migration are probably important selective pressures that have shaped the flight ability and morphology of raptors.

6

Flight Mechanics: Empirical Research

The preceding chapter introduced avian flight morphology and how it influences gliding performance and energetics. The present chapter builds on that theory, describing and discussing the empirical findings of researchers who have investigated the flight morphology, flight type, and aerodynamic performance of raptor migrants. Ideally, this review would examine dozens of species, but empirical studies are available for only a few. This review focuses on eight species of raptors from the Cathartidae, Pandionidae, Accipitridae, and Falconidae. The species vary in size (ranging from about 100 to 5,500 g), type of migration (partial and complete migrants), distance of migration, habitat type, and trophic habits. For heuristic purposes, the morphology and aerodynamic performance of several nonraptors are also presented: monarch butterfly (*Danaus plexippus*), dog-faced bat (*Rousettas aegyptiacus*), Rock Dove (*Columba livia*), Northern Fulmar (*Fulmarus glacialis*), Common Crane (*Grus grus*), Wandering Albatross (*Diomedea exulans*), a motorized sailplane, and a Rogallo-wing hang glider. By examining a diverse set of flying "species" in this review, I hope to illustrate the range of size and shape of "organisms" capable of gliding and soaring and to demonstrate how morphology is related to aerodynamic performance.

Flight Morphology

At least five body "types" are recognizable among the Falconiformes (fig. 6.1): kites, harriers, falcons, accipiters, and buteos. These "types" are not distinct and form a continuum from high to low aspect ratio. Falcons and kites have high aspect ratio wings, whereas buteos and accipiters have lower aspect ratio wings. These morphotypes correspond roughly to those described by Hankin (1913), Savile (1957), and Brown and Amadon (1968). Convergent evolution has resulted in distantly related species' having similar morphology. Cases of possible convergence include pointed wings shared by the Swainson's Hawk, falcons, and kites; dihedral wings (held slightly above the body in a shallow V during gliding) of Swainson's Hawks, Northern Harriers, Zone-tailed Hawks (Willis 1963), and Turkey Vultures; and the slotted primaries of buteos, vultures, eagles, accipiters, and even owls and grouse. These examples of convergent flight morphology are presum-

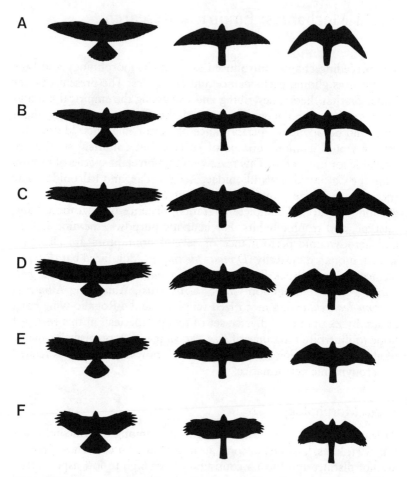

Figure 6.1 Continuum of "morphotypes" among the falconiforms as shown by silhouettes (not to scale) of (A) Peregrine Falcon, (B) Mississippi Kite, (C) Osprey, (D) Bald Eagle, (E) Red-shouldered Hawk, and (F) Sharp-shinned Hawk. Each species is depicted in a slow glide as in soaring with wings and tail fully open (left column), a moderate glide with wings and tail area reduced from slow glide (center column), and a fast glide with wing and tail area reduced dramatically as occurs when gliding between thermals. (Drawn by David A. Sibley.)

ably the result of similar selective pressures. After all, a behavior as strenuous and critical as flight has been intensely and constantly selected.

Morphological similarities among species and the types in figure 6.1 also result from behavioral modifications of wing and tail planform. Buteos, for example, pull their wings in and back during fast interthermal glides, reducing wing area and span. When they do their wings become pointed, similar to the wing of a Peregrine Falcon. Broad-winged Hawks gliding at high altitudes can be confused with Peregrine Falcons by inexperienced hawk watchers. This is only one example of how behavioral control of wing and tail planform change the appearance of a migrant.

The following aspects of flight morphology are examined in this chapter: wing span, wing area, tail area, mass, wing loading, aspect

Table 6.1 Summary of Flight Morphology of Raptors, Sailplanes, and Various Soaring Animals

Species	Mass (kg)	Span (m)	Wing Area (m²)	Wing Load (N/m²)	Aspect Ratio	Tail Area Open (N/m²)	Wingtip Shape*
RAPTORS							
Sharp-shinned Hawk[c]	0.14	0.51	0.057	24.05	4.57	0.023	3
Broad-winged Hawk[c]	0.46	0.81	0.118	35.42	5.61	0.029	2
Red-tailed Hawk[c]	1.36	1.08	0.222	49.68	5.20	0.068	3
Osprey[c]	1.68	1.49	0.297	54.75	7.20	0.682	2
Lanner Falcon[a]	0.57	1.01	0.132	42.39	7.72	NA	1
Black Vulture[b]	1.80	1.38	0.331	53.45	5.70	NA	3
White-backed Vulture[d]	5.39	2.18	0.690	76.50	6.90	NA	3
Andean Condor[e]	10.05	2.88	1.050	93.50	7.90	NA	3
NONRAPTORS							
Common Crane[f]	5.50	2.40	0.720	80.00	8.00	NA	3
Rock Dove[g]	0.40	0.67	0.063	52.20	6.50	0.010?	1
Fulmar[h]	0.73	1.10	0.121	58.93	10.02	NA	1
Dog-faced bat[i]	0.13	0.55	0.057	20.50	5.42	NA	2
Monarch butterfly[j]	0.001	0.01	0.0003	1.77	3.97	NA	3
Sailplane[k]	315	15.0	10.5	290	21.40	NA	1–2
Rogallo hang glider[k]	107	10.1	15.1	69.4	6.76	NA	2

Note: Not available or not applicable (NA).

[a]Tucker and Parrott 1970.
[b]Parrott 1970; Pennycuick 1983.
[c]Kerlinger 1982a.
[d]Pennycuick 1971a.
[e]McGahan 1973a.
[f]Pennycuick, Alerstam, and Larsson 1979.

[g]Pennycuick 1968b.
[h]Pennycuick 1960.
[i]Pennycuick 1968b.
[j]Gibo and Pallett 1979.
[k]Rogallo dealer.

*1 = pointed tip; 2 = intermediate;
3 = rounded tip.

A

B

C

Figure 6.2 Osprey gliding at (A) slow, (B) intermediate, and (C) fast air speeds. Slow glide is analogous to soaring in lift, whereas fast ("flexed") glides are used between thermals. (Photographs A and C by Clay Sutton, B by Frank Schleicher.)

ratio, and wingtips. The structure of wingtips will be considered after the discussion of the standard aspects of flight morphology.

Wing area, tail area, wing span, mass, wing load, and aspect ratio for eight species of raptors are presented in table 6.1. The measurements given are for birds with wings opened fully (100% of span), the way area is traditionally measured and reported (Poole 1938). Among migrating raptors, flight with wings extended fully is limited to soaring, turning, landing, and takeoff. When birds are gliding at faster speeds between thermals, the wings are swept back and closer to the body, such that functional wing span is less than 100% of the extended span (figs. 5.2, 6.2; also see figures in Pennycuick 1968b; Tucker and Parrott 1970; McGahan 1973a). The relations between wing span and wing area, wing load, and aspect ratio were examined by Kerlinger (1982a) for the Sharp-shinned Hawk, Broad-winged Hawk, Red-tailed Hawk, and Osprey by tracing the outline of the wings of dead specimens (one male and one female of each species) on a piece of paper. Four tracings of each specimen were drawn, corresponding approximately to 100%, 85%, 70%, and 60% of full span. These represent a continuum of slow to fast glides.

Hankin (1913) and Raspet (1960) noted that wing area increased with wing span. Pennycuick (1968b), Tucker and Parrott (1970), and Kerlinger (1982a) demonstrated that wing area increased linearly with span (fig. 6.3), at least within the normal range used during flight. Slopes varied from 0.09 for the Sharp-shinned Hawk, the smallest species, to 0.21 for the Red-tailed Hawk. Slopes for the two buteos were steeper than for either the Osprey or the Sharp-shinned Hawk.

Functional wing load increased when wings were changed from fully open (100% of span) to the flex-gliding position. The two buteos showed a greater rate of increase (steeper slope) of wing load (fig. 6.4) than did other species, as expected from the relation between wing span and wing area. The Osprey had the flattest slope. Functional wing loadings overlapped between adjacent species, but at different wing areas and spans. Red-tailed Hawks and Ospreys overlapped extensively, and Broad-winged Hawks overlapped somewhat with both. For this to occur during flight, the Broad-winged Hawk must be in a steep glide with its wings swept back and the others must be in shallow glides.

The pattern for aspect ratio was the opposite of the pattern evident with wing load (fig. 6.5). As wing span decreases so does wing area, but at a greater rate. Thus, at fast glide speeds the aspect ratio of raptors is considerably less than at slower glide speeds.

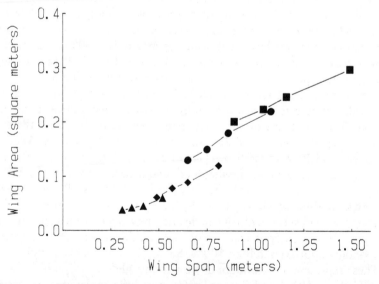

Figure 6.3 Relation between wing area and changing wing span (square = Osprey, circle = Red-tailed Hawk, diamond = Broad-winged Hawk, triangle = Sharp-shinned Hawk). Determined from the mean of one male and one female measured after death.

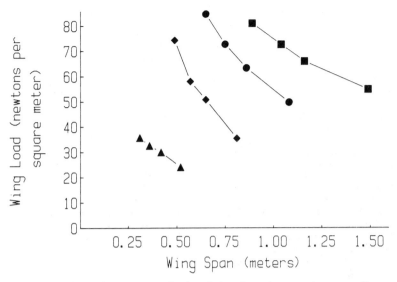

Figure 6.4 Relation between wing load and changing wing span (square = Osprey, circle = Red-tailed Hawk, diamond = Broad-winged Hawk, triangle = Sharp-shinned Hawk). Determined from the mean of one male and one female measured after death.

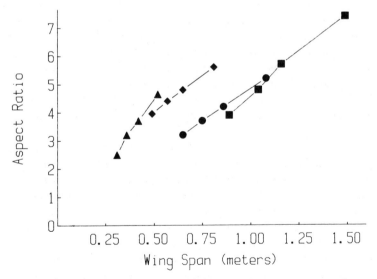

Figure 6.5 Relation between aspect ratio and changing wing span (square = Osprey, circle = Red-tailed Hawk, diamond = Broad-winged Hawk, triangle = Sharp-shinned Hawk. Determined from the mean of one male and one female measured after death.

Table 6.2 Summary of Wing Loadings of Various North American Raptors
with and without the Fully Spread Tail

	Wing Load (N/m²)		
	---	---	---
Species	Without Tail	With Fully Spread Tail	Percentage Reduction of Wing Load
Osprey	54.75	45.28	17.3
Sharp-shinned Hawk	24.05	17.07	29.3
Cooper's Hawk	44.71	32.14	28.1
Goshawk	59.87	43.87	26.7
Broad-winged Hawk	35.42	30.27	20.1
Red-tailed Hawk	49.68	39.84	19.7
Rough-legged Hawk[a]	47.03	37.91	19.3
Bald Eagle	80.94	69.36	14.3

Note: Measurements of wing and tail were from dead birds acquired from the New
York State Museum and New York State Department of Environmental Conservation
(W. Stone). Masses were taken from Snyder and Wiley (1976). One male and one female
of each species were pooled.

[a]Measurements were from one female.

In many raptors the tail is an important airfoil used to generate lift.
It is used for slow gliding (soaring) and maneuvering. Tail areas given
in table 6.1 are for tails held open (with the outer tail feathers sub-
tending about 90°) and closed (outer tail feathers held parallel). Tail
area varied greatly among species. By opening the tail the Sharp-
shinned Hawk reduces functional wing loading by 30% (table 6.2).
Other species realize lighter wing loadings by opening their tails (table
6.2), but not to as great a degree as Sharp-shinned Hawks, because
the tail area of this species is disproportionately greater in size. Bald
Eagles and Ospreys realized the least reduction of functional wing
loading. Is the tail less important as an airfoil in larger raptors than
among smaller birds?

The lower wing load realized by raptors when the tail is spread
permits flight at very slow speeds, and the tail serves as a brake.
(Many species such as the Martial Eagle and some African vultures
use their feet as brakes [Pennycuick 1972a; 1971b] by hanging them
below their bodies while gliding.) Accipiters use their long tails to
make turns while pursuing prey in the forest and during soaring flight.
Raptors of open country also spread their tails during slow flight and
when making turns (fig. 6.6). Kestrels spread their tails and lower
them when wind hovering (not true hovering, since air speeds are positive), presumably to generate more lift than is possible with the wings
alone.

Although the tail is important to most raptors during flight, a few have extremely short tails. The Bataleur Eagle has virtually no tail, yet it soars extensively and captures various prey (Brown and Amadon 1968). The Short-tailed Hawk also has a short tail, yet it soars and often captures avian prey (H. Darrow, personal communication). It is interesting that these species are agile and maneuverable even though they have such small tails. Most species that feed on birds, such as accipiters and falcons, have relatively larger tails than other raptors.

Values of morphological characteristics as reported in table 6.1 are meant for comparisons among species. There is variance within each species that can be attributed to population differences, sexual size dimorphism, and differences between adult and immature birds. Population differences have been demonstrated infrequently for raptors, although some northern populations are larger than more southerly populations, as is the case with Red-tailed Hawks (Geller and Temple 1983).

The detailed measurement of migrating Sharp-shinned Hawks by Mueller, Berger, and Allez (1981a) is an excellent example of the type of data that can be gathered during banding. They found differences between adult and immature hawks as well as between male and female hawks with regard to wing span, wing area, wing width, wing loading, aspect ratio, mass, and tail length and area. Adults of both sexes had longer wings, greater wing area, wing load, and mass, whereas immatures had longer tails and greater tail areas. Based on aerodynamic theory Mueller, Berger, and Allez (1981a) argued that smaller wings, lighter wing loadings, and larger tails of immature birds were adaptations for slower and more maneuverable flight, and that the flight of these individuals was less energetically expensive than that of adults. The slower and more maneuverable flight of younger birds may prevent them from crashing into obstacles while learning how to capture prey. Once they become adept at prey capture, they may be able to deal with faster speeds and the greater cost of flight that is a consequence of maturity. Not only is their argument logical, but their data are plentiful and robust. Mueller, Berger, and Allez (1981a) also present a formula for calculating area of the tail when held in various positions:

$$T_a + (\sin A)(T_1^2), \qquad (6.1)$$

where T_a is tail area, A is the angle subtended by the outer tail feathers, and T_1 is the tail length. Other papers by Mueller and his colleagues document intraspecific variation of flight morphology of other raptors (Mueller, Berger, and Allez 1979, 1981b).

A

Figure 6.6 Photograph of (A) a Merlin and (B) a Northern Harrier making turns as during soaring or maneuvering showing the fully opened tail and outstretched wings (100% of span). (Photographs by Frank Schleicher.)

Structure of Wingtips

A survey of eighteen North American falconiforms reveals diversity in the structure of wingtips. To compare species I examined the primary feathers of study skins (one male and one female from each species) from the Field Museum of Natural History in Chicago. I recorded the

B

number of the longest primary (measured from the wrist), the numbers of all primaries in which emargination (slotting) was visible, flat wing chord, and the length of the slot on each primary (measured from the point of inflexion to the distal end of the feather). A relative measure of slotting was obtained by dividing the length of the slot on a feather by the wing chord. (Measuring individual feathers is intrusive and damages specimens, so these feathers should be measured before a study skin specimen is prepared.)

Table 6.3 Structure of Wingtips for Selected North American Falconiform Migrants

Species	Number of Slotted Primaries[a]	Longest Primary ·(number)	Most Slotted Primary (%)[b]	Shape of Wingtip[c]
Turkey Vulture	3	7 (8, 9)	10 (42.0)	3
Northern Harrier	4	7 (8)	10 (32.4)	1–2
Osprey	4 (3)	8 (9)	9 (33.6)	1–2
Sharp-shinned Hawk	5	7 (6, 8)[d]	10 (38.4)	3
Cooper's Hawk	5	7	9–10 (33.2)	3
Goshawk	5	7	10– 9 (29.8)	3
Broad-winged Hawk	3	7 (8)	10 (29.0)	2
Red-shouldered Hawk	4	7	9 (32.7)	3
Red-tailed Hawk	4	7 (6)	10– 8 (33.3)	3
Rough-legged Hawk	4	7 (8)	10 (31.0)	2
Swainson's Hawk	3	8	10– 9 (27.9)	1–2
Bald Eagle	6	7 (8)	10 (39.2)	2–3
Golden Eagle	6	7 (8)	9 (39.2)	3
American Kestrel	2	9 (8)	10 (17.7)	1
Merlin	2	9	10 (16.3)	1
Prairie Falcon	2	9	10 (15.5)	1
Peregrine Falcon	1	9	10 (14.2)	1
Gyrfalcon	1	9	10 (16.0)	1

Note: Taken from two study skins (one male, one female) from the Field Museum of Natural History.

[a]Number of primaries in which slotting was evident to the naked eye starting with primary 10 and moving inward.

[b]Number of the primary with the greatest percentage of slotting—slot depth measured with calipers from the tip of the feather, percentage of slotting calculated by dividing the slot depth by the flat wing chord.

[c]Shape of wingtip: 1 = pointed tip, high aspect ratio; 2 = intermediate, intermediate aspect ratio; and 3 = rounded tip, low aspect ratio.

[d]Numbers in parentheses are the numbers of other primaries that are as long, or nearly as long, as the longest, measured from the wrist.

The relative length of primaries and the pattern of slotting was related to wing shape. Among falcons with pointed, high aspect ratio wings, the ninth primary was usually longest (table 6.3), whereas among the buteos and accipiters with rounded, lower aspect ratio wings, the seventh primary was usually longest. Primaries six through eight were similar in length in the latter group. The eighth primary was either the longest or equal to the longest (seventh) in Swainson's Hawks, Ospreys, Bald Eagles, and Broad-winged Hawks. The Swainson's Hawk seems to have a higher aspect ratio wing than other buteos, and the wings appear pointed in flight. To a lesser extent this is also true for the Broad-winged Hawk.

The number of primaries that were emarginate was least in species like the falcons with high aspect ratio wings. Slotting was distinct only on the tenth primary of Peregrine Falcons and Gyrfalcons, the largest North American falcons. The ninth primary of the smaller falcons was slightly emarginate, and their primaries do not appear slotted in flight (fig. 6.6). The percentage of the wing that was slotted (primaries nine and ten) never exceeded 20% of the wing chord among the falcons, and the slots were narrow and V-shaped (terminology of Savile 1957). The percentage of the wing that was slotted among the nonfalconid species ranged from 28% in Rough-legged Hawks to 42% in Turkey Vultures. These slots are deeper and shaped like a U. Even the degree of emargination on primaries five through eight of some nonfalconids extended to 20%. The eagles and Sharp-shinned Hawk showed the greatest number of slotted primaries (table 6.3; fig. 6.7). Aspect ratio is smaller among these species than among the falcons, and the tenth primary is slotted strongly (38%–39%). Primaries five and six from these species were also emarginate, ranging between 21% and 31% of wing chord. Buteos were intermediate between the highly slotted accipiter and eagle wings and the relatively unslotted falcon wing.

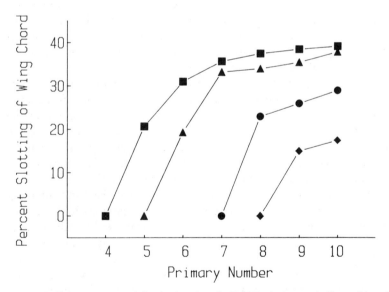

Figure 6.7 Slotting pattern of distal primaries of a Bald Eagle (squares), Sharp-shinned Hawk (triangles), Broad-winged Hawk (circles), and an American Kestrel (diamonds). Percentage of slotting for each primary was determined as the length of the slot (from inflection point to the primary tip) divided by the wing chord.

Among the buteos the relative size of slots ranged from a maximum of 28%–33% for primaries nine or ten (the longer of the two) to a minimum of 23%–29% of chord for primaries eight and seven. Swainson's Hawks and Broad-winged Hawks showed the least slotting among buteos. To show the difference in slotting patterns among species, the relative size of slots has been plotted in figure 6.7 for an eagle, a buteo, a falcon, and an accipiter. By plotting percentages, we can control for size differences (although probably not in an allometric sense).

Although I was unable to quantify the stiffness of primary feathers, it seemed that this characteristic varied with wing shape. Species with pointed wings and little slotting (high aspect ratio) had very stiff primaries, whereas species with rounded wings and extensive slotting had softer, more pliable feathers. Withers (1981) has discussed the aeroelastic properties of feathers, pointing out that they may reduce wingtip vortices, thereby permitting flight at slower air speeds. The flexibility and lack of flexibility of primaries between the slotted versus (relatively) unslotted wing can be seen in figure 6.6. The distal primaries of the Merlin in that photograph are barely bent, yet those of the Northern Harrier are greatly bent. Aeroelastic properties of the distal primaries may functionally enlarge existing slots, thereby increasing maneuverability at slow speeds. This may allow birds with slotted wings to have smaller turning radii than birds with stiff wings and may be one of the reasons raptors with slotted wings are not as capable of prolonged powered flights.

Pennycuick (1983) maintains that low aspect ratio wings are best for "high power at a low forward speed," as during takeoff from the ground or when making turns around vegetation. In addition, he states that "the tip slots may be seen as an adaptation for minimizing the deleterious effects of the low aspect ratio on gliding performance." As with "wingtip sails" on experimental aircraft (Spillman 1978), slots probably reduce induced drag (Pennycuick 1983).

In summary, there is a negative relation between the degree of wingtip slotting and aspect ratio and between how many feathers are slotted and the number of the longest primary. In species like falcons with pointed, high aspect ratio wings, the ninth primary is longest, and slotting is not developed inward from the ninth primary. Species with rounded, low aspect ratio wings have slotted feathers extending from primary five through primaries seven to ten, and the slotting is highly developed. Enigmatic species include the Northern Harrier (fig. 6.6) and Osprey, with relatively high aspect ratio wings and distinct slotting, and the Swainson's Hawk, a buteo with a pointed, high aspect ratio wing without prominent wingtip slots. These three species inhabitat open areas such as prairies, lakes, and oceans.

The variation of wingtip morphology among these eighteen species permits us to speculate on the function of slotting. Because wingtip slots occur most often in species with low aspect ratio wings and in large species, slots may compensate for the lack of long or pointed wings (Pennycuick 1983). Slots may permit slower flight (as in soaring) by reducing tip vortex phenomena or may promote faster acceleration than is possible without slots. The latter is less important in migrants but extremely important for species that need rapid acceleration during prey pursuit and capture or that must take off from the ground after gorging on carrion or prey. Falcons, with high aspect ratio wings, forage in open country and maintain high speeds until prey is captured. When taking off in open country, species with high aspect ratio wings can accelerate over a long distance, unlike forest birds. Slotting in low aspect ratio wings probably reduces induced drag during soaring flight, takeoff, and maneuvers.

Aerodynamic Performance

Methods

The gliding and soaring performance of the species examined in this chapter (table 6.4) was studied in four ways: flying captive birds in wind tunnels, "observing" migrants with tracking radar, photographing (cinematography) or observing (using a range finder and mirror) birds in free flight, and timing individuals as they glided in controlled situations (indoors). Wind tunnel studies are most desirable because they yield accurate measurements of air speed and sink rate and because the researcher is in control. Information about the morphology of individuals flown in the wind tunnel, along with repeated measurements, permits precise determination of flight capabilities. However, it is difficult to gain access to wind tunnels, and birds often will not fly readily in them.

Photographic techniques involve determining sink rate and ground speed from serial photographs of free-flying birds. Wind speed and direction must be measured simultaneously with the photographs so that air speed can be calculated. An ingenious variation of the photographic technique was devised by Pennycuick (1972a), who mounted a motor-driven 35 mm camera on a motorized sailplane. The aerodynamic performance of vultures was determined by photographing them during foraging flights. The difference in performance between the sailplane and the vulture was calculated from the photographic

Table 6.4 Summary of Aerodynamic Performance of Raptors, Sailing Craft, and Various Soaring Animals

Species	Turning Radius (m) 30° Bank Angle	Maximum (Lift:Drag) Glide	Air Speed at Best Glide (mps)	Sink Rate at Best Glide (mps)	Maximum Coefficient of Lift	Cruising Speed (mps) (Foraging or Migration)	Lift:Drag at Cruising Speed
RAPTORS							
Sharp-shinned Hawk[a]	7[b]	9.0	10.5	1.2	1.3	22.5	7.0
Broad-winged Hawk[a]	11[b]	10.5	11.6	1.1	1.5	24.2	8.1
Lanner Falcon[c]	11[b]	10.1	10–14	1.0	1.6	—	—
Red-tailed Hawk[a]	14[b]	9.1	14.5	1.6	1.4	23.9	7.2
Osprey[a]	16[b]	12.2	11.0	0.9	1.5	24.9	8.6
Black Vulture[c,d]	17 (25°)	11.6	13.9	1.2	1.3–1.6	16.8	7.3
White-backed Vulture[e]	19	15.0	13.5	1.1	1.6	16–20+	9–10
Andean Condor[f]	43	NA	NA	NA	0.7	15.0	14.0
NONRAPTORS							
Monarch Butterfly[g]	0.5	3.6	2.6	0.7	—	—	—
Dog-faced bat[c]	—	6.8	8.0	1.2	1.5	—	—
Rock Dove[c]	19[b]	5.5–6.0	18.0	3.0	1.3–1.5	—	—
Fulmar[f]	13[b]	8.5	10.7	1.3	1.8	18.7	—
Common Crane[b],	20[b]	15.6	14.0	0.9	1.6	20.0	12.0
Hang glider[h]	29[b]	11.0	12.0	1.0	0.9	15.0	7–8
Sailplane[h]	114[b]	38.0	23.7	0.6	1.0	51.0	17.0

Note: Source for species as in table 6.1
Methods used to determine aerodynamic performance:
[a]tracking radar
[b]calculation from standard equations
[c]wind tunnel
[d]ornithodolite
[e]photographic method from motor-glider
[f]cinematography
[g]flown in still air
[h]dealer specifications

record and the known performance of the sailplane. Analysis of film is tedious, and there is no guarantee that free-flying birds will use their full range of air speeds during filming.

Hankin (1913) studied the soaring behavior and performance of various raptors by combining a range finder, mirror, and metronome. Because his research was conducted before the principles of flight were formalized, his observations are remarkable (and seem to be precise). While observing the movements of a soaring bird in a mirror, Hankin made marks on the mirror at one-second or half-second intervals (indicated by a metronome). These measurements permitted estimation of how long it took a bird to complete one circle. The range finder was used to estimate the diameter of the circle.

Tracking radar measures ground speed and sink rate of migrants more rapidly than the other techniques. Furthermore, high-flying birds can be studied with ease. For gliding performance to be determined, wind speed and direction must be measured at the same altitude and time as the migrant being tracked. Kerlinger (1982a) measured wind speed and direction at the height of migrants by tracking helium-filled weather balloons that carried radar reflective material. There are two drawbacks to the system. First, tracking radar is expensive. Second, wind gusts (updrafts and downdrafts) that cannot be detected (or controlled) influence the accuracy of measurements. Tracking migrants for several kilometers resolves this problem to some extent. Third, migrants cannot be induced to fly at particular air speeds, especially slow speeds, that might be of interest to researchers.

The last method entails direct measurement of air speed and glide angle of migrants flying in a room where wind is absent (Gibo and Pallett 1979). This is possible only for small, slow-flying organisms such as insects. The methods used to determine the performance of the organisms described in this chapter are given in table 6.4.

General Flight Behavior

Characterizing general flight behavior of migrants is not easy and has been done by few researchers. Ideally, flight behavior of a migrant should be quantified as contingencies based upon atmospheric and topographic conditions from which a statistical analysis can be conducted.

Most raptors rely on thermal soaring and gliding flight during migration, infrequently resorting to flapping flight or ridge gliding. Broun (1949), Rudebeck (1950), Haugh and Cade (1966), Haugh (1972), and Heintzelman (1975) have briefly characterized the flight behavior of migrants, but these observations were made in ridge or coastal situations. During radar studies I observed that most migrants use thermals when they are present and that thermaling flight predominated among most species.

The studies of migrating hawks at Falsterbo in southern Sweden by Rudebeck (1950) document gross flight behavior as well as interspecific differences. He noted the different propensity of various species to continue migration on days or during hours of the day when thermal conditions were poor or absent. Most of the species he studied relied on thermals when they were present. He found that the Common Buzzard, Honey Buzzard, and European Sparrowhawk relied on soaring flight 90%, 60%–80%, and 15%–30% of the time, respectively. He further noted that species such as the sparrowhawk did not always soar when thermal conditions were good and that this species often migrates early in the morning before thermaling conditions are well developed. Rudebeck gives no further quantification of the different types or modes of flight used by the migrants he observed or of changes in behavior with weather conditions.

Dunne's observations are some of the few quantitative measurements available. During two spring migrations Dunne (1977, 1978) noted soaring flight among more than ten species of migrants. Because Dunne made his observations at a ridge (Raccoon Ridge, New Jersey), it is possible that migrants were influenced by updrafts from the ridge. Of interest is Dunne's finding that most raptors relied on thermaling flight and resorted to intermittent gliding or powered flight less frequently. Between 23% and 38% of Ospreys, Northern Harriers, Goshawks, Red-tailed Hawks, and American Kestrels were observed soaring, whereas 40% to 62% of Sharp-shinned Hawks, Cooper's Hawks, Red-shouldered Hawks, Broad-winged Hawks, and Rough-legged Hawks were seen soaring. Many of the migrants he observed were not following the ridge, so it seems they relied primarily on thermals.

Cochran's (1972, 1975, 1985) radiotelemetry studies of Sharp-shinned Hawks and Peregrine Falcons showed that many individual migrants use soaring flight throughout the day in their migration. For Peregrine Falcons migrating inland from the Atlantic Ocean, soaring and gliding flight, presumably in thermals, accounted for 90% of the time spent in flight. In the Midwest this species spent between 37% and 69% of their flight time soaring in thermals and between 31% and 63% of their time gliding between thermals. Individual Sharp-shinned Hawks, a species that has been characterized as not being thermal dependent (Heintzelman 1975), soared nearly continuously on some days (Cochran 1972). On other days only 25% of the time was spent soaring in thermals. Cochran was able to differentiate between time spent soaring and gliding from the changing strength of the radio signal from the transmitter. As the antenna attached to the migrant's tail changed orientation during soaring flight, the signal became stronger, then weaker. From this difference Cochran was also able to quantify the time it took for a bird to complete a single circle within a thermal.

Studies like Dunne's and Cochran's need to be conducted at a variety of topographic sites and with varying weather. Quantitative data can then be used to compare the general flight behavior of different species and to test hypotheses about the best type of flight given certain atmospheric conditions.

Gliding Performance

Empirically derived glide polars are available for only eight raptors and several nonraptor species. Few researchers have been able to obtain precise measurements of sink rate at varying air speeds (fig. 6.8). The most important aspects of the aerodynamic performance of these species are summarized in table 6.4. Comparing the gliding ability of birds, insects, and bats to that of sailplanes, it is surprising to see that

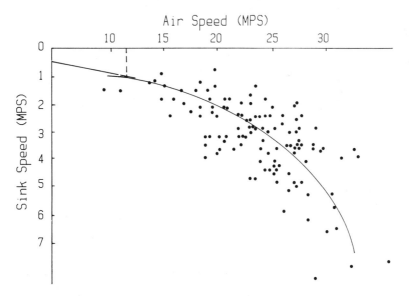

Figure 6.8 Glide polar (performance curve) of a Broad-winged Hawk as determined with tracking radar (from Kerlinger 1982a).

animals do not perform favorably (figs. 6.9, 6.10, table 6.4). Maximum glide ratios (max L/D) for raptors are greater than for pigeons, bats, and insects; about the same as for cranes, fulmars, and hang gliders; and lower than for sailplanes. The maximum glide ratios of raptors ranged from about 9–15:1, values one-half to one-fourth those reported for sailplanes (fig. 6.9). This means that in a straight glide from a given altitude sailplanes can glide three to four times as far as raptors. Even an albatross, with a glide ratio of 18–24:1 (Wood 1973; Pennycuick 1983), realizes only one-half the maximum glide ratio of a sailplane (fig. 6.10). In addition, air speed of raptors at best glide is only one-half to one-fourth that of sailplanes. A final difference between the performances of sailplanes and raptors during straight gliding flight is the shape of the glide polar (fig. 6.9). Polars of raptors and other flying animals drop off more steeply than those of sailplanes. Thus, flying animals fly slower, with lower glide ratios and steeper glide polars than do sailplanes.

The largest soaring birds, including Andean Condors and African vultures (*Gyps*), realize maximum glide ratios of 13–15:1 at about 13–15 mps. Ospreys, with a mass less than one-half that of large vultures, have best glides that are slightly lower, averaging 12:1 at speeds

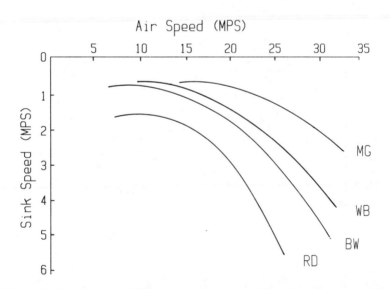

Figure 6.9 Comparison of gliding performance curves (glide polars) of a motor glider (MG, from Pennycuick 1971a), African White-backed Vulture (WB, from Pennycuick 1971a), Broad-winged Hawk (BW, from Kerlinger 1982a), and a Rock Dove (RD, from Pennycuick 1968b).

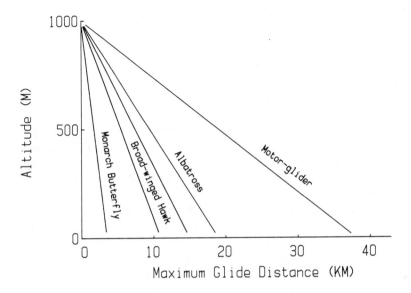

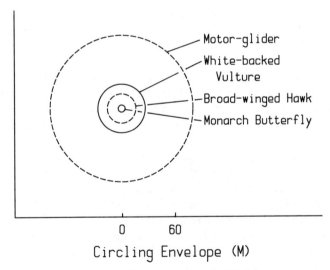

Figure 6.10 Comparison of maximum glide ratios and circling abilities (turning envelope or radius at a 30° bank angle) of a motor glider (from Pennycuick 1971a), African White-backed Vulture (from Pennycuick 1971a), Broad-winged Hawk (from Kerlinger 1982a), and Monarch butterfly (calculated from information given in Gibo and Pallett 1979).

of about 11 mps. Smaller soaring raptors, in the 400–1,300 g range, realize best glides of 8–11:1. From these values we may extrapolate to raptors of nearly all sizes and shapes. Species like the Bald Eagle and Golden Eagle probably realize maximum glide ratios of 13–14:1 at air speeds near 13 mps. Of course, eagles and large falcons can glide faster than vultures in steep glides. Although these values are not comparable to those of sailplanes, they are better than those of hang gliders. The first hang gliders, also called sky surfers, had only 4.5:1 glide ratios (Stong 1974). Modern hang gliders compete with some birds (maxL/D = 10–13:1), at least in the maximum glide ratio category, but overall the birds still have better performance (and safety records as well).

The values presented in table 6.4 are not all for gliding flight. Sharp-shinned Hawks, Common Cranes, and some other species often use power-assisted glides. Sharp-shinned Hawks sometimes incorporate short bursts of flapping flight (three to seven flaps per burst) during interthermal glides. Other raptors resort to this strategy, which increases glide ratios by increasing lift and thrust. Even Broad-winged Hawks, a species reported as an obligate gliding migrant, use power-assisted glides during short water crossings. These "glides" have been termed "partially powered glides" by Pennycuick, Alerstam, and Larsson (1979), so the data for species like Sharp-shinned Hawks should be interpreted accordingly. It is probable that small hawks perform so poorly during pure gliding that they must use powered glides, which are a compromise between energetically expensive powered flight and cheaper gliding flight. Powered glides flatten "glide ratio" and increase speeds between thermals. A species like the Common Crane uses partially powered glides when thermals are weak or undependable or to increase cross-country speed (Pennycuick, Alerstam, and Larsson 1979).

Maximum lift coefficients for the eight raptors ranged from 0.7 for the Andean Condor to 1.6 for the Lanner Falcon (table 6.4). The values in table 6.4 were calculated using only wing area. If tail area were included, the values would be lower. Because of the important role of the tail as a variable-surface airfoil, perhaps the influence of tail area should be investigated. If this was done for the birds in table 6.4, maximum lift coefficients would decline, approaching that of sailplanes and hang gliders, which do not have movable tails. The asymmetric structure of the outer tail feathers (rachis being off center toward the leading edge) of raptors and other birds is evidence that these feathers have an important aerodynamic function.

Because wing span and area are not available for the range of air speeds used by raptors, an estimation of lift coefficients at various air speeds is risky. Tucker and Parrott (1970) measured gliding performance of a Lanner Falcon in a wind tunnel and arrived at a maximum coefficient of lift of 1.6. The value of 0.7 reported by McGahan (1973a) for the Andean Condor seems rather small, although Mac-Cready (1976) reported values of 0.91 for Turkey Vultures and 0.85 for Black Vultures. Because Pennycuick (1983) included a massive data set in his study of maximum lift coefficients, his determination of 1.35 for the Black Vulture seems to be robust. These species were observed flying in thermals at air speeds of 7.8 mps. Coefficients of lift for four species were determined by Kerlinger (1982a) using equation 5.9 (Pennycuick 1975). Lift coefficients for these species at different air speeds are given in figure 6.11. Approximate wing span and area for these air speeds and species are available in figure 6.3. The relation between air speeds and lift coefficient is hyperbolic, lift coefficients decreasing with air speed. The curves appear to converge at faster air speeds. Curves were flattest for Ospreys and steepest for Sharp-shinned Hawks. It appears that a large wing span promotes a slower decline in lift coefficient with increasing air speed. Maximum coefficients of lift ranged from 1.3 at 6 mps for the Sharp-shinned Hawk to

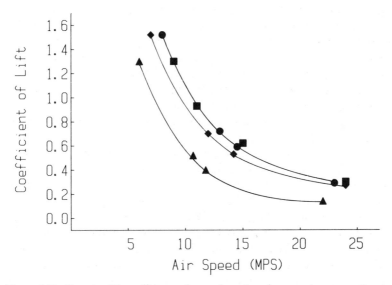

Figure 6.11 Changing lift coefficients of several species of raptors (square = Osprey, circle = Red-tailed Hawk, diamond = Broad-winged Hawk, triangle = Sharp-shinned Hawk) flying at a range of air speeds. Slowest speeds are the speeds used while soaring in thermals; faster speeds are used during interthermal glides.

1.5 at 7–8 mps for the Broad-winged Hawk. Minimum lift coefficients were 0.14 or more at air speeds used for migration (chap. 11).

Circling Performance: Turning Radius within Thermals

To compare the circling performances of raptors and other soaring "craft," I used equation 5.12 (Pennycuick 1975). Because a bank angle of 30° is normal for soaring birds (Pennycuick 1972a), sailplanes, and hawks (migrating in New York and Texas, personal observations), I used this angle to calculate the turning radii in table 6.4. The maximum lift coefficients, weights, and wing spans were taken from the literature (tables 6.1, 6.4), using 100% of wing span and wing area in the calculations. These equations should be reliable for predicting the circling performance of raptors, except for species such as the Sharp-shinned Hawk, which often uses powered glides, thereby generating more lift than during gliding flight. By calculating turning radius, we can "control" all variance resulting from behavior (motivation? and experience), thermal characteristics, and bank angle. Thus, the comparison is for "ideal" birds and may not be a precise representation of the performance of these species and aircraft in natural situations. Because the calculations are based on a powerful theoretical framework (Mises 1959; Welch, Welch, and Irving 1968; Pennycuick 1971a, 1975), they should be accurate predictors of performance. The turning radius of raptors ranged from 7 to 43 m. It is probable that these estimations bracket the extreme values for raptors, with most migratory species falling between 10 and 30 m.

A turning radius of 43 m for the Andean Condor may be large and, in theory, should be smaller than for the hang-glider. It was interesting to note that a Fulmar realizes the same turning radius as a Red-tailed Hawk, even though the hawk is 400 g heavier. If tail area were included in the calculation the discrepancy would be larger. The reason for this discrepancy may be differences of wing and tail structure. Fulmars have long, narrow, pointed wings and relatively small tails, whereas Red-tailed Hawks have wide, rounded wings with slotted primaries and relatively large tails. When wings and tail are spread, the Red-tailed Hawk can vary its functional wing loading more than the Fulmar, allowing the bird to make tight circles at very slow air speeds. Adaptive differences seem to be related to the environments these species inhabit. Buteos must use small thermals over land, whereas the Fulmar uses wind-generated updrafts and can use flapping flight for long distances. In addition, strength of wind over the ocean may provide more constant lift than is present over terrestrial habitats. Note also that the Rock Dove, even though it weighs less than a Red-tailed Hawk, cannot turn in as small a circle. Again, it is adapted more for flapping than gliding and resorts to the latter when circumstances permit.

The turning radius of soaring birds has been measured for few species. Because turning radius is dependent upon bank angle, the researcher must measure radius of turns at a range of bank angles to describe the performance of a given species and to compare species. The bank angle used by soaring birds varies with the strength and diameter of a thermal. In large, strong thermals a shallow angle of bank is used, with associated slow rate of sink. For most raptors the radius of circles at speeds near the minimum air speed and bank angles of about 30° ranged from about 6 m for the Sharp-shinned Hawk to over 40 m for the largest vultures (table 6.5), similar to values calculated for a 30° bank angle (table 6.4). These birds range in weight from 100 g to more than 5,000 g, and from a span of about 0.5 m to over 2 m—nearly the entire range of raptors. The range of angles of bank used was between 20° and 35°.

The measurements of Pennycuick (1971a, 1983) for vultures, frigatebirds, and pelicans are the most complete set available for soaring flight. Using an ornithodolite and a motorized sailplane, he has been able to measure the bank angle, air speed, and turning radius of

Table 6.5 Empirical Determinations of Circling Performance of Raptors, Other Birds, and Aircraft

Species	Circling Radius (m)	Bank Angle	Circle Time (sec)	Air Speed (mps)	Reference
Black Vulture	17.1	24.7	12.5	8.8	Pennycuick 1983
Brown Pelican	18.0	22.9	13.3	8.6	Pennycuick 1983
Magnificent Frigatebird	12.0	23.7	10.6	7.2	Pennycuick 1983
Black Vulture	24.4	20–30	14.1[a]	10.9	MacCready 1976
Turkey Vulture	12.5	20–30	10.1[a]	7.8	MacCready 1976
Magnificent Frigatebird	12.5	20–30	10.1[a]	7.8	MacCready 1976
Lappet-faced Vulture	15.0	35.0	9.4	10.0	Pennycuick 1971a
Indian White-backed Vulture	40–50	?	13–16	10.0	Hankin 1913
Kite (*Milvus* sp.)	12.0	?	7–9	5.0[a]	Hankin 1913
Broad-winged Hawk	15.4	24.5	13.9[a]	7.0[b]	Welch 1975
Sharp-shinned Hawk	5.7–11.5[a]	?	9–12	6.0[b]	Cochran 1972
Hang glider	24.4	30.0	16.8[a]	9.1[c]	MacCready 1976
Motor glider	61.0	35.0	18.3[a]	20.5	Pennycuick 1971a

[a]Calculated from values given by author.

[b]Minimum air speed from table 6.4.

[c]Approximate minimum air speed given by Stong (1974).

hundreds of free-flying birds. Some of the discrepancies between empirical (table 6.5) and calculated values (table 6.4) for circling performance may be reconcilable by differing bank angles and thermal conditions. Whatever the reason, the phenomenon needs to be studied more thoroughly.

Several conclusions can be made about circling performance from tables 6.4 and 6.5. Most important, the circling radii of smaller species are smaller than those of larger species. This enables smaller birds to use smaller thermals while flying at shallow bank angles, thereby reducing sink rate. These birds can increase their bank angles, realizing still smaller turning radii, while increasing their sink rate only slightly. Soaring birds are able to circle in smaller circles than sailplanes and hang gliders. Raptors up to the size of vultures have turning envelopes one-third to one-quarter that of sailplanes.

Discrepancies among the calculated values of circling radius can be explained by Pennycuick's (1983) recent demonstration that species with high aspect ratios can soar better than those with lower aspect ratios. Measurements of soaring performance of pelicans and Black Vultures, species with different aspect ratios but similar wing loadings, showed that the pelicans have similar turning envelopes (table 6.5). This means that, in addition to wing loading, wing shape is important for determining soaring performance (turning envelope).

The adaptive value of a small turning radius to a raptor should be obvious. In chapter 4 we saw that thermals vary in size and strength and that these characteristics vary with time of day. A small turning radius allows raptors to use small or weak thermals and the area with strongest updrafts at the core of a thermal (fig. 4.9). Sailplane pilots who depend on thermals rarely take off before 1000–1100 h and must land before 1600–1700 h. Small raptors are able to take off earlier (before 0800–0900 h; Pennycuick 1971a; Kerlinger and Gauthreaux 1985a) by using small, weak thermals and can continue flying after sailplanes are forced to land. They can also fly on days that sailplane pilots consider poor for flying. Thus, the ability of raptors and other birds to use small, weak thermals efficiently offsets their poor gliding performance and slow speeds.

Sailplane pilots (Pennycuick 1972a; personal observations) frequently are frustrated by not being able to compete with soaring birds in a thermal. By comparing the circling envelope of a motor glider with that of a vulture, Pennycuick (1971a) clarified the differences between birds and aircraft. Whereas the sink rates of vultures and sailplanes are about the same, the turning radius of a motor glider is more than two times that of a vulture (table 6.5). If thermals were large enough, both could soar easily. However, the motor glider requires thermals more than four times as large as the vulture needs, because area increases as the square of thermal radius. Pennycuick also points out that the "solar energy required to initiate the thermal is correspondingly greater."

The following incident illustrates the magnitude of the difference between soaring performance of sailplanes and soaring birds. On a May day in 1986 near Hattiesburg, Mississippi, I was the second passenger in a Grob-109 motorized sailplane. As we were soaring in a thermal at about 700 m, we sighted four or five Turkey Vultures some 300 m below. Despite a climb rate of about 1.3 mps, the vultures passed us within 2–4 mins. Their rate of climb had to exceed 3 mps (not unlike the climb rates of migrating hawks). Since they were on our inside wing (closest to the center of the thermal), they were probably flying in the strongest updrafts near the thermal's core. Because our turning radius exceeded theirs by a large margin, we could not utilize the region of strongest lift. Although adequate lift was available for vultures to fly cross country on that day, we could not do so because thermals were too small to be used by the Grob-109. This was a dramatic lesson as to the advantage of a small turning envelope. These birds must soar every day to forage, so it is essential that they be adapted to exploit weak or small thermals.

Summary and Conclusions

The morphology and aerodynamic performance of migratory raptors were examined and compared with those of insects, bats, other birds, hang gliders, and sailplanes. A continuum of morphotypes of raptors was described: falcons, kites, harriers, buteos, and accipiters. Species like falcons and kites have high aspect ratio and pointed wings with relatively unslotted tips. Buteos, vultures, eagles, and accipiters have lower aspect ratio and rounded wings with slotted tips. The wing area, tail area, wing span, mass, wing loading, aspect ratio, functional wing loading, functional aspect ratio, and wingtip structure of numerous raptors were compared with each other and with those of other flying objects. Quantitative data describing the flight morphology of raptors and most other birds are scarce, and the methods and units used to report flight morphology are variable. A standardized method for measuring avian flight morphology is needed. When reliable data are available, a thorough description and analysis of the flight morphology of raptors will be possible.

In their aerodynamic capabilities raptors are equivalent to other soaring birds but differ from sailplanes. Raptors are smaller ˉand weigh less than sailplanes, so they are slower and have steeper glide polars. The maximum glide ratios of raptors whose aerodynamic performance has been studied range from about 8–9:1 (at 10–12 mps), in the case of Sharp-shinned Hawks and possibly Red-tailed Hawks, to 12–14:1 (at 11–13 mps) for Ospreys and eagles, to 14–15:1 (at 13–15 mps) for large vultures. Sailplanes realize best glides of 37–50:1 at 20–24 mps. Maximum speeds during interthermal glides for most raptors are over 25 mps. During migration Red-tailed Hawks, Ospreys, and Broad-winged Hawks average glide ratios of 7–10:1. Turning radii of raptors and other birds are smaller than those of sailplanes, ranging from less than 10 m to about 40 m at bank angles of 20°–35°. Natural selection has favored the ability to use weak updrafts over fast speed and a flat glide polar. By virtue of their ability to vary wing and tail planform by behavioral means, raptors exploit a wider range of air speeds and glide ratios than would be possible if wing span, wing area, and tail area were fixed and wingtip configurations were constant. The ability to modify morphology behaviorally allows raptors to stay aloft (and migrate) in all but the weakest updrafts, whereas sailplanes can stay aloft only when updrafts are strong and abundant.

7

Flight Direction: The Roles of Wind, Topography, and Geography

Orientation and navigation have been studied more than any aspect of animal migration. How salmon return to the rivers where they were spawned, turtles home to oceanic islands, and birds return to breeding or non-breeding sites has fascinated and frustrated scientists for decades. Most studies have focused on the cues that animals use to orient. A recent review by Able (1980) has made it clear that animals orient by several cues, including celestial bodies (sun, moon, stars), olfactory (chemical) and geomagnetic input, sea currents, wind and polarized light. Furthermore, some animals may possess hierarchical systems in which multiple cues are utilized. These cues are usually not enough to permit an animal to find its way home or to some other destination. In addition to orientation cues (a compass), an animal must also "know" where it is and where it is going. To accomplish this it must have a "map" by which it can navigate.

During migration hawks experience varying wind, topographic, and geographic conditions. Wind can blow a bird off course, make progress impossible, or promote rapid flight. When migrants encounter a ridge, mountain range, or coastline they must "decide" whether to follow topographic features or ignore them. These same migrants also must decide whether to cross barriers such as large bodies of water, deserts, prairies, or mountain ranges. Thus, wind, topography, and geography interact to cause hawks to deviate from a direct flight (straight line) during migration.

Little research has focused on the orientation of hawks during migration, and few studies have examined navigation (but see Mueller and Berger 1969; Drost 1938). This chapter focuses on the flight direction of hawks as it is related to wind, topography, and geography—that is, it tests the wind drift hypothesis.

Evidence for Navigation

Demonstrating that an animal possesses navigation abilities is difficult. Only two studies have addressed the question of navigation for migrating raptors. Mueller and Berger (1969) concluded that the navigation system of migrating hawks was "crude" and thus resulted in large geographic "displacement" from migratory goals. In addition, they stated that "hawks wander widely while returning to their breeding areas." Evidence for their conclusions comes from recoveries of birds banded during spring migration and from the flight directions of migrants passing their banding station. Of 791 birds banded during spring migration at Cedar Grove, Wisconsin, 23 (2.9%) were recovered, 7 of them south of Cedar Grove. Mueller and Berger suggested that some birds had "overshot" their breeding area and had to migrate southward. Flight directions of migrants revealed that about 40% of all migrants were moving southward in "reverse migration." Unlike true reversed migration (Raveling 1976) Mueller and Berger saw birds going both north and south on the same day. This may support the hypothesis that some birds overshoot breeding sites during spring migration.

The "displacement" study of Drost (1938) is significant because it was one of the first field experiments with migrating birds and because it involved two age classes of migrants. After capturing European Sparrowhawks during autumn migration in Helgoland, off the northwestern coast of West Germany, Drost transported them 600 km east to western Poland, where they were released. Of 209 hawks released, 36 (17.2%) were subsequently recovered. Immature birds tended to move along a northeast to southwest axis (fig. 7.1), the same as most sparrowhawks recovered after banding on the coast of western Europe (and parallel to the axis of migration of birds captured at the Courland Spit along the Baltic Sea of the USSR as reported by Paevskii 1973). Adult birds, however, flew more toward the west, as if they possessed a map and were trying to find the axis along which they had migrated in previous years. These results suggest that immature sparrowhawks possess a "genetically" defined axis of migration along which they orient during their first migration. Adults seem to have true navigational abilities, that is, a map and a compass. Perdeck (1958) and others have replicated Drost's (1938) pioneering study using passerines. Radiotelemetry studies of displaced migrants might be particularly rewarding.

The conclusions of Mueller and Berger (1969) and Drost (1938) may or may not be reconcilable. Mueller and Berger's sample sizes are

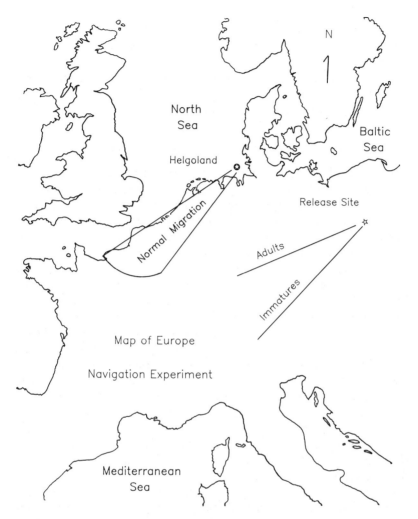

Figure 7.1 Summary of a displacement experiment by Drost (1938) of European Sparrowhawks trapped at Helgoland, Germany. Birds released in Poland were displaced from the capture site, whereas the remaining birds were released at Helgoland. Experienced (adult) migrants flew on more westerly courses toward the coast of Europe than unexperienced (immature, first year) migrants, whose migratory paths were nearly parallel to those of "normal" birds that were released at the coast where they were trapped.

small, and their conclusion should be considered tentative. It is not likely that raptors have poor navigational abilities, especially those that breed in the far north and must find their breeding sites rapidly after spring migration. It is probable that there is variation among and within species depending upon the intensity of selection on that particular population. Thus, the question is open.

Evidence for Wind Drift: Hawk Migration Counts

Two papers have been influential in shaping the way hawk migration has been viewed and studied. In 1936, Allen and Peterson presented an illustrated model of how raptors, specifically Sharp-shinned Hawks, arrived at Cape May Point, New Jersey, during autumn migration. They stated that these small hawks normally used an inland migration route in autumn but arrived at the Atlantic coast after being drifted there by north to west winds. Once at the coast, they were believed to follow it until they arrived at Cape May Point. Their figure (fig. 7.2) shows hypothesized flight directions of Sharp-shinned Hawks with north to west winds as opposed to other winds. The evidence for drift cited by Allen and Peterson was the large number of hawks counted at Cay May Point with winds from the north to west and the small number counted on other winds. The hypothesis of Allen and Peterson was reformulated and stated more formally by Mueller and Berger (1967a). They used the definition of leading line proposed by Geyer von Schweppenburg (1963; but see Rudebeck 1950) and generalized the hypothesis to include topographic situations such as tree lines and mountain ridges. These two papers probably have been cited more than any others in the hawk migration literature.

The evidence for drift is mostly from studies in which hawk migration counts were used as a primary source of data. Mueller and Berger (1961, 1967b) showed that the largest counts of migrating Sharp-shinned Hawks occurred at Cedar Grove, Wisconsin, along the Lake Michigan shore on days with west winds. Haugh (1972) analyzed data from several hawk lookouts, including Hawk Mountain on the Kittatinny Ridge and Derby Hill on the southeast shore of Lake Ontario (fig. 2.1). He demonstrated that more hawks were counted in autumn at Hawk Mountain on west to north winds than other winds and that more hawks were seen on south to east winds at Derby Hill in spring than on other winds. Although he states that many hawks would have arrived at the shore of Lake Ontario on any wind because of its east to west orientation, he concludes that wind drift is an important factor in creating large flights of hawks at both of these top-

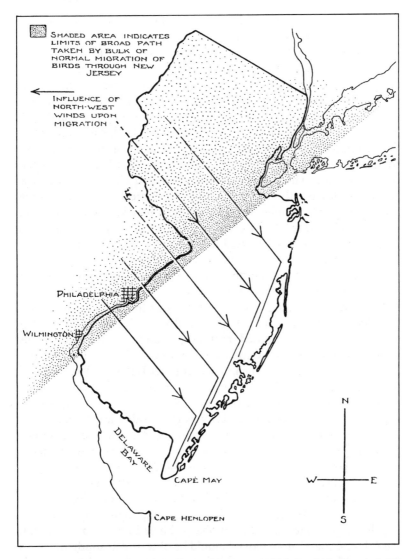

Figure 7.2 The classic figure by Allen and Peterson (1936) in which the wind drift hypothesis was proposed to explain large numbers of migrating Sharp-shinned Hawks at Cape May Point, New Jersey (courtesy of the American Ornithologists' Union). The idea has been generalized to explain large numbers of birds along other bodies of water and along ridges.

ographic situations. Heintzelman (1975) has also invoked wind drift to explain large numbers of hawks at ridge sites.

The idea that hawks follow leading lines and that leading lines are effective probably originated when early researchers noted the vast numbers of raptors seen migrating along ridges and coastlines. The number of some species counted at leading lines is great but may be small compared with the total population. For example, about 10,000 Broad-winged Hawks are counted annually at Hawk Mountain, along the Kittatinny Ridge. This number seems large, but it is small when compared with the total number of Broad-winged Hawks in eastern North America. The majority of Broad-winged Hawks are dispersed across eastern North America, where there are few ridges or other leading lines. The number of Sharp-shinned Hawks counted in an autumn at Cape May often exceeds 50,000. This may be a significant proportion of the hawks breeding in eastern Canada and New England, but we do not know how large. In parts of North America where there are no leading lines few migrating raptors are counted, but that does not mean that they do not migrate at those places.

One author has questioned the wind drift hypothesis as presented by Allen and Peterson (1936) and Mueller and Berger (1967a). Murray (1964, 1969) proposed that large numbers of Sharp-shinned Hawks migrate through Cape May regardless of wind, but that they descend to low altitudes when they encounter Delaware Bay only with north to west winds. When these birds descend they can be counted easily, as well as recounted (Murray 1964; chap. 8). Thus, Murray hypothesized that a counting bias was responsible for the differences in numbers seen. He does concede that the Atlantic coast may be a diversion line whereby migrants are diverted south after encountering the New Jersey shoreline. A diversion line differs from a leading line (sensu Mueller and Berger 1967b) in that wind drift need not be invoked to explain how birds arrived at the coastline or ridge.

Murray (1964) makes another interesting point that has been largely ignored. He states that if Sharp-shinned Hawks are drifted to the Atlantic coast, more hawks should be observed at southern coastal sites. As Murray points out, this is not the case. Few hawks can be counted along the New Jersey coast except at the end of the Cape May peninsula. The same argument may hold for ridges. The numbers of hawks counted along ridges should increase as one travels south (in autumn). There is no evidence, to date, that shows that hawks actually follow leading lines for "long distances." Counts do not demonstrate that hawks follow ridges because they show only how many hawks fly past a given point. Many of the largest flights of Broad-winged

Hawks recorded at Hawk Mountain have been observed crossing the ridge, not following it (A. Nagy, personal communication). Klem et al. (1985) examined the behavior of numerous species of hawks at a "gap" in the Kittatinny Ridge. They found that few hawks left the ridge. Obviously, hawks do follow ridges and coastlines for short distances.

Murray's hypothesis has been shown to be at least partially correct by Kerlinger (1984) and Kerlinger and Gauthreaux (1984). There are several other hypotheses that can be invoked to explain the larger numbers of hawks counted at ridges and coastlines in autumn with north to west winds and in spring with south to east winds. First, it is probable that more birds migrate when winds are favorable for migration; from south in spring and north in autumn. Second, birds fly faster (and farther) with following winds than with opposing winds, so that if the same number are aloft on days with opposing and following winds more will be counted with following winds. If any of the alternate explanations is correct, the evidence for wind drift becomes less convincing. Only radar or radiotelemetry studies can be used to test these alternate hypotheses and to determine the effectiveness of a leading or diversion line.

There is yet another reason why some raptors migrate along coastlines. Species seen along the Atlantic coast are mostly bird eaters such as Sharp-shinned Hawks, Cooper's Hawks, Merlins, and Peregrine Falcons. In addition, American Kestrels and Ospreys are present in coastal counts in larger numbers and proportions than along inland ridges. Broad-winged and Red-tailed hawks occur along the Atlantic coast, but not in large numbers. For Sharp-shinned Hawks, Cooper's Hawks, Peregrines, Merlins, and Ospreys, the coast is an enormous cafeteria. Millions of passerine and shorebird migrants frequent the coast during autumn. These migrants are fatigued or inexperienced and readily fall prey to raptors.

It is my opinion that hawk migration counts tell us little about migration strategies, especially regarding flight behavior. Hawk migration counts are an "indirect" means of studying flight behavior (Murray 1964; Mueller and Berger 1967b), and they should be used with caution. As Murray pointed out in his "heretical" paper, drift has not been demonstrated for any species that migrates in daylight (over land). Why should hawks be different? The only appropriate and acceptable means of studying orientation (other than laboratory tests, i.e., funnel or cage experiments) and testing the wind drift hypothesis is by measuring flight directions of birds during migration at situations where leading lines are absent.

Directional Data and Tests of the Wind Drift Hypothesis

In this section I summarize and critique studies in which the wind drift hypothesis was tested. These tests are crucial to understanding the orientation behavior of hawks during migration because they involve directional data. Most involved measuring wind and flight direction of individual migrants or flocks in topographic situations where leading lines were absent. It is of importance that several species were examined at varying topographic situations and geographic locations during both spring and autumn migration.

Theory and Definitions

Before testing the wind drift hypothesis I must present some theory. For a bird to migrate from point A to point B, it obviously must fly. If there is no wind, it can face directly toward its goal as shown in figure 7.3A. In this figure the terms heading and track are introduced. A bird's track is the direction it realizes over the ground, whereas the heading is the direction it faces. Track and heading are the same when there is no wind.

It is rare for the atmosphere to be still; there is almost always some wind. When wind is present track and heading are different unless the wind is parallel (following or opposing wind) to the flight path. If the wind is from the left or right of the flight path, track and heading differ as in figure 7.3B, 7.3C by some angle θ. In figure 7.3C, I have shown how a bird must fly in order to compensate for a wind blowing lateral to its preferred direction of flight. The greater the wind component perpendicular to the flight path, the greater θ must be to compensate for lateral wind or the faster a migrant must fly along the same heading.

If a bird does not compensate, or if it merely faces its preferred direction and flies along a fixed heading, it will drift from its course (fig. 7.3D). Also, as wind speed increases there is a point at which a bird can no longer compensate because it cannot fly fast enough or at a large enough angle (θ) to realize its destination. The point at which this occurs is a function of the maximum air speed the bird can maintain, which is determined by aerodynamic and energetic constraints. Evans (1966) and later Kerlinger and Gauthreaux (1984) termed this the threshold for drift.

Aircraft pilots refer to the angle between heading and track as the drift angle. Students of bird migration have defined several types of drift. If a bird flies along a fixed heading toward its destination, it will be drifted from that heading by lateral winds. This is the most common type of drift in the avian literature. A second type of drift occurs

A. No wind, track = heading

B. Vector Triangle

Wind

C. Wind ⟶, Drift

D. Wind ⟶, Compensation

E. Wind ⟶, Partial Compensation

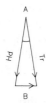

F. No wind, no displacement

Soaring

Gliding

G. Wind ⟶, Passive Displacement
with Drift

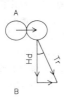

H. Wind ⟶, Passive Displacement
with Compensation

I. Wind ⟶, Passive Displacement
with Partial Compensation

J. Wind ⟶, Passive Displacement
with Overcompensation

Figure 7.3 Schematic diagram showing the relations among headings, tracks, wind direction, soaring direction, compensation, and drift.

when a bird flies downwind, even though that wind may not be in the appropriate direction for migration. This is called "downwind drift" (Williamson 1955; Gauthreaux and Able 1970). The third type of drift, called "partial drift," occurs when an animal compensates for lateral wind, but not completely (fig. 7.3E).

The models presented in figure 7.3A-E are used to test whether passerines or other powered migrants are drifted by wind. Because most raptors use soaring and gliding flight during migration, testing the wind drift hypothesis is not as straightforward. The models above can be used, but only for birds gliding between areas of thermal or other lift. To test whether raptors are drifted by wind from a preferred course, we must look at flight direction during both soaring and gliding flight.

In figure 7.3F-J are situations analogous to those presented in figure 7.3A-E. The difference is that soaring flight is included. As with straight flapping flight, no drift or displacement occurs when there is no wind. With following and opposing wind, lateral drift is not apparent, although a bird will be displaced either toward or away from its goal during soaring bouts. Things become complex when wind is lateral to the axis of migration. Obviously, a bird soaring in a thermal will be transported over the ground with the thermal. This displacement is necessary if the bird is to use the region with the strongest updrafts at the thermal core.

Mueller and Berger (1967a) maintained that the displacement that occurs while soaring is wind drift. They have cited the definitions of drift given by Williamson (1955) and Lack and Williamson (1959) and modified them. This may not be correct for two reasons. First, a bird soaring in a thermal is not orienting actively. Its flight direction is passive. Second, Williamson (1955) and Lack and Williamson (1959) were referring mostly to passerine migrants that used flapping, not soaring flight. Their definition of "drift" was downwind drift (fig. 7.3F).

The problem is not resolved easily because the displacement from a preferred course that occurs during soaring does move a bird from its axis of migration. Eventually, a soaring bird must recover (or compensate) for displacement incurred while soaring. It can do so by setting glide headings into the wind as in figure 7.3I.

The problem of wind drift borders on a semantic argument and is perhaps an intractable paradigm. Whatever terminology is used, researchers must examine the flight direction of migrants during both soaring and gliding bouts. They must then determine whether the directions realized during these bouts are related to wind direction. Students of avian migration have examined the relation between flight direction (with powered flight) and wind direction for several decades. The data and arguments presented below will not resolve whether wind drift occurs among migrating hawks, but it will yield information regarding the orientation behavior of these migrants.

Empirical Tests and Comparative Orientation Data

Sharp-shinned Hawks and American Kestrels Crossing Water

The simplest test of the wind drift hypothesis is the case of raptors using intermittent gliding flight over water. Kerlinger (1984) measured flight directions of migrating Sharp-shinned Hawks and American Kestrels at Cape May Point, New Jersey, and Whitefish Point, Michigan (fig. 7.4). Short distance water crossings are a unique opportunity for studying how hawks orient with different wind conditions because the researcher "knows" the bird's migratory goal, the bird can see its goal, and large sample sizes aid statistical analysis. Track and heading of migrants and wind speed and direction were measured during crossings with hand-held compass, binoculars, and hot-wire anemometer. Headings were the direction a bird faced when leaving the shore-

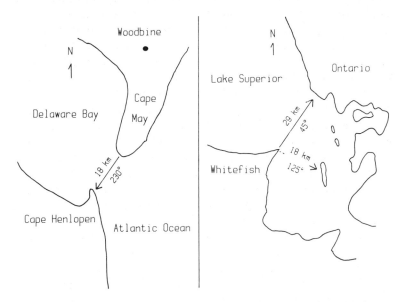

Figure 7.4 Map of study sites at Cape May Point, New Jersey, and an inland site at Woodbine, New Jersey, and at Whitefish Point, Michigan (redrawn from Kerlinger 1984 and Kerlinger and Gauthreaux 1984; courtesy of Tindall Ballierre).

line. Track was the direction the bird realized over the water. Testing for drift was accomplished by regressing tracks and headings on the wind component perpendicular to the crossing direction. A negative relation between track and wind is predicted, and headings will be constant (independent of wind) if wind drift occurs. Conversely, tracks will be constant (independent of wind) and headings will be positively related to wind direction if compensation occurs.

Sharp-shinned Hawks and American Kestrels crossing Delaware Bay at Cape May Point during autumn compensated for wind, as indicated by strong positive relations between headings and wind direction. Tracks were unrelated to wind direction (table 7.1). By adjusting headings into the wind, individuals of both species avoided drift. The test is not complete, because few hawks attempted crossings when lateral winds were greater than 5 mps (Kerlinger 1984; chap. 10). A few of the hawks that attempted crossings with stronger winds progressed slowly, and some appeared to be drifted over the Atlantic Ocean. Most of these hawks returned to the shore.

The situation was somewhat different at Whitefish Point during spring. Although headings of Sharp-shinned Hawks were positively related to wind, tracks were negatively related (table 7.1). These relations were interpreted as partial compensation (partial drift). American Kestrels compensated completely, as is evident from the lack of a significant regression for tracks and a strong regression for headings. Because the crossing at Whitefish Point is different from that at Cape May, Kerlinger invoked a post-hoc explanation. At Cape May the direction a bird must realize to make the shortest crossing subtends only about 30°, while at Whitefish Point a bird can achieve crossings that are similar in distance by flying in compass directions ranging from 125° to about 45°. By flying to the east at Whitefish Point migrants may "island hop" to the Canadian shore. Flight to the northeast (45°) is entirely over water. Partial drift by westerly winds allows a hawk to achieve a faster and less costly crossing by flying to the east and island hopping. Islands may be important during poor weather or when birds are fatigued. At Cape May such a strategy is not feasible, so birds must compensate completely. Interestingly, the mean tracks for both species were to the inland side of the crossings: Cape May Point, 236° for Sharp-shinned Hawks and 255° for American Kestrels; Whitefish Point, 72° for Sharp-shinned Hawks and 75° for American Kestrels. Tracks to the inland side of the most direct crossing may be a conservative strategy to avoid being blown out over Lake Superior or the Atlantic Ocean.

It appears that Sharp-shinned Hawks and American Kestrels compensate for lateral winds during intermittent gliding (mostly level) flight over water. There are two caveats, however. First, there is a threshold of wind speed above which these species cannot compensate. For species like the Merlin or Peregrine Falcon the threshold for wind drift is greater because these raptors are capable of faster flight than Sharp-shinned Hawks or American Kestrels. Second, strategies such as partial drift, pseudodrift, or optimal use of wind (Alerstam 1979a) may be involved that make the wind drift paradigm too simplistic.

Table 7.1 Regression of Headings and Tracks on Wind Direction for American Kestrels and Sharp-shinned Hawks during Water Crossings at Whitefish Point, Michigan, and Cape May Point, New Jersey

Crossing	Sharp-shinned Hawks	American Kestrels
CAPE MAY POINT (Crossing direction = 230°)		
Tracks		
Regression equation	—	—
r^2	0.002[b]	0.039[b]
N (observation periods)	24	17
Mean direction ± 99% confidence intervals	236° ± 9°	232° ± 11°
r (length of mean vector)	0.96**	0.95**
Headings		
Regression equation	y = 6.44x + 236	y = 7.81x + 240
r^2	0.826**	0.661**
N (observation periods)	24	17
Mean direction ± 99% confidence intervals	248° ± 12°	259° ± 19°
r (length of mean vector)	0.93**	0.89**
WHITEFISH POINT (Crossing direction = 45°)		
Tracks		
Regression equation	y = -3.34x + 62	—
r^2	0.408*	0.05[b]
N (observation periods)	19	44[a]
Mean direction ± 99% confidence intervals	72° ± 9°	75° ± 8°
r (length of mean vector)	0.96**	0.94**
Headings		
Regression equation	y = 6.43x + 68	y = 5.99x + 70
r^2	0.663**	0.692**
N (observation periods)	19	44[a]
Mean direction ± 99% confidence intervals	48° ± 13°	72° ± 13°
r (length of mean vector)	0.92**	0.86**

Source: Kerlinger 1984 and unpublished data.

[a]Sample size consists of individual migrants (or flocks) instead of observation periods because of the small sample size for American Kestrels at Whitefish Point.

[b]Not significant ($p > 0.10$).

$*p < 0.05; **p < 0.01$.

*Autumn Migration of Sharp-shinned Hawks Inland
from the New Jersey Coast*

Hawks seldom use soaring flight over water (Kerlinger 1984, 1985a; chap. 10), whereas they typically resort to soaring and interthermal gliding over land (Broun 1949; Mueller and Berger 1967a; Haugh 1972; Heintzelman 1975; Kerlinger and Gauthreaux 1984, 1985a, 1985b; Kerlinger, Bingman, and Able 1985). Kerlinger and Gauthreaux (1984) tested the wind drift hypothesis over land by measuring the directions of migrants during soaring and gliding at an inland site (Woodbine) about 35 km north of Cape May Point and over 12 km from the nearest coastline (fig. 7.4). Their intention was to test the original hypothesis of Allen and Peterson (1936; fig. 7.2).

Directions of soaring bouts were strongly correlated with wind direction (table 7.2), indicating a net downwind tendency. Headings of birds gliding between thermals were also related to wind direction. Mean gliding tracks on days with easterly winds did not differ significantly from those on days with westerly winds. The mean direction of tracks on days with east winds was only 2° different (not statistically significant) from days with west winds. The difference between mean headings with west winds and east winds was significant, however (table 7.2). Note that the mean headings of Sharp-shinned Hawks in south New Jersey would bring large numbers of these migrants to the coast, which is oriented toward about 220°. Also, the mean gliding track direction (194°) was within 1° of the mean track direction reported for this species from central New York by Kerlinger, Bingman, and Able (1985; see below). These results are inconsistent with predictions of the wind drift hypothesis.

Although downwind soaring flight by migrating hawks has been interpreted as drift by Mueller and Berger (1967a), Kerlinger and Gauthreaux (1984, 1985a) disputed use of the term drift to describe downwind soaring flight. They argued that the term as conventionally used is for a type of orientation by passerines, shorebirds, and other migrants that do not soar. They also stated that orientation during soaring flight was passive such that compensation cannot occur and that using the term drift in this way is inappropriate and confusing.

Because Sharp-shinned Hawks did adjust headings to realize a constant track over the ground during gliding bouts, Kerlinger and Gauthreaux concluded that these hawks are capable of compensating for some, but not all, winds. They state, as was the case with flight over water, that there is a threshold for drift. The major fault with the Kerlinger and Gauthreaux study was that they could observe only

Table 7.2 Flight Directions and Tests of the Wind Drift Hypothesis for
Sharp-shinned Hawks during Autumn 1982 at Woodbine, New Jersey
(Inland from Cape May Point, New Jersey)

Type of Flight	Number of Individuals (N)	Mean Direction ± 99% Confidence Intervals	Length of Mean Vector (r)
Tracks of glides			
All days[a]	15	194° ± 15°	0.95**
All migrants	172	194° ± 8°	0.86**
East winds	65	195° ± 10°	0.91**
West winds	90	193° ± 8°	0.84**
Calm winds	17	162° ± 36°	0.67**
Headings of glides			
All migrants	54	188° ± 23°	0.63**
East winds	19	148° ± 14°	0.93**
West winds	18	242° ± 12°	0.95**
Soaring (all migrants)	58	159° ± 25°	0.60**

Regression	N	Equation	r^2
Soaring direction on wind direction	54	$y = 1.099x - 8.84$	0.94**
Headings on wind direction	37	$y = -0.76x + 348$	0.68**
Daily mean tracks on wind direction	12	—	0.16[b]
Net direction when soaring and glides are combined on wind direction	23	—	0.08[b]

Source: Kerlinger and Gauthreaux 1984.
[a]Sample size is for days with daily means being the units of analysis.
**$p < 0.01$.
[b]Not significant ($p > 0.05$).

small segments of a migrant's flight. By measuring the net direction of
a sequence of several gliding and soaring bouts, a researcher could
determine whether the direction during glides compensated for displacement during soaring bouts. No relation between wind and resulting direction was evident for individual migrants that were observed in both glides and climbs (table 7.2).

Spring Migration at the Strait of Gibraltar

To my knowledge, studies by Evans and Lathbury (1973) and Houghton (1971) at the Strait of Gibraltar were the first to present quantitative information on flight direction of raptor migrants (other than
banding returns) and address the question of wind drift. They found
that the mean direction of daily mean tracks ($N = 9$ days with winds
from the west up to 3,000 m) was 16° ± 3.5° (SE). Headings were

estimated to be to the west of north. They also noted that migrants realized the same directions over land as they had over the sea and did not head for the Rock of Gibraltar. They concluded that raptors can and do compensate for wind, except when the wind component perpendicular to the desired flight path is strong. Evans (1966) stated earlier that such wind speeds (termed the threshold to drift) are close or equal to the air speed of the bird.

The data of Evans and Lathbury (1973) and Houghton (1971) are inadequate for a complete and comprehensive test of the wind drift hypothesis because many species were included in their data set and flight direction was presented only for flight with west winds. What direction do hawks realize when winds are from the east? However, their data suggest partial, if not complete, compensation.

Autumn Migration of Broad-winged Hawks in Ontario, Canada

Another early test of the wind drift hypothesis in which directional data were used is described by Richardson (1975). He used a surveillance radar (an AASR-1 radar, 550 kW) to monitor flocks of Broad-winged Hawks as they migrated through southern Ontario near London and Toronto. There is no doubt as to the identity of the radar targets because flocks of Broad-winged Hawks are distinct on a radar screen (PPI). Furthermore, visual confirmations were made at Hawk Cliff on the north shore of Lake Erie.

The Broad-winged Hawks Richardson "observed" formed long lines on the radar screen, sometimes representing thousands of birds oriented along an east-northeast to west-southwest flight path. Tracks of individual birds, however, were "somewhat south of WSW, since they moved WSW along the line and simultaneously S with the line." On some days these lines were stationary; that is, lateral movement was lacking, with birds moving along the axis of the lines. Also, the geographic position of the lines varied from day to day, showing that no location was consistently good for a migration count station.

Richardson reported that weather influenced the behavior and distribution of migrants in three ways: the relative numbers within a few miles of the north shore of Lake Ontario varied; the direction and amount of lateral movement of the flock lines varied; and the mean track (net) direction of the flocks changed. Richardson concluded by saying that "the results of the various analyses were less than perfectly consistent," but there was "strong evidence" that there were relatively more hawks close to the lake with northerly winds and more inland on southerly winds. This may or may not indicate drift because "flight direction of individual echoes showed no obvious relationship to any weather variable." Richardson's data seem to show that some drift did occur. Richardson's results are important with respect to the validity of using hawk migration counts as indicators of the numbers of hawks aloft.

Spring Migration of Broad-winged Hawks and Other Hawks through South Texas

The orientation of Broad-winged Hawks and several other hawks during spring migration through south Texas makes an interesting comparison with the work of Richardson from Ontario, because some Broad-winged Hawks were returning to Ontario. Kerlinger and Gauthreaux (1985a, 1985b) used marine surveillance radar with simultaneous visual observations and hand-held compass to measure flight direction of flocks and individuals during soaring and gliding flight. They found that the net tracks of Broad-winged Hawks soaring in thermals were northwest and highly correlated with wind direction (table 7.3). Tracks of gliding Broad-winged Hawks were unrelated to the wind component perpendicular to 360°, the presumed migratory direction or axis of migration. The mean track of gliding hawks flying from 20 to more than 1,400 m above the ground was to the east of north, and the variability of directions was small (table 7.3). From the mean track, the mean heading of these birds was estimated to be 16°.

Table 7.3 Flight Directions of Hawks Migrating during Spring in South Texas

| | | Direction of Flight | |
Species	N	Mean Track Direction ± 95% Confidence Intervals	Length of Mean Vector (r)
Turkey Vulture	12	4° ± 12°	0.95**
Mississippi Kite	10	5° ± 23°	0.87**
Sharp-shinned Hawk	33	349° ± 9°	0.92**
Broad-winged Hawk[a]	195	6° ± 3°	0.93**
Swainson's Hawk	31	360° ± 8°	0.94**
American Kestrel	25	349° ± 7°	0.96**
Net direction soaring flight in thermals			
Broad-winged Hawks[b]	120	312° ± 5°	0.88**
Other species pooled	27	323° ± 4°	0.97**

Source: Kerlinger and Gauthreaux 1985a,b.

[a]Regression of glide direction (tracks) on wind component perpendicular to 360° was not significant ($r^2 = 0.02$, $p > 0.10$, $N = 195$).

[b]Direction realized during a soaring bout was highly correlated with wind direction ($r = 0.88$, $p < 0.001$, $N = 120$).

**$p < 0.01$ by the Rayleigh statistic (directional data).

Because prevailing surface winds were from the southeast and south from the surface up to 1,000 m during spring migration (table 4.2), headings toward 16° should have compensated for the easterly component of the wind. Tracks of 5° might also be enough to compensate for displacement during soaring bouts.

Soaring climbs of the five other species of hawks studied in south Texas were nearly downwind when species were pooled and did not vary greatly in direction as indicated by the mean vector length (table 7.3). It was not possible to conduct statistical analyses with these data because of small samples. Mean tracks of glides for these species ranged from 349° to 5° and were well oriented (table 7.3). Differences among the mean gliding tracks for the six species are evident from confidence intervals. Although headings could not be determined (because of a lack of information on air speeds of these species), significant differences among species are probable. Because most measurements were taken when winds were from the southeast (hawks rarely migrated with north winds), the comparisons are probably valid. Different glide directions may indicate that the species (and perhaps individuals within a species) have different destinations. An alternative hypothesis is that Sharp-shinned Hawks and American Kestrels, whose track directions were to the west of north, were drifted to some extent to the west of north by prevailing southeasterly winds.

Findings for Broad-winged Hawks present a dilemma. Kerlinger and Gauthreaux (1985a) showed that Broad-winged Hawks migrating during spring in south Texas seem to fly mostly with prevailing south-southeast winds and need not fine tune their glide headings to wind direction. By setting headings of interthermal glides to the east of north (16°) as they do, migrants compensated for displacement during soaring flight in prevailing winds. Because gliding tracks were not related to wind direction, the wind drift hypothesis was not confirmed.

Comparative Orientation Behavior of Nine Species in Central New York during Autumn

The study of Kerlinger, Bingman, and Able (1985) from central New York is important because it was conducted north of known autumn concentrations in eastern North America where drift is hypothesized to occur; birds had been migrating for a short time (ca. 1–5 days) and distance before reaching the study site; the study was conducted at a site where potential leading lines were absent; and several species were examined so that a comparative analysis was feasible. The wind drift hypothesis was tested in two ways. First, a principal axis of migration

(PAM, sensu Gauthreaux 1978b) was determined for Sharp-shinned, Broad-winged, and Red-tailed hawks from the literature and banding returns. The PAM is the direct path (rhumb line) between breeding and wintering sites (or land bridges). For Broad-winged Hawks the PAM was 240°, a rhumb line from Albany, New York, to Houston, Texas. Most Broad-winged Hawks in North America are known to fly through Texas into Central and South America in autumn (Smith 1980). PAMs for Sharp-shinned and Red-tailed hawks were determined from banding data to be 215° (Clark 1985a; Holt and Frock 1980). By comparing the PAM with mean tracks and headings of each species (using 99% confidence intervals) we can determine whether individuals of these species attempted to realize a straight-line course during migration.

The second test of the wind drift hypothesis involved regressing headings and tracks on the wind component lateral (perpendicular) to the PAM of each species. A significant positive relation between track and wind indicates drift, whereas no relation indicates independence. A significant negative relation between heading and wind indicates that the bird is compensating for drift. The coefficient of determination (r^2) reveals the strength of the relation or how much variance of flight direction was explained by the wind, whereas the slope (regression coefficient) indicates the amount of drift or compensation.

Soaring flight for Red-tailed, Sharp-shinned, and Broad-winged hawks was toward the south-southeast and oriented significantly (table 7.4). Soaring climbs were positively correlated with wind direction showing a downwind tendency, although all three species showed a bias toward the south that was slightly away from the mean wind direction (table 7.4). This could have been a result of birds' beginning to move in the appropriate migratory direction before leaving a thermal but while still climbing. Overall, these results show that migrating hawks realize downwind tracks while soaring.

Mean track directions during gliding bouts for the nine species ranged from 180° to 206°, whereas mean headings ranged from 190° to 215°. A statistical comparison (Watson-Williams test; Zar 1984) of the directions of the three most numerous species shows that both headings and tracks of these species differed. Headings of Sharp-shinned Hawks were to the south of those of Red-tailed and Broad-winged hawks, as was the case with tracks. Mean headings show that the species are not homogeneous in the directions they attempt to migrate (table 7.5). Sharp-shinned Hawks, American Kestrels, Red-shouldered Hawks, Northern Harriers, and Cooper's Hawks headed more toward the south than Broad-winged and Red-tailed hawks,

Table 7.4 Net Direction of Hawks Soaring in Thermals in Central New York and the Relation between Wind and Net Direction

Season and Species	N	Direction of Flight		
		Mean ± 95% Confidence Intervals	Length of Mean Vector (r)	Correlation Coefficient, Wind vs. Direction
AUTUMN				
Sharp-shinned Hawk	35	170° ± 18°	0.61**	0.51**
Broad-winged Hawk	46	156° ± 24°	0.50**	0.74**
Red-tailed Hawk	59	173° ± 11°	0.76**	0.61**
Other species[a]	19	137° ± 27°	0.64**	0.25[b]
SPRING				
Sharp-shinned Hawk	22	66° ± 24°	0.67**	0.95**
Broad-winged Hawk	18	107° ± 62°	0.36[b]	0.88**
Osprey	3	143°	0.90[b]	—
Other species[c]	6	163°	0.23[b]	0.85*

Source: Kerlinger, Bingman, and Able 1985 and unpublished data.

[a]Includes Northern Harrier, Osprey, Cooper's Hawk, Goshawk, Red-shouldered Hawk, and American Kestrel.

[b]Not significant ($p > 0.05$), either no mean direction or no significant correlation.

[c]Includes Cooper's Hawk, Red-shouldered Hawk, Northern Harrier, and American Kestrel.

*$p < 0.05$; **$p < 0.01$, by the Rayleigh statistic or t-test in the case of correlation coefficients.

which headed to the southwest. If this difference among headings were to persist as these birds move southward, a greater proportion of American Kestrels, Sharp-shinned Hawks, and Northern Harriers would reach the Atlantic coast.

Comparing headings with PAMs shows that neither Sharp-shinned nor Broad-winged hawks attempted to fly directly toward their migratory goals, whereas Red-tailed Hawks did. Mean headings of glides for Sharp-shinned and Broad-winged hawks differed by 11° and 26° from the PAMs, and tracks of glides differed by 22° and 33°, respectively. The mean tracks of glides for Red-tailed Hawks were significantly to the south of the PAM, but the difference was not as great as for the other two species. Small sample sizes for the remaining six species precludes rigorous analysis, but it seems that some may not attempt to realize straight-line migrations to their wintering grounds. These data are not consistent with wind drift because a preferred heading to the south will bring birds to the Atlantic coast without drift and because headings show that some species do not attempt to fly directly toward their migratory goals.

The regressions of flight direction on lateral component of the wind also do not support the wind drift hypothesis (table 7.6). The positive relation between headings and wind for Red-tailed and Sharp-shinned hawks shows that some individuals compensated, at least partially, by facing into lateral winds. The weak relation between track and lateral wind component for Broad-winged and Sharp-shinned hawks shows that some drift (either some individuals or partial drift) occurred. Lateral winds in this study were a result of prevailing westerlies (table 4.1). Because winds explained such a small portion of the variance in flight direction, it appears that the wholesale drift proposed by Allen and Peterson (1936) and Mueller and Berger (1967a) does not occur.

Comparative Orientation Behavior of Nine Species in Central New York during Spring

Kerlinger and his colleagues studied the return flight of hawks in central New York using the same tracking radar at Albany during spring

Table 7.5 Direction of Interthermal Glides during Autumn and Spring Migration in Central New York as Measured with Tracking Radar during Autumn 1978 and 1979 and Spring 1978–1980

		Direction of Flight			
		Tracks		Headings	
Season and Species	N	Mean ± 95% Confidence Intervals	Length of Mean Vector (r)	Mean ± 95% Confidence Intervals	Length of Mean Vector (r)
AUTUMN					
Northern Harrier	9	190° ± 28°	0.84**	202° ± 18°	0.93**
Sharp-shinned Hawk	90	193° ± 3°	0.94**	204° ± 3°	0.94**
Cooper's Hawk	6	180°	0.96*	190°	0.97*
Goshawk	7	206°	0.68*	208°	0.65*
Red-shouldered Hawk	17	193° ± 22°	0.76**	201° ± 23°	0.74**
Broad-winged Hawk	131	203° ± 4°	0.88**	214° ± 4°	0.88**
Red-tailed Hawk	128	206° ± 5°	0.86**	214° ± 6°	0.83**
Osprey	8	204° ± 25°	0.90**	211° ± 46°	0.65*
American Kestrel	18	195° ± 20°	0.77**	203° ± 27°	0.66*
SPRING					
Sharp-shinned Hawk	34	18° ± 7°	0.92**	4° ± 10°	0.88**
Broad-winged Hawk	33	22° ± 16°	0.76**	18° ± 18°	0.71**
Osprey	17	27° ± 19°	0.80**	20° ± 22°	0.75**
Other species pooled	4	—	0.71[a]	—	—

Source: Autumn data from Kerlinger, Bingman, and Able 1985 and unpublished data.
[a]Not significant ($p > 0.05$); no mean direction.
*$p < 0.05$; **$p < 0.01$ by the Raleigh statistic.

Table 7.6 Regression Analyses of Wind Drift Hypothesis

	Broad-winged Hawk	Sharp-shinned Hawk	Red-tailed Hawk
Tracks of glides			
Regression equation	$y = -1.6x + 209$	$y = -1.1x + 196$	—
r^2	0.11*	0.04*	0.00[a]
N	131	90	128
Headings of glides			
Regression equation	—	$y = 1.4x + 199$	$y = 2.4x + 204$
r^2	0.02[a]	0.07*	0.08*
N	131	90	128

Source: Kerlinger, Bingman, and Able 1985.

Note: Tracks and headings of autumn migrants during 1978–1979 from central New York were regressed on wind component perpendicular to a species' PAM (principal axis of migration, as determined from band recovery data). Principal axis of migration was 240° for Broad-winged Hawks and 215° for Sharp-shinned and Red-tailed hawks.
[a]Not significant ($p > 0.10$).
*$p < 0.05$.

1978, 1979, and 1980. Similar to species in the studies cited above, Sharp-shinned Hawks, Broad-winged Hawks, Ospreys, and several other species realized downwind tracks during soaring flight, as is evident from the robust correlation coefficients in table 7.4. (Because of the large number of Red-tailed Hawks that breed near the radar, it was not possible to study the spring migration of this species.) Mean directions of soaring bouts were to the east and overlapped with mean directions of soaring flight during autumn migration for these species (table 7.4). This finding again reflects the strong westerly component of the prevailing winds during spring and autumn in central New York.

Mean track directions of Broad-winged Hawks, Sharp-shinned Hawks, Ospreys, and several other species during interthermal glides were to the northeast, ranging from 18° to 27° (table 7.5). The mean track directions for gliding flight of Broad-winged Hawks, Sharp-shinned Hawks, and Ospreys were within 5° of the back azimuth of the mean tracks during autumn. This means that the mean directions realized during interthermal glides were within 5° of being in the opposite direction (180°) of mean autumn tracks. Mean headings of glides were to the north of tracks. The difference was greatest for Sharp-shinned Hawks (about 14°) and least for Broad-winged Hawks (about 4°). Headings of glides were 4° to 16° to the north of back azimuths of autumn headings for these species and may reflect an attempt to face into prevailing westerly winds (table 4.1) to compensate for them. The paucity of data precludes a statistical test of the wind drift hypothesis for spring migration in central New York.

Autumn Migration in Central New England Studied from a Motor Glider

Using a motorized sailplane, the New England Hawk Patrol (Hopkins et al. 1979) determined the flight direction (gliding tracks, presumably) of hundreds of migrating (table 7.7) Sharp-shinned Hawks, Broad-winged Hawks, and Ospreys flying at high altitudes. Tracks of glides were more westerly than those of hawks flying near Albany, about 150 km to the west, perhaps because many migrants were flying from maritime Canada and northeastern New England or had previously encountered the Atlantic coast. To avoid the coast, individuals must set their headings to the west of south, especially when winds are from the west. Flight lines of Broad-winged Hawks published by the New England Hawk Watch (Hopkins 1975) also reflect the strong westerly trend of migrants flying through New England during autumn.

None of the studies from New England include analyses or information about wind speed and direction at the altitude of migrants, so regressions could not be used to determine whether flight directions were influenced by wind. Hopkins (1975) has noted that Broad-winged Hawks changed their headings as wind direction changed, appearing to compensate for potentially drifting winds. Also, the large variance in flight directions of flocks of Broad-winged Hawks on some days was related to the presence of other flocks. During interthermal glides migrants often changed their headings to join flocks that were climbing in thermals. It would be interesting to know the altitude of

Table 7.7 Summary of Flight Directions of Migrants during Autumn Migration in Southern New England as Observed by Motor Glider

Species	Number of Measurements	Number of Individuals	Mean Track Direction ± 95% Confidence Intervals	Length of Mean Vector (r)
Broad-winged Hawk	26	269	243° ± 5°	0.95**
Osprey	8	9	240° ± 13°	0.96**
Other species combined[a]	3	4	231°	0.87

Source: Based on data from Hopkins et al. 1979.

[a]Other species include Sharp-shinned Hawk, Turkey Vulture, and Merlin.

**$p < 0.05$.

migrants when flight directions changed. That is, did these birds deviate from their preferred migratory direction only when they needed to find a thermal?

Autumn Migration Studies from Switzerland Using Tracking Radar

Studies by Schmid, Steuri, and Bruderer (1986) from Switzerland summarized the flight direction of the European Sparrowhawk and Common Buzzard, the ecological equivalents of Sharp-shinned (or Cooper's) Hawks and Red-tailed Hawks from North America. Flight directions were measured with a tracking radar and compared with data from banding studies in addition to the topographic conditions at the tracking radar site. The "axis" of migration of these species was to the southwest as shown from banding returns and was not very different from the direction of birds measured with radar (approximately 220° to 240°). An interesting shift in the migratory direction of Common Buzzards noted between September (242°) and October (273°) was attributed to an "obstructing effect" of the border of the Alps. Because some European mountain ranges are oriented from east to west, they may have a very different effect on migrating hawks than mountain ranges in North America.

Radiotelemetry Studies

Orientation studies in which radiotelemetry is used may yield the best answers to the wind drift question. Data provided by telemetry include flight direction during soaring and gliding for an entire day's flight as well as directions during flights on successive days. The behavior of individual migrants can be monitored so that changes in behavior are detected instantly.

Although several researchers have used telemetry, few have analyzed their data with respect to wind drift. Cochran's (1972, 1975, 1985) tracks of Sharp-shinned Hawks and Peregrine Falcons would be ideal tests of the wind drift hypothesis because he was able to follow individuals for several hundred kilometers on successive days with varying wind conditions. Statements about the behavior of Peregrine Falcons (Cochran 1975) indicate that these birds adjust headings in a manner that approximates compensation. Harmata's (1984) work with Bald Eagles may also shed some light on the wind drift question. The entire spring migration of at least one Bald Eagle was monitored between Colorado and a breeding site in central Saskatchewan. One eagle was noted to be blown off course by very strong winds, after which it resumed a direct flight toward its nesting site. Another bird deviated from its course to follow a ridge, the Missouri

Coteau, for nearly 100 km. After leaving the ridge it resumed a direct course toward its breeding site. These observations and those by Cochran are tantalizing tidbits, showing how telemetry data can be used to study the orientation behavior of migrants.

Banding Returns, Recaptures, and Recoveries: A Neglected Source of Data

One of the problems with determining whether raptors are drifted by wind from a preferred pathway is that the researcher is rarely sure of the migratory goal of his subjects. Kerlinger and his colleagues have attempted to incorporate a presumed axis of migration (PAM) in their tests of the wind drift hypothesis. (The PAM was originally conceived by Gauthreaux [1978b] for determining whether passerines flying after dawn attempted to compensate for drift incurred during the previous night's migration.) In these studies the PAM was defined using "known" migratory pathways and banding data (Kerlinger, Bingman, and Able 1985). Banding data will be more important in future studies, allowing researchers to be confident about where migrants are going and where they originated.

In the past banding data have been used as evidence for (Mueller and Berger 1967b) as well as against wind drift (Kerlinger and Gauthreaux 1984; Clark 1985a). Recoveries of Sharp-shinned Hawks banded at Cedar Grove, Wisconsin, and recovered at different latitudes presumably showed that birds recovered south of Wisconsin and north of the Tennessee border (about 37° N) were mostly to the south and east of the banding site. Migrants recovered south of the Tennessee border were spread to the southeast and southwest of the banding site. Mueller and Berger interpreted this difference as drift by prevailing west and northwest winds. However, the number of returns was small, especially if age and sex classes are considered separately. Because few birds were recovered from the breeding grounds after banding at Cedar Grove, it is impossible to say with certainty where the birds originated. If birds were drifted to the Lake Michigan shore at Cedar Grove, as Mueller and Berger claimed, they would have to originate from breeding areas northwest of that location. The few spring returns do not support the wind drift hypothesis, although a summer return from Alberta may be consistent with wind drift.

Of 27,399 Sharp-shinned Hawks banded at Cape May Point by Clark (1985a) as of 1983, 337 (1.2%) were recovered or recaptured elsewhere. Eighty-three of these were recovered during subsequent breeding seasons in New England, Quebec, and the Canadian Maritime Provinces. Of those recovered late in the autumn and winter, most were from the coastal plain of New Jersey south to Florida, east of the Appalachian Mountains. Thus, to reach their wintering range most migrants must fly to the southwest. If birds were drifted by west to north winds, more of the migrants captured at Cape May should originate to the northwest of Cape May. These data together with behavioral data from central New York (Kerlinger, Bingman, and Able 1985) and from an inland site in south New Jersey (Kerlinger and Gauthreaux 1984) do not confirm the wind drift hypothesis for Sharp-shinned Hawks in eastern North America.

The two examples given above are only first attempts at using banding data for testing the wind drift hypothesis. Other studies have examined banding data (e.g., Brinker and Erdman 1985), but none have included information that sheds light on the orientation strategies of raptors. Banding data can play an important role in determining migratory pathways and studying orientation of migrants. Because the return rate from most banding operations is small (fewer than five birds returned for every one hundred banded), few banders will accumulate enough data for statistical analysis. A cooperative effort by banders and scientists is needed to resolve these intriguing questions.

Orientation Strategy and Prevailing Wind

After presenting several tests of the wind drift hypothesis and considering count and banding data, it is obvious that the orientation strategy of migrating hawks is complex. Furthermore, wind drift is an elusive phenomenon that has rarely been documented conclusively. Attempting to fit all orientation data into the narrow paradigm of wind drift versus compensation may be unrealistic and naive. I have chosen to do so because wind drift has been accepted as the central, organizing theory in hawk migration research (Heintzelman 1975) and has been the paradigm used in many field studies of bird orientation.

There is a more realistic and better means of evaluating the orientation strategies of migrating hawks. In 1979 Alerstam (1979a, 1979b) presented models that showed birds could migrate efficiently and quickly by using the wind in an optimal fashion. In one model Alerstam proposed that birds allow themselves to be drifted partially at high altitudes and correct (overcompensate) for that drift at low altitudes (fig. 7.5). Alerstam's rationale was that wind is stronger at high altitude than at lower altitude, making compensation during high altitude flight not profitable (fig. 7.5). By strong wind, Alerstam means strong in relation to the air speed of migrants. A second model showed how migrants should drift optimally with variable winds (winds varying from hour to hour or day to day). Finally, he also showed how optimal drift could be used where winds shifted predictably over a long migration pathway. Hereafter, these hypotheses will be referred to as the "optimal use of wind hypothesis." Although the three models are slightly different, they are similar enough to be considered within a common framework.

With the various forms of the hypothesis, birds do not fly directly to their goal but realize tracks that allow them to utilize following winds (or avoid opposing and lateral winds). Such a migration strategy ensures that birds move quickly and efficiently away from or through inhospitable regions.

Alerstam's optimal use of wind hypothesis is important because he used a mathematical approach that permitted quantitative predictions regarding flight behavior. Although Alerstam was the first to state the hypothesis mathematically, some of the basic ideas about winds changing predictably along a migrant's pathway were proposed in rudimentary form by Bellrose and Graber (1963), Rabol (1974), and Gauthreaux (1978b). Since Alerstam's paper, several authors have presented data or simulations (Stoddard, Marsden, and Williams 1983) showing elliptical flight paths of migrants. Bellrose and Graber (1963) probably said it best: "Birds may have evolved an elliptical,

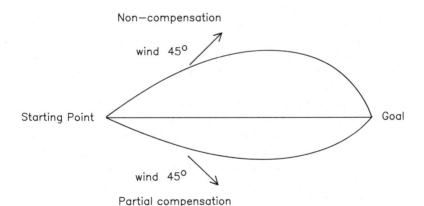

Figure 7.5 Alerstam's model showing optimal use of wind by a migrant (redrawn from Alerstam 1979a). Migrants are drifted by the wind during the initial stages of migration and later overcompensate to reach their goal. By invoking differential altitude use and varying prevailing winds, variations of this model can be used to test hypotheses.

clockwise migration route between spring and fall as an adaptation to the prevailing westerly and easterly winds." They were referring to eastern North America, where winds north of 35°–40° N are from the west and where winds south of 30°–35° N are from the east (table 4.1). Gauthreaux (1978b) presented a map showing prevailing winds in autumn and spring across North America (fig. 4.2) and stated that both phenology and orientation of migrants have been shaped by prevailing wind patterns in all parts of the world (Buskirk 1980; Curry-Lindahl 1981). What is most important about the Bellrose and Graber (1963) and Gauthreaux (1978b) papers is that they recognized the importance of prevailing wind as well as differences of prevailing winds between spring and autumn seasons.

The most cited example of an elliptical migration pathway is that of the Lesser Golden Plover (*Pluvialis dominica*) in North America (Bent 1937; Lincoln 1952; diagram in Welty 1982). A theoretical model by Stoddard, Marsden, and Williams (1983) shows a curvilinear migration pathway over the western North Atlantic between New England and South America. There are many examples of elliptical migration pathways from both the Old and New worlds. Curry-Lindahl (1981) calls elliptical migration pathways "loop migration." Elliptical migrations probably occur in Wahlberg's Eagle, the Amur Falcon, and possibly the Sooty Falcon and Lesser Kestrel, to name a few species. During autumn the falcons listed above undertake flights over different parts of the Indian Ocean, whereas during spring they may use routes over land (Brown and Amadon 1968; but not according to Curry-Lindahl 1981).

The example of a large water barrier shows that prevailing winds, topography, and geography may interact as selective factors to shape the orientation behavior of migrants. Obviously, the Indian Ocean and other bodies of water are strong selective factors, but so too may be prairies, mountain ranges, deserts, and forests. Many migrants avoid massive mountain ranges and deserts because they cannot forage adequately should they land, or because they are physiologically incapable of surviving such environments. The Great Plains of North America may be an obstacle to forest birds such as Sharp-shinned Hawks. If this is so, large numbers of this species that breed north of the prairies should fly eastward, eventually encountering the Great Lakes. The arrival of these birds at the Great Lakes may be caused by drift by prevailing west winds (summarized by Brinker and Erdman 1985), although these migrants may set headings to the east. After reaching the forests, these same birds may proceed southward. The result of avoiding the Great Plains would be a curvilinear migration.

The best example of low altitude "overcompensation" as proposed by Alerstam is the postdawn flight of nocturnally migrating passerines. Gauthreaux (1978c) noted that many passerine birds flew west or northwest shortly after dawn in the southeastern United States. This low altitude flight (<100 m) in a direction seemingly inappropriate for autumn migration was hypothesized to be a correction (redetermined migration direction, sensu Gauthreaux 1978c) for drift incurred during the previous night's flight.

There are meager data supporting both hypotheses for raptors. Raptors following ridges in the northeastern United States or elsewhere may be an example of low altitude overcompensation. Ridges such as the Kittatinny and others of the Blue Ridge chain, which hawks follow (for unknown distances), are not oriented in the appropriate direction for migration. A migrant that follows the Kittatinny Ridge in autumn will realize a course of 255°–260°, far to the west of the principal axis of migration (215°–240°; Kerlinger, Bingman, and Able 1985) for most species. These birds may have been drifted before reaching the ridges, or they could have chosen a heading to the south of their principal axis of migration, after which they followed the ridge to compensate for their movement from the PAM. Either way, the behavior corresponds to Alerstam's hypothesis.

Kerlinger, Bingman, and Able (1985) proposed an elliptical migration pathway for Broad-winged Hawks in eastern North America. Directional data presented above are consistent with the hypothesis. Because Broad-winged Hawks set headings to the south ($\bar{x} = 214°$) of their PAM (240°), they truly were not drifted from the PAM. The weak relation between wind and track direction suggests that a slight, partial drift may occur. Kerlinger, Bingman, and Able (1985) suggested that by not compensating for prevailing westerlies, Broad-winged Hawks realize a faster and more energy efficient migration out of eastern Canada to latitudes where east winds prevail (south of 35° N). Data from hawk migration counts may support this contention. In autumn many Broad-winged and other hawks are counted along the Kittatinny Ridge and Atlantic coast, but few are counted at these locations in spring. Large spring counts come from sites farther inland such as Derby Hill (Haugh and Cade 1966) and Braddock Bay (Moon and Moon 1985) on Lake Ontario. Proposed elliptical migrations of Broad-winged and other hawks from eastern North America are presented in figure 7.6 along with prevailing wind directions.

The flight directions reported in this chapter from tracking radar studies are from a fragment of the migration pathway of individual birds. The answer to whether birds are drifted or whether they compensate fully or partially for displacement incurred during soaring flight can be determined only when many birds are followed for hundreds of kilometers over country that is devoid of leading lines. These data are not available at this time, but with the plethora of radio-tracking studies that are now in progress they may emerge in the near future.

The future of orientation research is dependent upon several factors. For meaningful studies to be conducted, researchers should attempt to test the optimal use of wind hypothesis or a modification of it. To do so they must consider prevailing winds as well as flight direction, geographic origin, and destination of migrants. Radiotelemetry, radar, and banding data will all be helpful. I cannot emphasize enough the importance of prevailing winds. In several of my studies quantitative summaries of prevailing winds have been presented to show the

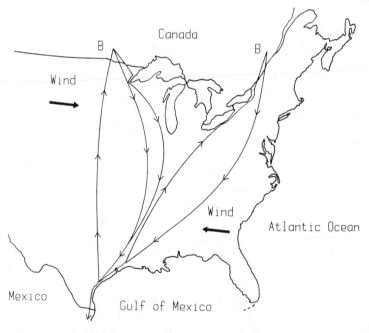

Figure 7.6 Hypothesized elliptical pathways of Broad-winged Hawks in the eastern and midwestern United States during their migration to Central and South America. Several different pathways are proposed to show that migrants may change their courses from year to year. Note how the Great Lakes modify the migration pathways of the two populations. B = breeding sites in eastern and central Canada.

prevailing direction of wind at a given site as well as how predictable or unpredictable wind direction is at a site at different times of the year. A knowledge of direction, strength, and predictability of winds (length of mean vector, *r*) along a migratory pathway will help researchers understand how a bird must fly to complete a safe and efficient migration. Wind, both horizontal and vertical, is undoubtedly the most important evolutionary factor (selective force) shaping the migratory strategy of soaring raptors. Without a quantitative knowledge of prevailing winds, meaningful studies cannot be undertaken.

Summary and Conclusions

Until recently the unifying theory in hawk migration research stated that migrating hawks were drifted by wind until they encountered and followed a leading or diversion line. Evidence supporting the wind drift hypothesis is indirect, mainly from the large numbers of hawks counted with west winds at ridge and coastal leading lines in eastern North America during autumn migration. Hawk migration count data cannot distinguish among alternative explanations of large counts. Only directional data are acceptable for determining whether wind influences the direction of migration. Directional data from radar and visual studies conducted in New York, New Jersey, Texas, Michigan, Gibraltar, and perhaps Ontario show that Broad-winged, Red-tailed, Sharp-shinned, and possibly other hawks realize downwind flight paths when soaring in thermals but are not influenced greatly by wind during interthermal gliding flight. The weak relations between wind and gliding tracks for a few species (Broad-winged and Sharp-shinned hawks in New York; Broad-winged Hawks in Ontario; Sharp-shinned Hawks at Whitefish Point) may indicate partial drift or incomplete compensation. Several species adjusted headings according to the strength of lateral winds in a manner consistent with compensation for wind. Banding data also do not support the wind drift hypothesis.

An alternative means of studying orientation behavior, proposed by Alerstam and others, incorporates partial drift, prevailing winds, and overcompensation. The resulting flight between breeding and nonbreeding ranges is a curved line, ellipse, or loop migration. With this strategy the optimal flight direction is determined from wind direction along the migration pathway and the migrant's air speed. The actual flight directions migrants use can then be compared with the flight directions predicted by the model. In eastern North America a clockwise, elliptical round-trip migration pathway is predicted. By flying with prevailing winds, migrants realize faster flights that are energetically less costly. The data for some migrants conform to this hypothesis, and some species do not attempt to fly a straight path between breeding and non-breeding ranges. Data from individuals flying for hundreds or thousands of kilometers are needed to test this hypothesis. I recommend that the term wind drift be used more carefully in the ornithological literature or not be used at all.

8

Altitude of Flight and Visibility of Migrants

A question asked frequently by hawk watchers and birders is, How high do migrants fly? The question has attracted researchers for two reasons. Behaviorists address the question from the perspective of decision making by the animal. That is, do migrants select altitudes where winds are favorable (Bruderer and Steidinger 1972), where temperatures allow cooling during strenuous flight (Torre-Bueno 1978), or where sufficient oxygen is available (Tucker 1970)? The altitude of bird migration is also of practical importance, especially with respect to aircraft safety (Gauthreaux 1974). Much of the research funding for bird migration studies has been allocated to determine the potential birds represent for damage to aircraft and loss of life (Leshem 1987). Conversely, conservationists are interested in the altitude of bird flight because television towers, skyscrapers, industrial smokestacks, high-tension wires, and lighthouses are hazards to birds, killing hundreds of thousands of migrants each year.

Determining the altitude of migrating birds is not easy. First, a researcher must have an apparatus to measure the altitude of individual migrants or flocks. Second, a sampling system must be devised that ensures that measurements are representative of migrants aloft at a given time. Neither task is trivial, although researchers in Europe and North America have made ingenious use of tracking and weather radar. A problem with radar ornithology is that species cannot be identified from radar echoes. By noting the relative size and speed of a radar echo, researchers can distinguish passerines, shorebirds, waterfowl, and soaring birds. Most radar studies of bird migration were conducted before 1980, and today the use of radar is rare.

From radar studies we know that the height of bird migration varies. Some factors that influence altitude of migration are geographic location, species (or species group), topographic situation (over water, over land, along ridges or coasts), time of day (night vs. day, or hour of the day or night), weather, and type of flight (gliding vs. powered). A comprehensive review of the altitude of avian migration is needed.

185

Among the interesting results of radar ornithology are those of Able (1970), who found that over 75% of passerine migration occurred below 900 m above ground level (hereafter all measurements are given as above ground level); Gauthreaux (1971), who demonstrated that passerines migrating over the Gulf of Mexico fly at less than 490 m at night and climb to over 1,000 m in the day; and Bruderer and Steidinger (1972), who showed that passerines adjusted altitude to fly in winds that were favorable for migration. There are other interesting findings. I have chosen three that help us understand what types of questions have been addressed by researchers working with radar.

The altitude at which raptors migrate has not been available until recently. Three topics that pertain to the altitude of raptor migration are addressed in this chapter: a review of studies published before 1980 in which data were limited or qualitative; a review of empirical studies involving radar and simultaneous visual observations; and an examination of the visibility of migrants to ground-based observers.

Results of Early Studies of Migration Altitude

Before about 1980, published altitudes of migrating hawks were mostly visual estimates (summarized in table 8.1). There are three problems with these studies. First, there is no way to determine the reliability (accuracy) of visual estimates and whether low flying birds were disproportionately represented. Second, few studies report sample sizes, means, or information about variance of altitude with weather, time of day, season, or species. Third, most studies were conducted in topographic situations such as ridges and coastlines (table 8.1) that occupy a small portion of the migration of hawks.

The estimates of altitude in table 8.1 vary greatly. The lowest is less than 10 m, whereas the highest is 6,400 m. It is safe to say that migrants flying along ridges fly at low altitudes, usually below 50 m. Several authors report this migration as "very low" (Broun 1935; common knowledge to hawk watchers). If migrants were not this low, thousands would not have succumbed to gunners along the Kittatinny Ridge in Pennsylvania, Cape May, New Jersey, and elsewhere. How often migration occurs at higher altitudes is difficult to say because migrants become difficult to see as they fly higher. They also do not depend on ridge lift when flying at high altitudes, so they may not follow ridges. Remarks by Allen and Peterson (1936), Stone (1937), Broun (1949), Haugh (1972), Cochran (1975), and others make it clear that some birds fly very high and are difficult to see and count.

Table 8.1 Summary of Selected References in Which "Qualitative" Information on Altitude of Hawk Migration Was Presented

Species	Altitude (m AGL)[a]	Method and Topographic Situation	Reference
Broad-winged Hawk, Sharp-shinned Hawk	180–600, "Lower"	VE,[b] ridge, New Jersey	Lengerke 1908
"Buteos"	2,100	VE, mountains, New Hampshire	Forbes and Forbes 1927
Many species	Maximum = 600	VE, Hawk Mountain, ridge, Pennsylvania	Broun 1935, 1949
Many species	"Very high," "Moderate," "low"	VE, Hawk Mountain, ridge, Pennsylvania	Broun and Goodwin 1943
Sharp-shinned Hawk	150 to limit of vision	VE, peninsula, Cape May Point, New Jersey	Allen and Peterson 1936
Sharp-shinned Hawk	Beyond limit of human vision	VE, peninsula, Cape May Point, New Jersey	Stone 1937
Broad-winged Hawk	360–800	Blimp, ridge, New Jersey	Stearns 1949
Many species	152–1,500	VE, shore of Lake Ontario, Derby Hill, New York	Haugh 1972
Honey Buzzard and other species	60–450, "varied considerably," "only visible through binoculars"	VE, Island of Malta	Beaman and Galea 1974
Broad-winged Hawk	At least 900 to 1,275	VE, ridge, Pennsylvania	Heintzelman 1975
Peregrine Falcon	9–90, >>900 at times, seldom <150	VE, aircraft, radiotelemetry, midwestern United States	Cochran 1975
Broad-winged Hawk	Variable to 800	Aircraft, Connecticut	Welch 1975
Broad-winged and Swainson's hawk	3,000–4,000, 6,400	VE, pilot reports, photographs, Panama	Smith 1980
Broad-winged and Swainson's hawk	1,800, 4,000–5,000	telescope, Panama	Smith 1985a
Broad-winged and Swainson's hawk	375–2,650	Motor glider, wave lift, Panama	Smith 1985b
Peregrine Falcon	<30	Radiotelemetry, over Atlantic Ocean off U.S.	Cochran 1985
Bald Eagle	<50 – >1,000	Radiotelemetry, VE, mountains and prairie, United States and Canada	Harmata 1984; Harmata, Toepfer, and Gerrard 1985
Lesser Spotted Eagle, Honey Buzzard, and Levant Sparrowhawk	1,300	Motor glider, desert, Israel	Leshem 1987

Note: "Qualitative" in this case means that some numbers were given but sample sizes, means, and variances were not reported.
[a] AGL indicates above ground level, although some studies do not specify if altitudes were reported as above ground or above sea level.
[b] VE indicates visual estimate.

I know of few studies before 1980 in which an effort was made to *measure* the altitude of migrating raptors. Hopkins et al. (1979) used the altimeter of a motor glider to measure the altitudes of Ospreys, Broad-winged Hawks, Sharp-shinned Hawks, Red-tailed Hawks, and two other hawk species during autumn migration over Connecticut and Massachusetts. Their report includes dozens of measurements with mean altitudes for six species between 457 and 855 m (table 8.2). This study is particularly significant because the birds Hopkins and his colleagues observed used thermals and did not follow ridges or coastlines (table 8.2). Leshem (1987; personal communication) reports soaring over the Israeli desert with Levant Sparrowhawks, Honey Buzzards, Lesser Spotted Eagles, and other species at slightly higher altitudes than those reported by Hopkins and his colleagues. Pennycuick (1972a) also used a motor glider to follow vultures and other soaring birds over the Serengeti in Kenya. Maximum midday altitudes ranged to 1,300–1,700 m for vultures during foraging flights. Although most of the birds he followed were not migrating, migrants were observed at similar altitudes. His findings for the African vultures are of interest because migrants also used thermals at that location.

The work of Houghton (1971, 1974) and of Evans and Lathbury (1973) probably constituted the first systematic study of the altitude of raptor migration. Houghton used radar during spring and autumn at the Strait of Gibraltar. To "identify" "types" of birds (large soaring vs. small flapping), wingbeat frequency of targets was recorded from the radar. Large targets that flapped infrequently were judged to be soaring birds such as hawks and cranes. In addition, direct observations of radar targets were made through a telescope mounted on the radar. Visual observations and counts by Evans and Lathbury (1973)

Table 8.2 Summary of Altitude Measurements Made When a Motor Glider Was Used to Follow Migrating Hawks

Species	Sample Size (Instances and Individuals)	Mean Altitude ± SD (m)
Sharp-shinned Hawk	9 (12)	610 ± 220
Broad-winged Hawk	35 (842)	855 ± 272
Red-tailed Hawk	4 (5)	457 ± 163
Osprey	13 (14)	818 ± 343
American Kestrel	1 (1)	640
Merlin	1 (1)	457

Source: Based on data from Hopkins et al. 1979.

at Gibraltar during the same season confirmed that raptors made up a large proportion of the migrants Houghton tracked.

The data presented by Houghton (1971) are in two histograms. Virtually all raptor migration over Gibraltar occurred between 300 and 2,000 m. Approximately one-half of the birds observed were between 610 and 1,066 m, and about 25% were above 1,066 m during spring and autumn. Evans and Lathbury (1973) noted that fewer raptors were seen at Gibraltar with east winds than with west winds, even though radar showed that many birds were moving when winds were from the east. They concluded that raptors used standing waves on east winds to soar to "several thousand m over the top of the Rock, well above visible range." Also, visibility of migrants to observers was reduced in midmorning as thermals promoted flight at high altitude. They stated that "visible movements are but a proportion of the total migration on each day" and that "counts of raptors migrating through Gibraltar cannot be used to give quantitative estimates of western European populations of birds of prey. The same reservation might apply to other focal points for raptor migration."

It was unfortunate that Houghton's studies were published in obscure volumes that are not available in most libraries. The research was conducted to learn about hazards to air traffic at Gibraltar and did not achieve what it could have if undertaken by an animal behaviorist.

Radar Studies and Other Studies of Altitude

This section presents findings of radar and other studies from southern New Jersey (Kerlinger and Gauthreaux 1984 and unpublished data), southern Texas (Kerlinger and Gauthreaux 1985a, 1985b, and unpublished data), central New York (Kerlinger, Bingman, and Able 1985 and unpublished data), a mountain pass in Switzerland (Schmid, Steuri, and Bruderer 1986), and several other locations. These studies are limited with respect to topographic and geographic situations as well as confined to a few seasons and species. However, they illustrate relations between altitude of migration and time of day, convective depth, wind, and topography, as well as differences among species, the height band, climb rates of migrants, and flight in and above clouds. These studies serve as baseline data for future research.

Altitude of Flight in Relation to Time of Day and Convective Depth

That the altitude of raptor migration varies with time of day is not surprising. As shown in chapter 4, convective depth varies with time

of day. Thermal convection is absent in early morning but develops
rapidly, reaching its maximum between 1100 and 1600 h. The altitude
of raptor migration follows a similar sequence because soaring hawks
are dependent upon convection. After takeoff, the altitude of Broad-
winged Hawks migrating during spring in south Texas increased rapidly (fig. 8.1), and within minutes migrants climbed to over 100 m
(Kerlinger and Gauthreaux 1985a). The rate at which migration altitude increased varied from day to day, presumably reflecting a variable
rate of thermal formation. With a thermal inversion at dawn, takeoff
was delayed and thermal formation was slow (31 March 1982; fig.
8.1). Another example of this rapid increase in height during early
morning comes from Sharp-shinned Hawks migrating through southern New Jersey in autumn (Kerlinger and Gauthreaux 1984). Although altitude was not measured with a radar in this study, an obvious difference was observed between the altitudes of hawks flying

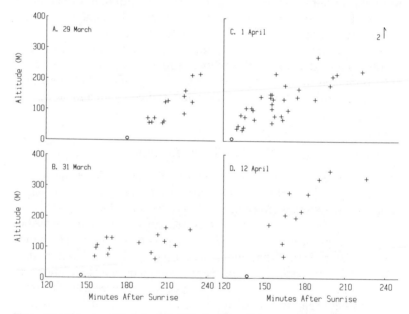

Figure 8.1 Altitudes of Broad-winged Hawks plotted against minutes after sunrise following takeoff (circles) on four mornings during spring migration at Santa Ana National
Wildlife Refuge (now Rio Grande Valley National Wildlife Refuge). Regression equa-
tions: (A) $y = 3.60x - 631$, $r^2 = 0.76$, $p < 0.01$; (B) $y = 0.89x - 63$, $r^2 = 0.34$;
$p < 0.05$; (C) $y = 2.68x - 301$, $r^2 = 0.73$, $p < 0.01$; (D) $y = 3.53x - 405$,
$r^2 = 0.72$, $p < 0.01$. Regression coefficients (slopes) mirror closely the rate of develop-
ment of the convective field. Each point represents 1 to over 1,000 migrants. (From
Kerlinger and Gauthreaux 1985a; courtesy of the American Ornithologists' Union.)

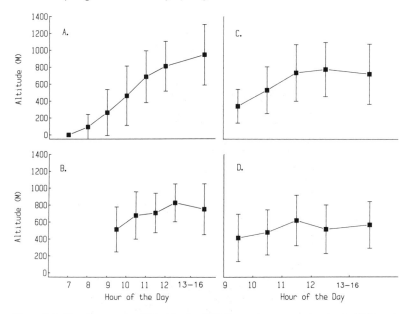

Figure 8.2 Hourly mean (± SD) altitude of (A) September convective depth, (B) Broad-winged (N = 220), (C) Red-tailed (N = 268), and (D) Sharp-shinned (N = 217) hawks during autumn migration in central New York. (From Kerlinger, Bingman, and Able 1985; courtesy of the *Canadian Journal of Zoology*.)

before and after 0900 h. Before 0900 h only 3 of 69 (4%) hawks were rated as difficult to see without binoculars (when directly overhead), whereas between 0900 and 1100 h 46 of 140 (33%) were rated as difficult to detect $\chi^2 = 19.38$, df = 1, $p < 0.01$).

The altitude of migration increases from early morning into the afternoon. Radar studies in New York (fig. 8.2) and south Texas (fig. 8.3) reveal maximum hourly mean altitudes between 1100 h and 1300 h. The pattern is consistent for Sharp-shinned, Broad-winged, and Red-tailed hawks in New York, showing that in situations where ridge or coastal lift is absent most hawks use thermal soaring as their prime source of lift.

From early morning (before 1000 h) until midmorning (1000–1059 h) Broad-winged Hawks gained, on average, 239 m during spring migration in south Texas (fig. 8.3). By midday (between 1100 and 1600 h) mean altitude increased by an additional 247 m. In New York during autumn Sharp-shinned Hawks, Broad-winged Hawks, and Red-tailed Hawks gained altitude between 0900 h when radar operations commenced and midday ($F = 2.31$–10.76, df = 4, N = 212–263,

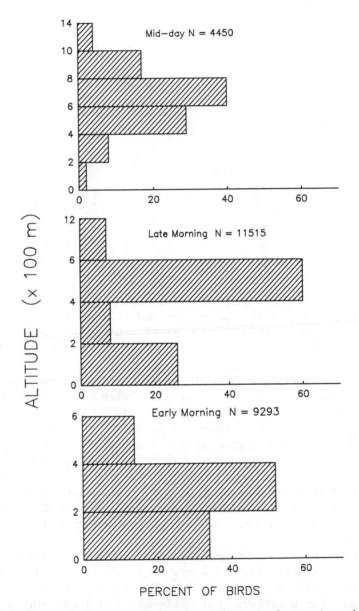

Figure 8.3 Altitudes of Broad-winged Hawks in early morning, late morning, and mid-day during spring migration in south Texas. N = total number of hawks represented during each time period. (From Kerlinger and Gauthreaux 1985a; courtesy of the American Ornithologists' Union.)

$p < 0.01–0.06$), but altitude decreased after 1300–1400 h. The rate of altitude gain between 0900 and 1200 h was nearly linear (1.7 to 4.2 mpm) in both Texas and New York.

The altitude of hawk migration is correlated with the convective depth (fig. 8.2). A significant, though weak, correlation ($r = 0.31$, $N = 51$ h, $p < 0.05$) between these variables was found for Broad-winged Hawks during autumn in New York. The lack of a stronger correlation probably reflects the late time radar operations commenced (0900 h) and the fact that migrants could not be tracked below about 300–400 m. If the few migrants that were airborne earlier had been tracked, a stronger relation would have emerged. Two other sources of variance may explain the poor relation. First, hawks change altitude constantly as they soar in thermals and glide between thermals. Second, the method of estimating convective depth (chap. 4) can be inaccurate. The method uses the 0700 h sounding and assumes level ground. The radar station near Albany was in hills, where thermals are stronger than over flat ground. More than 10% of all hawks flew 100–300 m higher than the estimated convective layer, suggesting that thermals were influenced by the terrain.

A strong relation was evident between convective depth and altitude of migrating Broad-winged Hawks in south Texas (fig. 8.3). Unlike birds in the New York study, few raptors exceeded the estimated convective depth. There are two reasons why a stronger relation was found in Texas than in New York. Two radars (MSR for low altitude flocks early in the day and VFBR for higher flying birds at midday) permitted measurement of birds from takeoff to high altitudes, which was not possible with the tracking radar in New York. Also, the flat terrain in south Texas may promote more accurate estimation of height of the convective layer.

There are striking similarities between the rate of altitude gain and the rate of convective field development. Figure 8.1 shows that the rate at which hawks gain altitude ranged between 0.9 and 3.6 mpm (regression coefficients). After the early morning period the rate ranged between 1.7 and 4.2 mpm (figs. 8.2 and 8.3). The rates of convective field development reported in the atmospheric literature (chap. 4) are between 2 and 6.1 mpm. Thus, the altitude of migrating raptors tracks the developing convective field.

Altitude of Flight in Relation to Wind and Cloud Cover

Cloud cover and wind should influence the altitude of migrants. In Texas, takeoff time of migrants was inhibited by cloud cover ($r = -0.35$, $N = 15$ days, $p > 0.10$), as was the altitude of migrants at various times of the day (Kerlinger and Gauthreaux 1985a). In early morning, before 1000 h, the relation between cloud cover and altitude was negative ($r = -0.34$, $N = 5$ days, $p > 0.10$). The relation increased during the periods 1000–1059 h ($r = -0.80$, $N = 6$ days, $p < 0.05$) and 1100–1600 h ($r = -0.71$, $N = 11$ days, $p < 0.05$). Clouds often inhibit the formation of thermals. With thick clouds the sun cannot heat the earth's surface, and thermals will be weak or nonexistent.

In New York during the September migration of Broad-winged Hawks no relation was evident between cloud cover and altitude of migrants ($r = 0.04$, $N = 51$ h, $p > 0.10$). On several days with thin but complete (nearly 100%) cloud cover, flight proceeded at "normal" altitudes (fig. 8.2). Some of these times followed clear skies and excellent conditions for thermaling flight. At these times complete cloud cover may have been a result of overdevelopment of cumulus clouds. Thus, there were still thermals present and the cloud cover broke up a short time later. Migrants did fly lower when stratus clouds covered the sky, but few migrants were aloft at those times, making it difficult to measure altitude of migration. There are two other reasons why the relation was so weak in New York. First, absence of cloud cover is not always indicative of good soaring conditions. Clear skies sometimes occur when warm air overrides cooler air during inversions (fig. 4.6B). Inversions make for miserable soaring. Second, as I pointed out above, the radar could not track low migrants.

The influence of wind was more dramatic than that of cloud cover. Studies at Cape May Point (fig. 7.4) by Kerlinger and Gauthreaux (1984) confirmed Murray's (1964) hypothesis that Sharp-shinned Hawks descend to low altitudes when winds are from the north and west during autumn. With winds over 2 mps from the west or north, Sharp-shinned Hawks almost always flew below 300 m, whereas with winds from other directions or weak winds they flew above 300 m. The results are robust because there was almost no overlap in altitude between the two weather conditions (Mann-Whitney $U = 2$, $N = 9$, 10 days, $p < 0.01$). The highest daily mean altitude for west wind conditions was only 283 ± 116 (SD) m, whereas for other winds it was 737 ± 179 m with few birds exceeding 1,000 m. Migrants descended or ascended within minutes of wind changes. For example, altitude of migrants increased from 282 ± 138 (SD; $N = 20$) m at about 1300 h on 28 September 1982 to 396 ± 126 ($N = 30$) m at about 1445 h ($t = 2.95$, $p < 0.01$). This change occurred when westerly winds abated and winds became calm.

A change of the dispersion of migrants within the Cape May peninsula coincided with altitude changes. With west winds, hawk counters see many migrants along the Atlantic coast near the tip of the peninsula at the hawk counting station (Dunne and Clark 1977). With other or calm winds migrants are seen in small numbers at the count station. At these times migrants fly (at high altitudes) near the center of the peninsula. Sea-breeze updrafts may form at these times. Because the peninsula is long and narrow, sea-breeze updrafts may be augmented by airflow from Delaware Bay and the Atlantic Ocean (fig. 7.4). West winds may inhibit these updrafts or push them out over the Atlantic. Hawks may be reluctant to soar over the ocean, especially if winds could push them offshore. Hence they descend to lower altitudes. This finding confirms the earlier claims of Rudebeck (1950), who found a similar behavior among European Sparrowhawks and Common Buzzards migrating at Falsterbo in southern Sweden.

Other migrants fly at lower altitudes through the Cape May peninsula when winds are from the west (Kerlinger and Gauthreaux, unpublished data). With west winds the species listed in table 8.3 averaged 264 ± 145 m, with many flying too low to detect on radar. These same birds averaged 594 ± 193 m when winds were calm or from directions other than west to north ($F = 55.86$, df = 1, 61, $p < 0.01$). Thus, they respond to wind like Sharp-shinned Hawks by descending to low altitudes with west winds. In Sweden migrant buteos and accipiters often climbed to 800–1,000 m with some winds, according to Rudebeck (1950), although altitude was estimated visually. Either migrants are afraid of being blown offshore with strong

Table 8.3 Summary of Altitude and Visibility of Hawks at Cape May Point, New Jersey, during Autumn 1982 as Determined by Vertical Fixed-Beam Radar

Species[a]	Sample Size	Mean Altitude ± SD (m)	Critical Altitude (m)[b]
Turkey Vulture	6	655 ± 285	700
Northern Harrier	7	626 ± 132	550
Cooper's Hawk	5	406 ± 269	560
Broad-winged Hawk	12	476 ± 187	625
Red-tailed Hawk	1	499	—
Osprey	11	470 ± 256	700
American Kestrel	21	314 ± 206	400
Merlin	1	257	—
Peregrine Falcon	2	483	560

Note: Samples represent days when soaring conditions were good.

[a]Data for Sharp-shinned Hawks are given elsewhere.

[b]Critical altitude is the altitude at which the species was judged to be "difficult" to see with the naked eye when directly overhead against a cloudless sky.

west winds, or thermals may not be available. From a behavioral perspective this is an important distinction. If this behavior is caused by a paucity of convection, then no behavioral decision occurred. Whether descent by migrants at Cape May is obligatory or facultative is an intriguing question.

Inland from Cape May Point at Woodbine (fig. 7.4) Sharp-shinned Hawks did not change altitude when winds changed (Kerlinger and Gauthreaux 1984). Although no radar measurements of altitude were taken, a visual system (calibrated with radar, fig. 8.4) revealed no statistical difference between the altitude of migrants with west winds as opposed to other or calm winds (Mann-Whitney $U = 58$, $N = 7, 8$ days, $p > 0.10$). The altitudes at Woodbine averaged 350 ± 136 m ($N = 8$ days) with west winds and 375 ± 106 ($N = 7$ days) with east winds. The reason for the low average altitude of flight at Woodbine was that observations were halted by 1100 h.

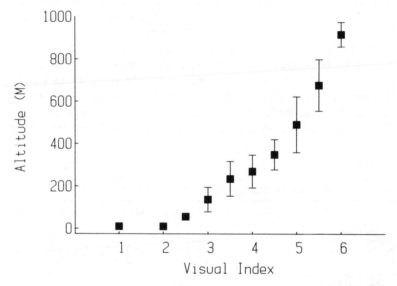

Figure 8.4 Comparison of a visual altitudinal estimation technique with simultaneous radar measurements (from Cape May Point, New Jersey; redrawn from Kerlinger and Gauthreaux 1984; courtesy of Tindall Ballierre): 1 = within 3 m of surface, 2 = 3–10 m, 3 = 11–30 m, 4 = 31–100 m, 5 = difficult to see without binoculars, 6 = not visible to the unaided eye (intermediate categories were used when the observer was unsure of the appropriate category). The total number of measurements used in the figure was 250 (class 1 was not included). A similar set of measurements was derived for Broad-winged Hawks migrating during spring in south Texas by Kerlinger and Gauthreaux (unpublished).

No obvious relation between wind and altitude of migration was reported in radar studies from New York and Texas, but Evans and Lathbury (1973) did note changes at Gibraltar. In Texas most migrants flew with south-southeasterly winds (Kerlinger and Gauthreaux 1985a). The absence of variation in wind precluded statistical analysis. In New York, migrants flew at high altitudes with most winds. The Gibraltar study by Evans and Lathbury (1973) relies upon radar data from Houghton (unpublished) showing that flight above 1,500 m occurred with east winds. Migrants flew below 1,500 m with west winds. They hypothesized that east winds generate powerful lee waves (chap. 4) on which migrants fly at great heights.

Although neither quantitative nor qualitative data are available from ridges and coasts (other than Cape May and Gibraltar), it is likely that wind and cloud cover influence the altitude of migrants in these topographic situations. At Hawk Mountain on the Kittatinny Ridge hawks are counted most often with west winds (Broun 1949; Haugh 1972). It is possible that raptors descend at these times to use ridge lift or to avoid being drifted (chap. 7). Millsap and Zook (1983) hypothesized that cold air wedges associated with the leading edges of cold fronts may be the reason why many hawks are counted on west and north winds at ridges in North America. Because warm air can override cold, dense air at the front, thermals may be inhibited at a time when ridge lift is abundant. The Millsap and Zook hypothesis should be tested. Thick cloud cover associated with the passage of a cold front also inhibits thermal formation. After a cold front passes, thermals are no longer inhibited but are promoted by cooler air and clear skies. When this happens raptors can soar to great heights.

Geographical, Seasonal, and Species Differences in Altitude of Flight

Differences in the altitude of flight among species, seasons, and geographical locations were less than anticipated. Statements by Heintzelman (1975) that the Red-tailed Hawk "seldom uses thermals" and that the Sharp-shinned Hawk is "not really dependent upon thermals" were found to be incorrect. Radar and visual observations by Kerlinger, Bingman, and Able (1985 and unpublished) in New York, Kerlinger and Gauthreaux (1984) in New Jersey, and by Kerlinger and Gauthreaux (1985a, 1985b) showed that all of the raptors studied used thermals as their primary source of lift. Radar data from the midday period confirm that these species and others soared extensively in thermals. What Heintzelman may have meant was that there are times when these and other species abandon thermal soaring and resort to ridge lift or powered flight.

Table 8.4 Summary of Radar Measurements of the Altitude of Raptor Migrants during Autumn Migration 1978–1979 in Central New York (Berne, Albany County)

Species	Number of Radar Tracks	Number of Hawks	Mean Altitude ± SD	Mean Track Length ± SD[a]	Percentage Too Low to Track	Mean Difference ± SD of Altitude[b]
Northern Harrier	11	16	774 ± 172	91 ± 47	31.3	184 ± 104
Osprey	14	19	831 ± 340	83 ± 105	26.3	261 ± 252
Sharp-shinned Hawk	126	263	755 ± 208	66 ± 51	45.6	227 ± 113
Cooper's Hawk	6	9	792 ± 203	88 ± 72	33.3	158 ± 113
Goshawk	7	12	803 ± 267	81 ± 61	41.7	168 ± 133
Broad-winged Hawk	187	791	791 ± 217	64 ± 59	9.5	237 ± 133
Red-shouldered Hawk	15	22	749 ± 276	117 ± 69	22.7	223 ± 161
Red-tailed Hawk	196	382	839 ± 250	77 ± 64	33.9	255 ± 141
American Kestrel	22	84	746 ± 236	58 ± 49	72.6	176 ± 140

Source: Kerlinger, Bingman, and Able 1985.

[a]Time in seconds that migrants were tracked.

[b]Differences were between the highest and lowest altitude during single radar tracks.

Considering each site separately, it appears that differences among species exist. Sharp-shinned Hawks and American Kestrels migrating during autumn in central New York flew lower than Red-shouldered Hawks, Broad-winged Hawks, and Ospreys, as indicated by the percentages of the species that were too low to track with the radar (table 8.4). Consequently, a smaller proportion of Sharp-shinned Hawks (7%) flew above 1,000 m than did Broad-winged Hawks and Red-tailed Hawks (about 16% of both species). Mean altitudes of flight as measured by radar (table 8.4) were similar, suggesting that all species were constrained by similar atmospheric phenomena. Data from spring migration in central New York (table 8.5) show minor differences among these species, although the data are not sufficient for statistical analysis. Autumn data show that Sharp-shinned Hawks and American Kestrels migrate at lower altitudes than some other species and suggest that these species resort to alternative forms of lift.

During flight over water, some species fly lower than others (Kerlinger 1985a). Raptors that regularly made water crossings at Cape May Point and Whitefish Point flew lower than other species ($r = 0.74$, $N = 10$ species, $p < 0.01$; chap. 10). Whereas no Turkey Vultures, Broad-winged Hawks, Rough-legged Hawks, or Red-tailed Hawks flew within 5 m of the waves, other species did so regularly. These included species that are capable of strong powered flight such as the Northern Harrier (22% of all individuals), Merlin (34%), Peregrine Falcon (23%), and Osprey (9%). Low altitude flight was often in adverse winds. Individuals that fly within one wingspan of the water may take advantage of ground effect (Withers 1979).

The midday mean altitudes of six species of migrants studied in south Texas during spring migration ranged from 531 m to 745 m

Table 8.5 Summary of Radar Measurements of the Altitude of Raptor Migrants during Spring Migration 1978–1980 in Central New York (Berne, Albany County)

Species	Number of Radar Tracks	Number of Hawks	Mean Altitude ± SD (m)	Percentage Too Low to Track	Percentage > 1,000 m
Northern Harrier	2	2	243–880[a]	—	—
Osprey	20	20	880 ± 243	6.3	25.0
Sharp-shinned Hawk	47	55	775 ± 209	12.0	12.7
Cooper's Hawk	1	1	924	—	—
Goshawk	2	2	421–779[a]	—	—
Broad-winged Hawk	54	72	936 ± 285	16.1	33.3
Red-shouldered Hawk	2	2	496–1,309	—	—
American Kestrel	3	3	665 ± 187	—	—

[a]Range given instead of standard deviation because of small sample size.

Table 8.6 Summary of Mean Altitude of Migrants during Spring Migration in South Texas Studied with Marine Surveillance and Vertical Fixed-Beam Radars

Species	Number of Radar Tracks	Number of Hawks	Mean ± SD[a]	Critical Altitude[b]
Turkey Vulture	12	26	552 ± 245	Unknown[c]
Sharp-shinned Hawk	14	18	745 ± 205	500
Cooper's Hawk	4	5	726 ± 403	600 +
Broad-winged Hawk	79	4,450	652 ± 258	600 +
Swainson's Hawk	15	154	531 ± 161	700 +
American Kestrel	11	13	551 ± 232	500
Summary of all species except Broad-winged Hawk	56	216	585 ± 227	

Source: Kerlinger and Gauthreaux 1985a, b.

[a]Measurements are from 0900–1500 h for all species except Broad-winged Hawk, whose measurements were from midday (1100–1630 h).

[b]Altitude at which species were judged as difficult to see with unaided eye when directly overhead against a cloudless sky.

[c]Individuals of this species were not judged difficult to see at the altitudes observed.

(table 8.6; Kerlinger and Gauthreaux 1985a, 1985b). Sharp-shinned Hawks flew higher than other species, although samples are small. Again, Sharp-shinned Hawks relied on thermals for lift, as did other species. I conclude that there was little difference among species in the altitude of migration at that site.

The meager radar data available from Cape May Point for species other than Sharp-shinned Hawks do not allow quantitative comparison among species. The means for six species for which at least five altitude measurements are available range from 406 to 655m (table 8.3); 31% of the measurements exceeded 600 m. A second problem with attempting comparisons at Cape May is that with west to north winds Northern Harriers, Sharp-shinned Hawks, American Kestrels, Merlins, and other migrants fly a few meters above the trees or dunes, making them too low to detect with the vertical radar (Kerlinger and Gauthreaux, unpublished data). Species like Broad-winged Hawks and Ospreys rarely descended to such low altitudes. When sharp-shinned Hawks and American Kestrels are not flying at treetop level, there is a large overlap in the altitudes used by all species because of a reliance on thermals. Comparing altitudes of the same species in New York, New Jersey, and Texas reveals no major differences among sites.

A possible difference was noted between altitudes in spring and autumn in central New York (tables 8.4 and 8.5). Mean altitudes were slightly higher for spring migration. A smaller proportion of Sharp-shinned Hawks, Ospreys, and American Kestrels flew at altitudes that were too low to track with the radar (Kerlinger, Bingman, and Able 1985, unpublished data). Convective activity is more pronounced in spring (fig. 4.8), which may account for some of the difference. The difference also may be attributable to the greater experience of hawks migrating in spring. During autumn a large proportion of birds are migrating for the first time (hatch-year birds). These migrants may not be adept at using thermals and consequently may fly at lower altitudes. The small numbers of low flying Broad-winged Hawks observed during autumn may be a result of immatures flocking with adults who are experienced at soaring. It would be interesting to follow individual migrants from the breeding ground to the wintering ground to see if thermaling efficiency improves with experience. The time spent flying at low altitude would decrease as birds moved south.

Flight at higher altitudes in spring may be of consequence to hawk counters. Combined with the fact that there are fewer birds migrating in spring (a result of winter mortality), high altitude flight may explain why so few hawks are counted during spring migration. Better or more efficient use of thermals may also explain why so few migrants are observed south of Pennsylvania during autumn, even though there are more migrants to observe.

The altitude of migrating hawks flying over land in North America is considerably higher than that reported from a tracking radar study in the Swiss Alps. Mean altitudes from Switzerland were lower than 350 m for Common Buzzards and European Sparrowhawks that encountered the Alps during autumn migration (Schmid, Steuri, and Bruderer 1986). Lower altitude of flight at this site is probably specific to the study site and similar sites. Studies of migrating cranes in Sweden (Pennycuick, Alerstam, and Larsson 1979) and of soaring species at Gibraltar (Houghton1971) and in Israel (Leshem 1987, personal communication) report altitudes similar to those from North America. It is possible that atmospheric conditions at the site used by Schmid and his colleagues are similar to those near ridges and mountain ranges in North America where raptors migrate at rather low altitudes (see above).

Height Band and Climb Rate

The data used in the descriptions and analyses above represent only one altitude measurement for each individual or a flock of migrants. The migrants from which measurements were taken change altitude continually, passing through several hundred meters of airspace as they soar up and then glide down. The vertical airspace used by a sailplane is called the "height band." Pennycuick (1972a) was probably the first to use the term in reference to soaring birds. The height band used by several raptor species during autumn migration through central New York (Kerlinger, Bingman, and Able 1985) ranged from 158 to 261 m (table 8.4). These values are the mean differences between the highest and lowest points of tracks more than 50 sec duration measured with a tracking radar. The values in the table underestimate the actual height band. When tracks more than 100 sec are examined, the height band for Sharp-shinned, Red-tailed, and Broad-winged hawks increased to 267–328 m (N = 36–65). A height band of 488 ± 177 m (SD; range 258–724 m, N = 11 days) was calculated for Broad-winged Hawks during spring migration in south Texas by subtracting the minimum altitude from the maximum altitude measured between 1100 and 1600 h on a given day (Kerlinger and Gauthreaux 1985a). Thus, most species climb and glide through a vertical space of at least 300 m during thermaling migration.

It would be interesting to know if the height band of migration changes with convective depth in a manner that allows birds to maximize distance traveled. Sailplane pilots adjust their height band in a way that allows them to travel farther than if they use the entire convective layer. Because the strength of thermals is less near the top of a thermal than at lower altitudes, pilots frequently leave thermals before reaching the top. By doing so, they do not waste time in poor lift. This strategy (referred to as distance maximization or updraft optimization) is similar to that of adjusting air speed and sink rate to convective conditions (chap. 11). When thermals are weak, pilots must use all possible lift by flying to the top of thermals. When thermals are strong, pilots leave them after using only the strongest lift, thereby spending less time thermaling and more time gliding cross-country. There is no doubt that migrating raptors use a greater band of height when thermals rise to high altitudes than when they are lower, but do they leave thermals when they could still soar to greater heights? The question is difficult to answer because of technical difficulties associated with tracking hawks and measuring lift in single convective elements simultaneously. Most raptors commence glides before they reach cloud base and therefore end climbs before they reach the top of a soarable thermal.

While moving through the height band, raptors can adjust the rate at which they sink by changing air speed and glide ratio. To maximize distance, climb rate in thermals should be as fast as possible so that the bird can spend more time gliding toward its migratory goal. Assuming that a migrant can locate the region of best lift within a thermal, climb rate is constrained by the vertical rate of air within the thermal and the aerodynamic performance of the migrant.

Hawks migrating through central New York used at least four means of gaining altitude: powered flight, soaring in wind deflected by hills and vegetation, soaring in thermals, and gliding through thermals. Powered flight was used primarily at low altitudes (<100m) in early morning. These migrants presumably were initiating migration for the day. Lift from wind off hills was also used at low altitudes.

Soaring in thermals was more common than other means of gaining altitude. Thermaling flight occurred from a few meters above the trees to more than 1,500 m. The rate of climb in thermals for birds that were between 350 and 1,700 m averaged nearly 3 mps for Sharp-shinned, Broad-winged, and Red-tailed hawks and for Ospreys (table 8.7). Smaller numbers of radar tracks of Northern Harriers, American

Table 8.7 Climb Rates of Thermal Soaring and Gliding Migrants during Autumn and Spring Migration Measured with a Tracking Radar in Central New York

Species	Number of Radar Tracks	Mean ± SD Climb Rate (mps)	Mean ± SD Time (sec)
SOARING			
Northern Harrier	6	3.0 ± 1.1	42 ± 25
Osprey	11	3.1 ± 1.1	106 ± 79
Sharp-shinned Hawk			
Spring	9	2.8 ± 0.7	59 ± 29
Autumn	35	2.7 ± 0.9	54 ± 34
Cooper's Hawk	5	2.2 ± 1.2	68 ± 43
Goshawk	3	2.7 ± 0.8	95 ± 22
Broad-winged Hawk			
Spring	15	3.4 ± 1.4	54 ± 22
Autumn	46	3.2 ± 1.4	58 ± 33
Red-tailed Hawk			
Spring	9	2.8 ± 1.4	53 ± 19
Autumn	59	2.9 ± 1.1	78 ± 50
Red-shouldered Hawk	9	2.7 ± 1.4	74 ± 51
American Kestrel	8	2.8 ± 1.1	63 ± 34
GLIDING CLIMBS			
Osprey	6	1.3 ± 0.8	48 ± 24
Sharp-shinned Hawk	15	1.7 ± 1.2	49 ± 20
Broad-winged Hawk	31	1.8 ± 1.1	41 ± 21
Red-tailed Hawk	12	1.3 ± 1.1	43 ± 22
Other species[a]	9	1.5 ± 0.8	36 ± 18

Source: Kerlinger, Bingman, and Able 1985 and unpublished data.

Note: Radar tracks > 20 sec; autumn and spring data pooled unless noted otherwise.

[a]Species include Northern Harrier, Cooper's Hawk, Red-shouldered Hawk, and American Kestrel.

Kestrels, and Red-shouldered Hawks revealed similar climb rates. Hopkins (1975), Welch (1975), and Hopkins et al. (1979) reported similar climb rates for Broad-winged Hawks, Sharp-shinned Hawks, and Ospreys during autumn migration in New England. Thus, climb rates of autumn migrants in eastern North America are about the same as those reported for a sailplane and vultures from the Serengeti of Africa (Pennycuick 1972a) and greater than those of Frigatebirds (*Fregata magnificens*), Brown Pelican (*Pelecanus occidentalis*), and Black Vultures on an island in Panama. The last three species realized climb rates of 0.4–0.6 mps. These birds were engaged in foraging flights or were milling about and may not have been attempting to maximize climb rate. Migrants, on the other hand, probably do attempt to maximize climb rate.

Slower climb rates have been reported for the European Sparrowhawk and Common Buzzard. Tracking radar studies by Schmid, Steuri, and Bruderer (1986) near the Alps in Switzerland show climb rates of less than 1 mps (means of 0.76 for sparrowhawk and 0.86 for buzzard). These birds were flying at rather low altitudes (means less than 350 m) when compared with migrants reported in North American studies, which may account for the slower climb rate. When confronted by the Alps these birds resorted to soaring in ridge lift to gain altitude before moving through mountain passes.

Other authors have also noted that soaring ability differs among species. Hankin (1913) working in India and Pennycuick (1972a) in Africa both noted that larger vultures tend to take off later in the morning than smaller vultures and hawks. This difference is related to the different soaring envelopes among the species. Keen observations by Hankin, who was probably the first to make systematic observations of the soaring behavior of vultures, allowed him to hypothesize that soaring ability was related to size. (His volume is mandatory reading for those interested in soaring birds, although it is difficult to find.)

Finally, by reducing glide speed when flying through updrafts hawks were able to move toward their migratory goal while gaining altitude. Air speeds during these glides averaged 5–6 mps slower than normal interthermal glides. This behavior has been reported for vultures making foraging flights over the Serengeti by Pennycuick (1972a) and is called "dolphin flying" by sailplane pilots (Pennycuick 1972a; Reichmann 1978). These vultures travel long distances by gliding through strong lift without soaring or flapping. One vulture observed by Pennycuick (1972a) realized a "glide ratio" of 60:1 (with respect to the ground, not the air) by dolphin flying. Migrants of many species gained altitude while gliding in a straight line in central New York (table 8.7). Climb rates were one-half to one-third the rates realized when soaring in thermals. These migrants were gliding either in large thermals (>1 km) or in thermal streets (chap. 4). Thermals over 1 km in diameter are known from East Africa (Pennycuick 1972a) and temperate regions (Hardy and Ottersten 1969; Konrad 1970). Smith (1985b) states that buteos migrating through Panama in autumn use linear updrafts created by waves (ca. 1,800 m). Gliding through thermals, thermal streets, or waves results in a fast cross-country speed. Whether migrants use them for long distances is not known, although Smith reports that the lenticular clouds in Panama created by waves are 60–80 km in length.

Flight in or above Clouds

Although radar studies do not document migrants flying in or above clouds, numerous reports of this phenomenon abound in the literature. Flight in opaque, cumulus, fog, and other types of clouds has been reported by Forbes and Forbes (1927, Broad-winged Hawk), Hopkins (1975, Broad-winged Hawk), Welch (1975), Heintzelman (1975, 1986, Sharp-shinned Hawk, Osprey, Turkey Vulture, Broad-winged Hawk, Red-tailed Hawk), Servheen (1976, Bald Eagles), Borneman (1976, California Condor), and Smith (1980, 1985a, Turkey Vulture, Black Vulture, Swainson's Hawk, Broad-winged Hawk). Some of the raptors reported by these authors were not migrating. The altitude at which these birds flew into clouds ranged from below 100 m (Heintzelman 1975, in fog/mist) to more than 6,000 m. I have seen three species (Broad-winged Hawk, Red-tailed Hawk, and Golden Eagle) soar into clouds, but all three observations were at altitudes less than 700 m.

Observations of hawks flying in clouds are too numerous and well documented to dismiss. But how often do raptors fly in clouds, and how far do they fly? To answer these questions we must examine the structure of convection in and around clouds, and we need better quantitative data on flight behavior of migrants when they are confronted by clouds.

The pattern of updrafts in and around clouds is documented in the atmospheric literature. The level at which clouds form is dependent upon the temperature profile and humidity of the atmosphere. Briefly, when warm, moist, rising air reaches the altitude at which condensation occurs, a cloud forms (for details see Stull 1988). Despite cloud formation, the warm air continues to rise. Warner and Telford (1967) showed that updrafts continue into and above cumulus clouds. Thus, a migrant can soar into cumulus cloud and continue to soar through some clouds. The same is true of lenticular (lens shaped) clouds that form in lee or mountain waves.

The numbers of migrants that soar into clouds seems to be small compared with the total number observed. It is significant that radar studies have not documented this phenomenon and that many of the citations above do not document flight in thick, solid, or opaque clouds. The observations of Hopkins et al. (1979) and others reveal that most raptors leave thermals before entering cumulus clouds or shortly after entering such clouds. Smith's (1985a) observations of Broad-winged Hawks and other migrants gliding in the "bottom 3 m" of lenticular clouds confirm my suspicion that raptors rarely fly within opaque clouds where their visibility is completely obscured. It is probably easier for raptors to see out of the edge of clouds than for observers to see the raptors soaring or gliding just within clouds. Although Welch (1975) "definitely established that Broad-winged Hawks do fly in solid cloud," it probably does not occur often.

Why raptors do not soar into opaque clouds more often is not known. Passerine migrants tracked with radar in opaque cloud at night maintained straight courses in a direction appropriate for migration (Griffin 1973). This observation and others (reviewed by Able 1980) suggest that some migrants utilize nonvisual cues for orienting at night. There are two reasons raptors might choose to avoid flight in opaque clouds. Lift may decrease near cloud base, although lift in cumulus clouds can be stronger than below cloud base (Warner and Telford 1967). More likely, raptors may not want to enter a medium in which they have limited visibility. Unlike nocturnal migrants, raptors are used to having abundant visual cues available to them.

Visibility and Detectability of Migrating Hawks

A common complaint among hawk watchers is that making accurate counts is "neck breaking and eye straining" (Allen and Peterson 1936). Other authors have also stated that raptors sometimes fly at heights that make them difficult to see and count. Allen and Peterson (1936), Stone (1937), and P. Dunne (personal communication) noted that Sharp-shinned Hawks and other species sometimes migrate at very high altitudes at Cape May Point. At times counters have noted that even larger hawks such as Honey Buzzards are "only visible through binoculars" (Beaman and Galea 1974). Cochran (1975), Christensen et al. (1981), and others have stated that larger migrants such as Peregrine Falcons (900 m, Cochran 1975) sometimes fly at altitudes at which they cannot be seen with the unaided eye. If migrants flew below 100 m at all times, there would be no problem counting them, but high altitude flight is common. What does high altitude flight mean for the accuracy of hawk migration counts?

Murray (1964) seems to be the first to state formally that hawk migration counts were biased to low altitude flight. He proposed that Sharp-shinned Hawks descended to low altitudes at Cape May Point only when winds were from west to north and that migration counts did not reflect the numbers of migrants aloft. His hypothesis was confirmed (Kerlinger and Gauthreaux 1984). Because much of what has been published about the behavior of migrating hawks has been based on hawk migration counts, it is important to determine if they are reliable and valid indicators of the numbers of hawks aloft.

Flight at high altitude has also been invoked to explain noonday lulls and the daily rhythm of migration (Heintzelman 1975; Kerlinger 1985b). Mueller and Berger (1973) demonstrated distinct peaks in the numbers of hawks counted at specific times of the day (fig. 8.5). For accipters (mostly Sharp-shinned Hawks) the peak in the number counted was evident at 0800–0900 h, whereas the peak for buteos other than "aberrant" Broad-winged Hawks was at 0900–1000 h. Two peaks were noted for falcons; one at 0800–0900 h and a larger one at 1300–1400 h. Allen and Peterson (1936) also reported a morning peak in numbers of Sharp-shinned Hawks. Not all of the peaks reported for a given species occur at the same time. Peak Sharp-shinned Hawk numbers at Hawk Mountain were in midafternoon (Broun 1949). Peaks for several other species are presented by Heintzelman (1975) and Hoffman (1985). Mueller and Berger (1973) hypothesized that peaks were indicative of the numbers of hawks aloft or a daily rhythm of migration. They stated that peaks or rhythms may be indicative of endogenous behavioral patterns, presumably like the migratory restlessness (*Zugenruhe*) of passerines that migrate at night.

If peaks in migration counts indicate greater numbers of migrants aloft, do troughs or lulls represent fewer? This is not necessarily so. Heintzelman (1975) showed lulls in single-day migration counts for Broad-winged, Sharp-shinned, and Red-tailed hawks. These lulls, termed noonday lulls, were near midday. Similar lulls have been reported in midday at Malta (Beaman and Galea 1974) and in midmorning from Gibraltar (Evans and Lathbury 1973). There are similarities among studies that report lulls and peaks. All relied on hawk migration counts, and both reported more migrants in the early morning (0800–1000 h) than at midday. Might these phenomena be an

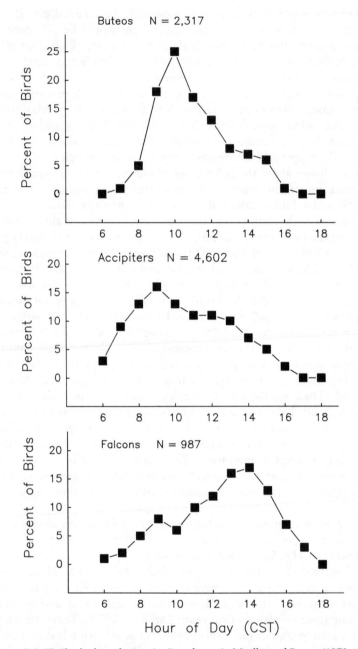

Figure 8.5 "Daily rhythm of migration" as shown in Mueller and Berger (1973; courtesy of the American Ornithologists' Union).

artifact of the count technique related to Murray's high altitude hypothesis?

To answer this query five pieces of information are needed: the altitude of hawk migration; if or how the altitude of migration varies with time of day; how high a migrant must fly before it is no longer visible to a ground-based observer; whether raptors migrate in or above clouds; and an accurate estimate of the number of hawks aloft. In previous sections data were presented documenting all but the last. I now present theoretical considerations and empirical findings that pertain to the visibility and detectability of migrating raptors.

At least five variables influence the visibility of a migrating raptor to a ground-based observer: size of the bird; distance from the observer; surface area presented to the observer; background (blue sky, clouds, vegetation) against which a bird is viewed; and aerosols or other atmospheric particles. It is difficult to evaluate the relative importance of these variables. Only the roles of bird size, distance from an observer, and surface area presented to the observer are considered here.

The most important factor influencing visibility of a migrant is its distance from the observer. Consider a bird flying at 400 m above the ground, lower than the mean midday altitude of many species. When directly overhead the bird is only 400 m from the observer. As the bird moves away the angle between the line of sight and the ground decreases. When the angle reaches 60° the distance between observer and bird increases to 462m. At 30° above the horizon, the distance increases to 800 m, and at 15° the distance is over 1,500 m (fig. 8.6). Thus, a bird flying at a low altitude can be far from an observer. This is important because most hawk watchers and raptor banders seldom scan higher than 30° above the horizon. How high above the horizon do hawk watchers scan, and is there a difference among observers and sites?

The surface area of a migrant that is visible to a ground-based observer also influences visibility. A migrant is most visible when overhead with its entire ventral surface presented to the observer (fig. 8.6). A bird flying at the same altitude that is viewed at 15° to 30° above the horizon is not as easy to see because less ventral surface is presented. When a gliding migrant is viewed at less than 10° above the horizon, only a "side-on," "head-on," or "tail-on" silhouette is presented. At these times migrants are difficult to detect. Most hawk watchers are familiar with the phenomenon called "blinking out," where a bird is visible one moment and disappears the next. Birds that blink out are often viewed when soaring in thermals at low angles above the horizon. During each circle, they present varying ventral, side, tail, and head views due to the angle of bank. Only when the ventral surface is fully exposed is the bird visible, unless light is reflected by its dorsal surface. Thus, there is a strong interaction between surface area visible to the observer and distance between bird and observer such that closer birds present greater surface area (if altitude is constant).

Empirical studies of the detectability of migrants by ground-based observers are scarce. Studies by Sattler and Bart (1985a, 1985b) showed that observers underestimate the numbers of hawks migrating past a given point. Their results also demonstrated that some species are more detectable than others. For most species the counter detected more than 70% of the raptors passing a site, although the percentage varied from under 50% to over 90% depending upon the species. They concluded that birds were missed because observers became fatigued and because of differing flight behavior among species. Again, count studies were shown to be biased to low flying birds.

Kochenberger and Dunne (1985) also examined the reliability of hawk migration counts. They investigated two questions: Does the number of hawks counted change when a count station is moved short distances (<300 m)? Is the number of hawks observed correlated with

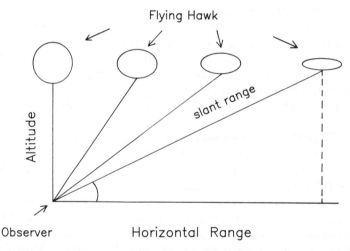

Figure 8.6 Schematic depiction of the relation between distance between an observer and a migrant, elevation of the migrant above the horizon, and surface area of a migrant visible to an observer. The circles and ellipses represent the ventral surface of a migrant flying at 400 m above ground level.

the number of counters? To answer these questions Kochenberger and Dunne designed an elegant field test. By placing one to three observers (assigned randomly each day) to one of four locations near the end of the Cape May peninsula, they created a blockwise design. They found that the position of the hawk lookout was important and that the number of counters made a slight difference in the number of hawks counted. A greater number of some species was seen at some locations than others, although daily counts from all locations were correlated highly.

The Sattler and Bart (1985a, 1985b) and Kochenberger and Dunne (1985) studies are important because they were the first to examine the "reliability" of counts made by ground-based observers. By reliability, Sattler and Bart referred to the accuracy of a count made by one individual. Because their system relied on a second observer to monitor migrants moving through a specified area of the sky above a hawk counter, their study did not test interobserver reliability, nor did they address the complete reliability picture. As with any count study, theirs was biased to low flying birds. A thorough test of observer reliability would be one in which a hawk counter was compared with a machine such as radar that detects birds at great distances and accurately measures the distance between counter and migrant. Such a system would estimate the volume of airspace that can be *sampled* accurately by a counter. Reliability coefficients determined by Sattler and Bart could then be used to correct for observer fatigue and variability of behavior among species. More well-designed studies like the two described above are needed.

Although no studies have attempted to compare a human counter with radar, radar and simultaneous visual observations have been used to determine altitude and visibility of migrants. The combined radar and visual efforts of Houghton (1971) and Evans and Lathbury (1973) showed that many migrants passed over counters without being detected.

More recently, simultaneous radar and visual studies by Kerlinger, Bingman, and Able (1985 and unpublished) and Kerlinger and Gauthreaux (1984, 1985a, 1985b) have determined how visible migrants are to observers when flying at known altitudes. They showed that few North American raptors are difficult to see with the naked eye when less than 300 m directly overhead. Sharp-shinned Hawks and American Kestrels, the smallest North American falconiforms studied, became "difficult to see" against cloudless skies at between 400 m (Kerlinger and Gauthreaux 1984, unpublished) and 500 m (Kerlinger and Gauthreaux 1985b), although some disappeared below 400 m when skies were hazy. These species disappeared often above 700 m. Larger species like Broad-winged Hawks were readily visible to the unaided eye below 550 m ($N = 10$ observations of flocks or individuals; Kerlinger 1985a), but at altitudes between 563 and 869 m ($N = 10$) "they were somewhat difficult to see without 7x binoculars." Above 870 m (879–1,368 m, $N = 10$) these hawks were not visible without binoculars. Two single Broad-winged Hawks were difficult to detect at 1,100 m even with 7x binoculars. Some of the findings for Broad-winged Hawks were confirmed by Kerlinger and Gauthreaux at Cape May Point (table 8.3).

Species that are larger than Broad-winged Hawks become difficult to see without binoculars at altitudes above 600 m (Kerlinger and Gauthreaux 1985b). In New Jersey, Kerlinger and Gauthreaux found that birds the size of the Northern Harrier, Cooper's Hawk, and Peregrine Falcon were difficult to see at altitudes slightly greater than 550 m, whereas the larger Turkey Vulture and Osprey became difficult to see at 700 m (table 8.3). Above 900 m many species either are difficult to detect without binoculars or disappear to the naked eye. During radar observations and simultaneous human observations in central New York, Texas, and New Jersey, dozens of strong echoes could not be identified visually even with binoculars or spotting scope. Most of these echoes were less than 1,500 m away, and some probably were hawks.

Unpublished data from Texas show that even flocks of hundreds of Broad-winged Hawks were difficult to detect at about 2,000 m and were not visible at the same distance on some occasions. These flocks were flying below 500 m but were less than 20° above the horizon. Even with 20x binoculars mounted on a tripod, large flocks were sometimes not visible at 2,500–3,000 m ($N = >5$ flocks of >500 individuals). The visibility in south Texas was influenced by hazy skies and high relative humidity.

Throughout my radar studies I have been surprised by how often birds went unobserved until located with radar. Many hawks passing at altitudes below 400 m were seen only after an alert observer was cued by the radar. Before radar use in central New York, I did not expect that so many migrating raptors could be seen away from ridges or coastlines. The radar demonstrated the prevalence of broad-front migration, as was shown earlier by Richardson (1975) in southern Ontario.

The results presented in the previous section regarding the relation between altitude and time of day must now be considered in light of visibility and detectability data. It is obvious from radar studies that a large and varying proportion of migrating raptors fly above the range of unaided vision. The data from combined radar and visual studies demonstrate that counting raptors that are flying at normal midday altitudes is difficult and that counts are probably not accurate. In addition, systematic differences in altitude that result from wind speed and direction, convective depth, percentage of cloud cover, and time of day bias counts of migrating hawks.

It is my opinion that noonday lulls and daily rhythms of migration are sampling biases that result from differential visibility (countability) of migrants. Consider the daily rhythms presented by Mueller and Berger (1973) in which peaks for buteos and accipiters occurred before 1030 h. Is it likely that more birds migrate early in the morning before the development of the convective layer than in late morning and early afternoon when thermals are strong and abundant? Noonday lulls coincide with the best soaring conditions of the day, when migration requires the least energy; they probably begin before noon and last until late afternoon. Note that noonday lulls and low counts in late morning previously have been hypothesized to be a result of high altitude flight (Heintzelman 1975; Evans and Lathbury 1973).

The results of empirical studies of the altitude and detectability of migrants have profound implications for the reliability and validity of the hawk migration count method and for the conclusions of count studies. Because counts are biased, studies in which differences in raptor counts have been used to infer migratory behavior are suspect. It is likely that the hawk migration count method is not valid for many scientific purposes. Before more time and money are invested in counting raptors, the questions of reliability and validity should be addressed.

Summary and Conclusions

Few quantitative or systematic measurements of the altitude of raptor migration were made before 1980. Radar and sailplane studies in the 1970s and 1980s provide an abundance of information about the altitudes of migrating hawks. Thermals are the predominant source of lift most migrants use, and the mean climb rate of raptors soaring in thermals was about 3 mps, with some exceptions. Altitude increased from early morning into midday and early afternoon as the convective layer developed. During flight over land devoid of topographic leading lines, most raptors fly between 300 and 1,000 m and move through a 250 to more than 500 m height band. Relatively few migrants (<<5%) flew higher than 1,500 m, and flight in opaque clouds is infrequent. Differences among species and geographic locations were minimal with some exceptions. Spring migration was slightly higher than autumn migration. Altitude of migrants was often lower, with strong winds and thick clouds. Smaller hawks such as Sharp-shinned Hawks were difficult to see without binoculars when more than 400 m overhead and were usually not visible without binoculars at over 800 m. Larger hawks were difficult to see without binoculars at about 800 m. Because migrants often fly at over 500 m at midday, a large proportion of hawks are not readily visible to ground-based observers. Factors influencing the visibility of migrants include size of migrant, distance from an observer, and the surface area presented to an observer. Migrants viewed at or near the horizon present less surface area (tail-on, head-on, side-on views) to an observer and cannot be seen at as great a distance as when the migrant is overhead. Noonday lulls and daily rhythms of migration are count biases resulting from differential altitude of flight that varies with convective depth. Hawk migration counts are biased to low altitude flight and are not reliable indicators of the numbers of birds aloft.

9

Flocking Behavior

Flocking can be observed among most groups of migrating birds. Among passerines that migrate at night flocking is rare (Balcomb 1977), occurring only after dawn or when birds are forced to fly during daylight hours over water barriers such as the Gulf of Mexico (Gauthreaux 1971). Flocks that form at dawn following a night of migration are the same mixed-species foraging flocks familiar to birders. The adaptive significance of flocking has been the subject of considerable speculation, and it is not clear whether the flocking behavior of different species is the result of similar selective pressures (Crook 1965; Pulliam 1973; Ward and Zahavi 1973; Caraco 1981; Pulliam and Millikan 1982).

Flocking implies social function. Tinbergen (1951) has defined an animal as being social when it strives to be in the neighborhood of fellow members of its species when performing some or all of its instinctive activities." Thompson (1926) uses the term to refer to "positive social behavior of individual birds that results in their congregating into groups (flocks)." He goes on to state that there are three characteristics of flocks: synchronous activities, cohesive movement, and maintenance of a constant dispersion of individuals within the group. As we will see, the behavior of some migrating hawks conforms to these stipulations.

When referring to groups of migrating raptors some authors refrain from using the term "flock." Instead they use "aggregation" (Brown and Amadon 1968) or "kettle" (Haugh 1972; Heintzelman 1975) to describe these groups. Hawk counters in North America also use the term "kettle." Other authors, however, refer to groups of migrating raptors as "flocks" (Broun 1949; Hamilton 1962; Brown and Amadon 1968; Newton 1979; Smith 1980). Henceforth, I use the term flock to refer to groups of migrating raptors except ephemeral and random aggregations. I recommend that "kettle" not be used.

The reluctance of some authors to use the term "flock" or "flocking" when referring to raptors is a result of the attitude that raptors are too aggressive to be social. Cade (1982) discusses this with respect to colonial nesting falcons. Raptors are not asocial animals, and even aggressive species have evolved social systems (sensu Brown 1974). Some maintain exclusive territories (Brown and Amadon 1968; Newton 1979), whereas others, mostly the "kites" and smaller falcons, sometimes breed in colonies. Social groups of raptors are most common during the migration and non-breeding seasons. Dozens of species form large flocks, sometimes exceeding 1,000 individuals, during these seasons that are characterized by a lack of aggression and territoriality.

Flocking behavior of migrants is relatively unstudied. A few studies have documented that it occurs, when it occurs, what species flock, and even that it occurs rarely (Balcomb 1977). The function of flocking has been questioned (Hamilton 1962; Lissaman and Schollenberger 1970; Keeton 1970; Wallraff 1978; Thake 1980; Kerlinger and Gauthreaux 1985a; Kerlinger, Bingman, and Able 1985), although raptors largely have been ignored. Even hawk counters have rarely noted flock sizes and species composition. Instead they record the numbers of hawks passing per hour. Many authors (Broun 1949; Brown and Amadon 1968; Haugh 1972; Heintzelman 1975; Cade 1982) give rough estimates of flock sizes for some species, and a few researchers have recorded flock size and species composition (Thake 1980; Kerlinger and Gauthreaux 1985a, 1985b; Kerlinger, Bingman, and Able 1985; S. Benz and B. Shelley, personal communication; P. Roberts, personal communication).

This chapter reviews the hypotheses proposed to explain flocking, presents three models that show how flocking facilitates the location and use of thermals, reviews empirical studies of flocking behavior, and compares existing data with predictions of the models.

Explanations of Flocking Behavior among Migrating Hawks

Several hypotheses have been proposed to explain the function of flocking among birds. Implicit in the explanations is that survival, and ultimately reproductive success, is greater for individuals that join flocks than for those who remain alone. Flocking and other social behaviors have evolved in such a large number of birds that a unitary cause is unlikely. Flocking is probably a result of multiple selective pressures. Furthermore, it may not serve the same function when the researcher studies it as when it evolved.

Flocking has been hypothesized to facilitate detection of predators (Lack 1954; Moynihan 1962); improve foraging efficiency (Pulliam

and Millikan 1982) or provide an information center (Ward and Zahavi 1973); improve navigation and orientation of migrants (Hamilton 1962; Keeton 1970; Thake 1980); and help birds find and use
thermals (Pennycuick 1972a, 1975; O'Malley and Evans 1982; Kerlinger and Gauthreaux 1985a, 1985b; Kerlinger, Bingman, and Able
1985). Only the last two hypotheses have been invoked for migrating
hawks, but the second may also be important. Only the last three will
be considered, because they are the most reasonable explanations.

The Foraging Efficiency Hypothesis

Foraging by groups of organisms may facilitate the location of patchy
food resources (Pulliam 1973; Pulliam and Millikan 1982). These
patches may be limited in both space and time. The mechanism by
which foraging efficiency increases may be immediate or delayed. By
foraging with others, a flock can search a much larger area than a lone
individual (immediate rewards). When a flock member locates a food
patch, all individuals in the group use the resource. Thus, there is a
degree of cooperation. Although the individual that finds the patch
will have to share the food, it will benefit in future foraging bouts
when it is not the first to locate a resource. This mechanism works
best when food resources are ephemeral, yet superabundant and too
much for one individual.

Foraging efficiency may also be improved even when individuals of
a flock do not forage in proximity to each other (delayed mechanism).
Ward and Zahavi (1973) proposed that a roosting flock may serve as
an information center. Individuals who have foraged successfully may
provide information (consciously or unconsciously) to other members
of the flock about food location and abundance (or lack thereof).
Honeybees do this by dancing to nest mates in their hives (Frisch
1967). If members of a roosting flock follow members who appear
well fed (e.g., a full crop), they may find abundant food. Ward and
Zahavi (1973) suggested that roosting flocks of small falcons that feed
on insects in East Africa serve as information centers. Knight and
Knight (1983) observed that immature Bald Eagles followed adults
away from winter roosts in Washington State and concluded that immatures gained information regarding the location of good forage.
Because eagles feed on locally abundant spent salmon (carrion), the
resource is patchy and ephemeral. Rabenold (1986) also concluded
that roosts serve as information centers for Turkey Vultures, another
carrion-dependent species. For an information center to be viable, a
resource must be available for several days so that the information is
meaningful.

I am reluctant to speculate that hawks migrating in flocks forage better as groups than as individuals. For the most part we know little about foraging by hawks during migration.

Many species that migrate in flocks remain in flocks through the non-breeding season. The most prominent of these are the insect followers that breed at temperate latitudes and migrate to central Africa. These include Amur Falcon, Red-footed Falcon, Lesser Kestrel, European Kestrel, European Hobby, Sooty Falcon, Eleonora's Falcon, and some others. Incredible roosting concentrations of more than 50,000 Amur and other falcons have been reported from Africa (Moreau 1972), although most flocks are smaller. Some of the larger hawks such as Lesser Spotted Eagle, Greater Spotted Eagle, Tawny Eagle, and the smaller Pallid Harrier also remain in flocks after migration, feeding occasionally on insects or other locally abundant prey (Brown and Amadon 1968; Brown 1970; Walter 1979). These species depend to a large extent on emerging (alate) termites, grasshopper plagues, and migratory locust swarms. In South America some Swainson's Hawks spend the non-breeding season in flocks moving from area to area where insects or other prey are abundant. Because the abundance and location of insects are unpredictable, these species may be nomadic. Thus, no stable "home range" is ever established, so that migration does not really end until the following breeding season.

Recent reports from the Sahel desert of Africa by J. M. Thiollay (in ICBP–World Working Group on Birds of Prey newsletter, 1985) indicate that with drought and increased insect control, flocks of many raptors are smaller or no longer present. If the trend continues, breeding populations in Europe, Asia, and Africa will suffer. The change of habitat in the Sahel is analogous to the changes in South America that may be contributing to declining populations of passerines in eastern North America (Keast and Morton 1980).

There are several species of raptors that migrate and spend the non-breeding season in flocks but do not feed on insects. These include Turkey Vultures, other vultures, and some harriers. Some of these species feed on carrion or rodent infestations, which are superabundant and ephemeral resources. When a carcass is located there is frequently too much food for one individual, so many exploit the resource together.

The Orientation/Navigation Hypothesis

Several authors have proposed that groups of migrants may be able to navigate or orient better than individuals. Hamilton (1962) suggested that individuals may convey directional information to flock members, especially in species that migrate long distances, such as Broad-winged and Swainson's hawks. He also hypothesized that some information on migratory pathways might be transmitted from generation to generation. There is a strong theoretical basis for the idea that better orientation is possible by a group than by an individual, but few published studies have demonstrated such a relation (Keeton 1970; Rabol and Noer 1973; Wallraff 1978; Helbig and Laske 1986).

Besides Hamilton (1962), Thake (1980) is the only author to suggest that flocking among hawks improves orientation or navigation. Thake used the size and spacing of flocks of Honey Buzzards flying over the island of Malta during autumn migration to test this hypothesis. Spacing of individuals was such that flocks may have been in visual contact with each other, thereby giving information on flight direction. Unfortunately, Thake reported no directional data.

A problem with the orientation/navigation hypothesis relates to the difference in seasonal timing of the age groups. If adults show immatures which way to fly, they would necessarily need to fly at the same time as adults, and flocks of migrants should not be homogeneous with respect to age. Among species like Broad-winged Hawks adults precede immatures by some one to four weeks in spring (Kerlinger and Gauthreaux 1985a) and could not show them the way. Similarly, different seasonal timing during autumn (see appendix 2) would present the same problem. The hypothesis does not depend upon cultural transmission, at least as presented by Thake (1980).

The Thermal Location and Utilization Hypothesis

Because powered flight is so costly compared with gliding and soaring, a migrant should avoid powered flight if possible. Migrants that find and use thermals efficiently will rarely resort to powered flight. Pennycuick (1972a) and other researchers (O'Malley and Evans 1982; Kerlinger 1985c; Kerlinger and Gauthreaux 1985a) have proposed that flocks of vultures and other soaring birds can locate thermals more efficiently than individuals.

There are several mechanisms by which individuals in flocks might locate and use thermals. Kerlinger (1985c) has presented three models: local enhancement, random encounter, and climb-rate feedback. The simplest is local enhancement. That is, if a migrant sees another using a thermal, it flies to that thermal (fig. 9.1). Remember, only two individuals are necessary here. In this situation the migrant attracted to the thermal was provided with information by the migrant soaring in the thermal. This may be the way some flocks form, but these may be ephemeral.

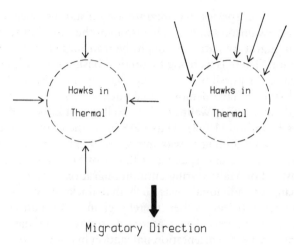

Migratory Direction

Figure 9.1 Two versions of the local enhancement model of thermal location by flocks of migrating raptors: flight of migrants from varying (random) directions into flocks of migrants already in a thermal (left), and flight of migrants in the migratory direction into flocks of migrants already in a thermal (right). Lines with arrows represent flight paths of individual migrants.

A second means by which a flock may locate thermals more efficiently than an individual migrant is random encounter. Migrants that travel long distances should not deviate greatly from their migratory course. By continually looking for migrants who have found thermals and flying to them (as in the local enhancement model), a raptor would fly along a circuitous pathway (fig. 9.1). If a migrant could encounter thermals randomly while gliding in the appropriate migratory direction, there would be no need to deviate from its course. By spacing themselves perpendicular to the migratory pathway (fig. 9.2) during interthermal glides, a flock of migrants would encounter thermals randomly more often than individual migrants. The probability of an individual's encountering a thermal is a function of the airspace it "samples" during a glide, which is equal to the wingspan of the bird times the length of its glide. For a flock in which n individuals space themselves to maximize the chance of encountering a thermal, the amount of airspace or frontal area sampled is the product of the glide length times the sum of the wing spans and lateral distance between individuals (fig. 9.2). Thus, the probability of encountering a thermal randomly is a function of the area sampled by the flock, the size of thermals, and the spacing of thermals. As flocks become wider, thermals become larger, or thermals are closer together, the probability of randomly encountering a thermal increases (fig. 9.3). At some point flock width will be such that thermals will always be encountered. The functions in figure 9.3 are hypothetical and can be tested when more is known about thermal size, spacing, and vertical extent.

The efficacy of this model depends upon the shape of a flock and the spacing of individuals within it. A prediction is that individuals in small flocks should disperse themselves more than migrants in larger flocks to "sample" an equivalent airspace. However, there will be a time when interindividual spacing becomes too great, analogous to a fisherman using a net with a mesh size that allows fish to get through. Furthermore, as the number of birds in a flock increases interindividual spacing becomes less important.

It should be remembered that the probability of random encounter is also dependent upon the amount and strength of convection on a given day. On days with little convection the probability of randomly encountering thermals may be so low that migration, even in flocks, is not feasible. There are also days when thermals are so strong and abundant that individual migrants can hardly fall out of the sky. Most of the time thermals are not strong and superabundant. In the early morning and late afternoon thermals are often weak and scarce. Flocking may be most important at these times. By flocking at these times, a migrant may avoid having to wait for good thermals or use powered flight.

When more is known about the dispersion of thermal elements within the boundary layer, quantitative predictions can be made that

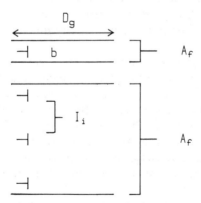

Figure 9.2 Spacing of individual migrants during interthermal glides showing the frontal area (A_f) covered by an individual bird (top) and by three birds (bottom). (D_g is distance of a glide, b is wing span of a migrant, and I_i is interindividual distance.) Such spacing may promote the random encounter of thermals by migrants in flocks.

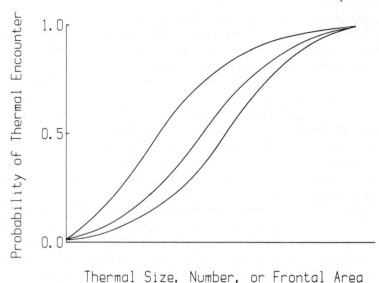

Figure 9.3 Hypothetical probabilities of encountering thermals randomly as the frontal area of a flock changes and as the size or number of thermals increases. Empirical determinations of parameters along the *x*-axis are necessary before reasonable probability functions can be ascertained.

will help us test the "models" described above. By using the known configuration and spacing patterns of migrating raptor flocks, simulations of gliding through thermal fields of varying dispersion will be possible. These will permit true probability functions to be generated as in figure 9.3.

Finally, the climb-rate feedback model is concerned with the utilization of updrafts within a thermal. On many occasions sailplane pilots find it difficult to locate or stay in strong lift even though they know they are in a thermal. This is because thermals move and because lift is not uniform within a thermal (chap. 4). The vertical ascent rate of air in a thermal varies, being strongest near the core and weakest near the edge. Obviously, the best place for a soaring hawk or sailplane to be is at the center or core of a thermal. Finding that core and using it is not easy.

Sailplane pilots locate the region of best lift within a thermal by "feeling" their way around it. The difference in lift between left and right wing sometimes yields information that can be used to stay "centered" in a thermal. If the thermal is small or weak, the region of best

Figure 9.4 Photograph of a flock of Broad-winged Hawks soaring in a thermal showing how individuals are spread over a larger area than could be covered by a single migrant. By dispersing itself over a larger area a flock may be able to locate the area within thermals with the strongest lift, thereby using thermals more efficiently than single migrants. (Photograph by Clay Sutton.)

lift, or even the thermal itself, may be lost. Because a flock of hawks is dispersed within a thermal over a larger area than either an individual migrant (fig. 9.4) or a sailplane, some individuals realize better climb rates. This variation among individuals potentially provides members of the flock with information about updraft strength. By moving away from migrants that are climbing slower or moving toward those climbing faster, individuals may find and use the best updrafts in a thermal. The potential advantage to any individual is related to the variance of climb rate among individuals. The feedback system continues until the flock leaves the thermal and begins when a new thermal is located—hence the name climb-rate feedback model. The model lends itself to mathematical presentation, although empirical tests will be most difficult.

Survey of Flocking Behavior of Migrating Hawks

As with most of the migratory behaviors examined in this book, little research has focused specifically on flocking, and few quantitative data were available. For this section I reviewed the literature, relying on Brown and Amadon (1968), Cade (1982), and other sources including accounts of migration at dozens of locations and faunal accounts. Each migratory species was assigned to a category according to flocking tendency: nonflocking—species in which no flocking is known during migration; occasionally (facultative) flocking—species that form small flocks, although many migrate alone; and regularly (obligate) flocking—species that usually migrate in flocks. In a few cases I was forced to make decisions, some of which may be disputed. Only field data will confirm my classifications. Thus, the review and analysis are tentative and serve as an initial step in future research.

Flocking occurs in one-half (50.3%) of the 133 migratory raptors (fig. 9.5) with only sixteen species being judged as "obligate flockers." Sizes of flocks of "obligate flockers" range from two to several thousand individuals. Flocking behavior is most pronounced among complete migrants, with ten of eighteen (55.6%) species regularly occurring in flocks during migration. Seven of the eight remaining complete migrants flock on some occasions, although the Gray Frog Hawk is not known to flock. Interestingly, six of the ten obligate flocking, complete migrants (Swainson's Hawk, Lesser Spotted Eagle, Lesser Kestrel, Amur Falcon, Red-footed Falcon, and Hobby) are insectivorous

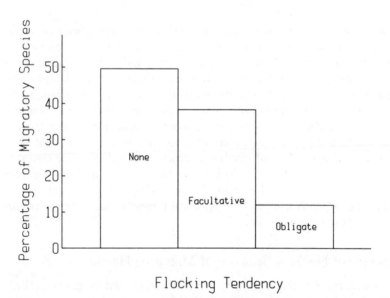

Figure 9.5 Flocking tendencies among the 133 species of migratory falconiforms.

to some degree during the non-breeding season and remain in flocks during that period. Three other complete migrants that fly in smaller flocks (Eleonora's Falcon, Sooty Falcon, and Greater Spotted Eagle) are also insectivorous during the non-breeding season and sometimes roost communally. Only four of ninety-one (4.4%) partial migrants form flocks of over one hundred birds (Turkey Vulture, Black Kite, European Kestrel, and Levant Sparrowhawk—perhaps also Brown Hawk). Three of these species (Turkey Vulture, Black Kite, European Kestrel) migrate long distances, and three (Black Kite, Levant Sparrowhawk, European Kestrel) follow insects in Africa during the non-breeding season. The European Kestrel is similar to the insectivorous complete migrants of its genus, often roosting and foraging with those species. Levant Sparrowhawk is a rather odd species because it is the only accipiter to form large flocks regularly.

Among partial migrants there is less of a tendency to fly in flocks than among complete migrants ($\chi^2 = 10.80$, df $= 2$, $p < 0.01$). Only 6.6% (six of ninety-one species) of partial migrants are known to regularly migrate in flocks, whereas another forty species (44.0%) occasionally fly in flocks. That leaves 49.5% (forty-five of ninety-one species) of the partial migrants that are not known to flock. Of the twenty-four species of irruptive/local migrants, four species (16.7%) sometimes migrate in flocks.

Examination of the sixteen species of migrants in which flocking occurs regularly reveals several shared characteristics (table 9.1). A greater proportion of these migrants tend to be long distance migrants, have insectivorous habits, occur in winter flocks, and make long distance water crossings than among other migrant raptors. Most prominent of these characteristics are long distance migration and insectivorous habits (at least during the non-breeding season). A greater number of flocking species tended to be both long distance migrants and insectivorous during the non-breeding season than expected by chance (tables 9.1, 9.2, and 9.3). Carrion also represents an ephemerally and locally abundant food resource similar to the emergences of termites and infestations of locusts. Species that use such resources might also be expected to use a flocking strategy. To determine how these characteristics are related to flocking and to each other, a log-linear analysis was employed (BMDP program 4f; see Kennedy 1983 for an explanation of log-linear models).

The log-linear analysis tests specified models. After specifying all possible models, the analysis rejected all but two, confirming the two independent chi-square analyses in table 9.3. Insectivory and distance are both independent predictors of flocking behavior among migrat-

Table 9.1 Characteristics of the Sixteen Species of Falconiforms That Flocked Regularly (Obligate Flockers) during Migration

Characteristics	Number and Percentage of Obligate Flockers	Number and Percentage of 117 Other Migrant Species	Chi-Square[a]
Migration distance > 3,000 km one-way	13 (81.3)	21 (17.9)	26.41**
Migrates from temperate zone into tropics	15 (93.8)	Unknown	
Flocks often >100 individuals	11 (68.8)	3 (2.6)	58.62**
Insect followers during non-breeding season	11–12 (68.8)	22 (18.8)	16.24**
Water crossings >100 km	8 (50.0)	8 (6.8)	23.66**
Forages or roosts in flocks during non-breeding season	10 (62.5)	Unknown	
Colonial breeders	3 (18.8)	Unknown	

[a]Chi-square corrected for continuity (2 × 2 table). Because expected values were <10 in one cell of all four analyses, a second (normal) chi-square was conducted in which the cell with the small expected value was excluded (reducing chi-square value). Results showed that all analyses contained deviations that differed from expected values at $p < 0.01$.
**$p < 0.01$.

Table 9.2 Summary of Flocking Tendencies of Migrating Falconiforms (133 Species) as Related to Distance of Migration and Insect Following during the Nonbreeding Season

| Flocking Tendency and Distance of Migration | Insect Following | | |
	No	Yes	Total
No flocking			
Short	30	3	33
Middle	21	5	26
Long	7	0	7
Facultative flocking			
Short	5	4	9
Middle	22	6	28
Long	9	5	14
Obligate flocking			
Short	1	0	1
Middle	2	0	2
Long	5	8	13

Note: Numbers represent the number of species in each cell.

Table 9.3 Chi-Square and Log-Linear Analyses of Data in Table 9.2

Analysis	Chi-Square	df	C[a]
Flocking, distance of migration	43.7 ($p < 0.01$)	6	0.50
Flocking tendency, insect following	12.2 ($p < 0.01$)	2	0.30
Distance of migration, insect following	10.8 ($p < 0.05$)	2	
Models specified in log-linear analyses[b] Flocking tendency, distance of migration; Flocking tendency, insect following	6.5 ($p = 0.37$)	6	—
Insect following, distance of migration; Flocking tendency, insect following; Flocking tendency, distance of migration	5.7 ($p = 0.23$)	4	
All other models	>12.9 ($p < 0.04$)	6+	

[a]Coefficient of contingency (analogous to coefficient of correlation).
[b]Nonsignificant chi-square value indicates data fit specified model (greatest standard normal deviate for a cell was 0.9, not significant). All other models were rejected.

ing hawks and are not significant by virtue of their relation to each other (conditional independence model). In birds that are long distance migrants and insectivorous during the non-breeding season, flocking evolved more often than in species that are shorter distance migrants and not insectivorous. I will speculate that flocking behavior evolved with trophic habits and greater distance of migration. Before closing this section, a caveat is in order. Because flocking has evolved independently among different raptor species, hypotheses invoking a single cause are likely flawed.

Empirical Studies of Flocking Behavior

It is a simple matter to describe quantitatively the flocking behavior of migrating raptors. During field studies I noted how many birds were seen together in a "cohesive" flock, when flocks merged or disintegrated, the time of day flocks were seen, the shape of the flocks, and "odd" behaviors. Using data collected at several locations permits a comparison of the flocking behavior of different species and at different times during migration. Some of these data can be used to test predictions of the hypotheses regarding the function of flocking. Determining flock size is difficult when topography promotes large aggregations of birds. It is not always possible to determine flock size at locations such as Cape May Point or Hawk Mountain, because many birds encountering the coast or following the ridge fly together for limited distances. In these cases it is difficult for migration counters to determine if hawks are in flocks. While studying water crossing behavior at Cape May Point and Whitefish Point, I noted many "flocks" of Sharp-shinned Hawks and American Kestrels (>10–25) that were larger than any I had seen at sites where "leading lines" were absent. The erratic behavior of these species at these locations precluded determining flock sizes, although crossing of water barriers may be facilitated by social context.

The tendency of raptors to migrate in flocks varies among North American species. Broad-winged Hawks, Swainson's Hawks, Turkey Vultures, and Mississippi Kites formed flocks more often (Kerlinger and Gauthreaux 1985a, 1985b; Kerlinger, Bingman, and Able 1985, unpublished) and formed larger flocks than other species. The largest flocks of these species are reported from Panama, where Broad-winged Hawks, Swainson's Hawks, and Turkey Vultures migrate in flocks of over 10,000 individuals at times (photographs in Smith 1973, 1980). From Panama north into south Texas flocks of these species numbering in the hundreds and often thousands are seen in both spring and autumn (Thiollay 1980; Kerlinger and Gauthreaux 1985a, 1985b). Some of these and other large aggregations of hawks may be a result of many flocks converging along coasts and in narrow land bridges such as the isthmus of Panama and Israel. In addition, some of these aggregations may result from use of a narrow band of updrafts by many flocks of hundreds or thousands of individuals.

The flocking behavior of Broad-winged Hawks is better known than that of any other species. Flocks of these hawks form shortly after the initiation of autumn migration. In central New York flocks of over 100 individuals occur on rare occasions (Kerlinger, Bingman, and Able 1985; Kerlinger, unpublished data from three locations, 1976–1980). Most flocks reported from New York did not exceed 30 individuals, although only 12% of all Broad-winged Hawks migrated alone during autumn (fig. 9.6). Flocks are organized loosely, often as skeins or echelons (sensu Heppner 1974). Hawks migrating through central New York probably had been migrating for only two to four days and had not yet formed large flocks. Interestingly, flocks of hundreds and sometimes thousands of hawks are reported east of New York from Mount Watchusett, Massachusetts (P. Roberts, personal communication), as well as elsewhere in New England, southern New York (personal observation, Hook Mountain, Rockland County, autumn 1976, 1977), and southern Ontario (Richardson 1975). By the time flocks reach southern Pennsylvania, most consist of more than

100 individuals (S. Benz, personal communication) and flocks of over 1,000 are reported from the northeastern United States every year. Photographs of the PPI screen of a surveillance radar published by Richardson (1975) show flocks of over 1,000 individuals migrating in south-central Ontario, north of Lake Ontario and Lake Erie. It is probable that these migrants travel in such large flocks for much of their migration.

Flocks of over 5,000 Broad-winged Hawks are not common in the northeastern United States and southern Canada. Every few years counts of over 15,000, called a "river of hawks" by one author (Welch 1975), are seen at such lookouts as Mount Watchussett, Massachusetts (P. Roberts, personal communication), and Hawk Mountain Sanctuary, Pennsylvania (A. Nagy, personal communication). Is it possible that these flocks occur in more years but are missed because they do not fly over a hawk lookout or because they fly high and fast? Observers on the ground and in motor gliders also report parallel lines of migrants in New England. These flight lines are separated by only 15–20 km (Hopkins et al. 1979). Therefore, a large portion of the Broad-winged Hawks from eastern North America move through the

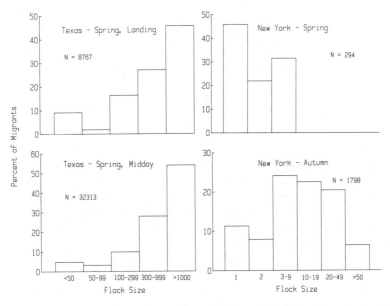

Figure 9.6 Flock sizes of Broad-winged Hawks observed during spring migration in south Texas before landing and at midday, and during spring and autumn migration in central New York.

New England and Pennsylvania area in one or three days each autumn.

The formation of flocks during autumn migration through New England has been observed from the ground and from motor gliders. Hopkins (1975) and Hopkins et al. (1979) noted that the headings of many flocks of Broad-winged Hawks during interthermal glides are extremely variable, ranging over 130°. They attribute this to flocks' being attracted to birds that are in thermals. This finding is interesting in light of the local enhancement model because it shows that a circuitous route can result from hawks' being attracted to other hawks using thermals. Just how much deviations slow migrants is not known. It is tempting to speculate that although these birds may lose time and distance during flock formation, they gain more than they lose.

The formation of flocks of Swainson's Hawks and Turkey Vultures is probably similar to that of Broad-winged Hawks. Swainson's Hawks seldom breed north of 53°N, yet during autumn migration they form flocks of 50 to more than 100 individuals by the time they reach 50° N (Woffinden 1986). Flocks of 50–100 birds are known from Regina, Saskatchewan, and larger flocks are known from Montana and North Dakota (reports in *American Birds*). Flocks of more than 700 birds are known from Nebraska and by the time the birds reach Texas flocks often exceed 1,000 migrants (reports in *American Birds*). A report in the HMANA newsletter (Autumn 1985), documents flocks of over 1,000 individuals near the Falcon Dam along the Rio Grande River in south Texas. Flocks of several hundred Swainson's Hawks and Turkey Vultures are known from southern California, where these raptors are not common.

Flocking by these species during spring is similar to autumn. Flocks of over 1,000 Broad-winged Hawks are not uncommon in south Texas, and the largest flocks exceed 3,000 (Kerlinger and Gauthreaux 1985a). As Broad-winged Hawks move north from Texas to breeding sites in northeastern and midwestern North America, flock sizes are smaller but can number in the hundreds. Although Haugh (1972) did not report flock sizes, he indicated that flocks of several hundred to over 1,000 migrate along the southeastern shore of Lake Ontario. Flocks of tens to hundreds of immature Broad-winged Hawks can be seen on the Door County peninsula of Wisconsin in early June (U. Petersen, personal communication). Flock sizes for Broad-winged Hawks from central New York in spring were smaller than those seen along the Great Lakes, with a maximum of about 10 birds (unpublished observations from five years of observations at three sites). It is probable that large flocks seen along the Great Lakes are, in part, a result of the leading or diversion line.

The picture for Swainson's Hawks in spring is similar to that of Broad-winged Hawks. Moving north from Texas, flocks are smaller than the flocks of thousands from south Texas. Reports on spring migration show flocks of 10–40 migrants in Oklahoma (reports in *American Birds*). An astounding 2,000 or more Swainson's Hawks were reported from Montana by Cameron (1907) during April of 1890, and a flock of about 2,100 birds was noted in Wyoming in April 1979 (Martin and McEneaney 1984). The latter group was divided into subgroups of 50–200 individuals. How often do flocks this large pass unobserved in the vastness of the Great Plains?

Observations of flocking migrants during roosting and takeoff activities have been reported by several authors. Smith (1980) reports landing and roosting by flocks of more than 1,000 Turkey Vultures, Broad-winged Hawks, and Swainson's Hawks from Panama. He reports up to 35 Broad-winged Hawks roosting in one tree. Other reports confirm that Swainson's Hawks roost close to each other in spring shortly before the nesting season. Cameron (1907) reports up to 50 roosting in one tree. Large numbers of Turkey Vultures also roost in the same tree during migration.

The only detailed study of flock formation following takeoff and breakup before landing is that of Kerlinger and Gauthreaux (1985a). In south Texas there are limited sites suitable for Broad-winged Hawks to roost. Remnant forests at Bentsen State Park and Rio Grande Valley National Wildlife Refuge (formerly Santa Ana National Wildlife Refuge) are ideal for observing migrants landing and taking off, because hundreds of thousands of hawks and vultures are attracted to them. The forest at Santa Ana, where Kerlinger and Gauthreaux (1985a) worked, is a 2,000 ha "oasis" of riparian and mesic forest surrounded by cabbage, onion, and sugarcane fields.

During their study (28 March–16 April 1982) Kerlinger and Gauthreaux observed more than 85,000 Broad-winged Hawks, of which 8,500 were seen landing and about 16,000 were seen taking off. The discrepancy between the number taking off and those landing is actually only 4,000 birds, because 4,000 birds apparently landed before Kerlinger and Gauthreaux arrived at the site. These 4,000 hawks were seen taking off on the first morning of the study. The difference of more than 3,000 between the number seen landing and those seen taking off was attributed to difficulties of counting migrants accurately. Even with the inaccuracies, the number of birds counted taking off was highly correlated with the number seen landing the previous evening ($r = 0.90$, $N = 18$ days, $p < 0.01$; data were log transformed). When Broad-winged Hawks were seen taking off, few remained in the refuge (confirmed by birders, refuge personnel, and our own trips through the refuge).

Takeoff from the refuge occurred 93–271 min ($\bar{x} = 139 \pm 45$ SD, $N = 15$ days) after sunrise (ca. 0815–1100 CST). Timing was related to wind direction and cloud cover. Not all migrants took off together, so takeoff time was defined as 30 min after the first two birds were seen aloft. Takeoff was earlier ($\bar{x} = 116 \pm 16$ min after sunrise, $N = 10$ days) when winds were favorable (from the south) than when they were unfavorable ($\bar{x} = 203 \pm 61$ min, $N = 3$ days, wind from the north). On the remaining mornings surface winds were light and variable ($\bar{x} = 156$ min, $N = 2$ days). There was also a slight but not significant correlation between takeoff time and percentage of cloud cover ($r = 0.35$, $N = 15$, ns).

Takeoffs generally lasted an hour or more. Individuals emerged from the forest canopy by themselves and quickly joined other migrants in flocks of 2 to 5 or more individuals. These flocks grew and moved rapidly to the north-northwest. Early morning flocks were unorganized and strung out in skeins or loose echelons. Within 20 min of takeoff, flocks of several hundred birds were evident. The earliest a flock of 500 was noted was 25 min after takeoff (or nearly 1 hr after the first birds were seen aloft). Maximum flock size in the first 60 min of flight was strongly correlated with the number seen taking off in that period ($r = 0.91$, $N = 15$ days, $p < 0.01$). Unlike a report by Smith (1980) from autumn in Panama, the migrants observed by Kerlinger and Gauthreaux (1985a) did not take off as cohesive flocks. It seemed that flocks formed almost at random and that the first birds to take off were more than 20 km to the north when the last migrants took off. Furthermore, birds taking off later may have joined flocks moving in from Mexico that had taken off earlier.

From 1100 h until after 1600 h flocks seemed to maintain integrity. During the period 1100–1600 h, (more than 90% of migrants were in flocks of over 100, 54% were in flocks of over 1,000 birds, and about 1% were alone (fig. 9.6). Midday flocks were well organized, with flight formations corresponding to extended clusters (sensu Heppner 1974). Smaller flocks (<200–400) spaced themselves across a front of about 30–50 m (fig. 9.7) when gliding between thermals, whereas flocks of over 1,000 birds usually covered a front of more than 50–100 m. Flocks of Broad-winged Hawks numbering 1,000–3,000 individuals formed extended clusters of 1 km to more than 3 km long during interthermal glides. Small flocks were shorter when gliding be-

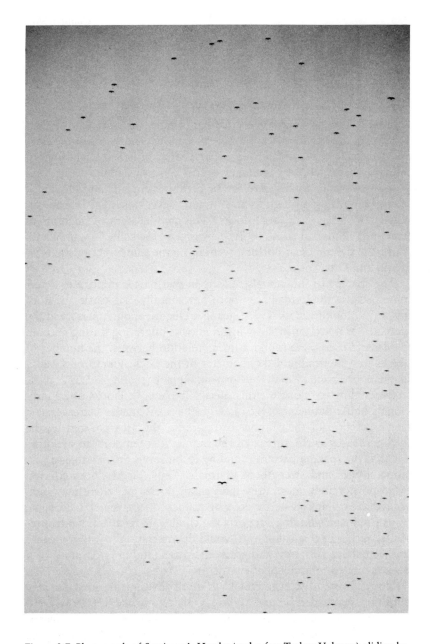

Figure 9.7 Photograph of Swainson's Hawks (and a few Turkey Vultures) gliding between thermals during autumn migration over Panama. Spacing of individuals perpendicular to the migratory direction may help migrants randomly encounter thermals without deviating from the preferred migratory direction. (Photograph by William S. Clark.)

tween thermals. The length of flocks was measured from the screen of
a marine surveillance radar. Interindividual spacing of migrants passing directly overhead was estimated visually to be about 5–15 m and
varied among flocks.

Gliding flocks of Broad-winged Hawks in south Texas averaged
about 1,000 migrants for every 1.6 km (1 mile) of radar echo. This is
similar to the number per kilometer of Honey Buzzards and Lesser
Spotted Eagles observed on surveillance radar in Israel. Flight lines of
migrants in Israel are sometimes over 70 km in length, being composed of smaller flocks. Are flocks of Swainson's or Broad-winged
Hawks migrating through Central America ever this long? Is it advan-
tageous to be in such large flocks? The parallel lines of migrants in
Israel are similar to those of Broad-winged Hawks in New England.
Thus, the dynamics of flocking seem to be the same in both the New
World and Old World.

The spacing of individuals soaring in thermals is difficult to determine because they are spaced both horizontally and vertically. Robinson (in Hopkins et al. 1979), using a cine-theodolite, measured the
diameter of flocks of Broad-winged Hawks migrating during autumn
in Massachusetts. He found a linear relation between the number of
hawks in a thermal and the diameter of the flock. Flocks of 3 had a
diameter of about 11 m in the horizontal plane (radius = 5.5m), close
to the circling envelope for this species (table 6.4). Flocks of 20 individuals had a diameter of 42 m. Whereas the diameter increases as a
linear function of the number of migrants, the area covered by the
flock increases as the square of the radius. It is tempting to hypothesize that the spacing pattern found by Robinson is attributable to variation in size and strength of thermals. Thus, larger flocks covered
disproportionately more area than small flocks and there may be a
limited area that can be covered by a flock of a given size. In some
ways Robinson's finding supports the climb-rate feedback model presented earlier in this chapter. It would be interesting to know if other
species behave this way and if the diameter (and area covered) of a
flock is related to atmospheric structure.

Flocks of Broad-winged Hawks in south Texas maintained integrity
until they descended before landing. Only two broke up before 1630
h. On these occasions about 50 and 100 migrants were seen descending into the refuge after leaving larger flocks (>300). Landing of
flocks occurred 16–195 min before sunset ($\bar{x}$ = 82 ± 64 min, SD,
N = 11 days). Of the 8,567 migrants seen landing, 91% were in
flocks of over 40 individuals (fig. 9.6). Landing time was earlier with
unfavorable winds ($\bar{x}$ = 171 ± 28 min, N = 3) than favorable winds

($\bar{x} = 48 \pm 33$, $N = 7$ days). On the remaining day winds were light and variable. There was also a significant positive correlation between landing time and percentage of cloud cover ($r = 0.87$, $N = 11$, $p < 0.01$), a result of inhibited convection.

Before landing most flocks descended several hundred meters. The altitude of flight at the beginning of such descents was more than 400–500 m, even as late as 1800 h. Descents were into the wind, and flocks remained together until less than 50 m above the trees. At that time individual movement seemed to become random. When birds reached a height of 5–15 m above the trees they dived into the forest canopy and did not emerge until morning. Birds were always alone when flying into the forest canopy. Casual observations in the refuge after landings or the next morning indicated that birds roosted alone, not in the same tree with other migrants. It would be interesting to observe landings of large flocks in forest "islands" where there are fewer trees than birds. Would migrants roost together in the same tree, and how closely would they roost? Also, are birds in spring (mostly adults in south Texas before 12 April) more reluctant to perch with conspecifics than in autumn? The discrepancy between Smith's (1980) observations and those of Kerlinger and Gauthreaux (1985a) may be due to the difference in social behavior between these seasons. Upon arrival at the breeding grounds these birds defend exclusive territories, and it is possible that a minimum roosting distance between individuals is similar to territorial defense.

In addition to documenting flocking behavior, Kerlinger and Gauthreaux (1985a) determined the number of hours flown per day during spring migration. By knowing takeoff and landing time they determined that Broad-winged Hawks migrate on average 8.2 ± 1.9 SD, hours per day ($N = 10$ days). Although landing and takeoff times are not for the same migrants, the large number of migrants in the calculation suggests a robust conclusion.

Flocking among other North American falconiform migrants tends to be less developed than among Broad-winged Hawks, Swainson's Hawks, Turkey Vultures, and Mississippi Kites (tables 9.4, 9.5, 9.6). Most other species form flocks, but these flocks are not as cohesive or as large. Kerlinger, Bingman, and Able (1985) showed that over 50% of the individuals from eight of the nine species they examined were alone (table 9.4) during autumn in central New York. Species such as Goshawks, Cooper's Hawks, and Northern Harriers were almost always alone, although elsewhere the latter two species fly with conspecifics or larger flocks of Broad-winged Hawks (table 9.4). Although the site of the Kerlinger et al. study was close to the origin of migra-

Table 9.4 Flocking Tendencies of Raptors Migrating during
Autumn 1978–1979 in Central New York

Species	Percentage Alone	Maximum Flock Size	Percentage with Other Species (Species)	Sample Size
Sharp-shinned Hawk	65.7	3	10.3 (Broad-winged Hawk)	271
Cooper's Hawk	100.0	—	—	10
Goshawk	93.3	—	6.7 (Sharp-shinned Hawk)	15
Broad-winged Hawk	12.1	115	5.8 (Sharp-shinned Hawk)	1,130
Red-shouldered Hawk	55.6	3	22.2 (Red-tailed Hawk)	27
Red-tailed Hawk	56.5	9	2.4 (Sharp-shinned Hawk)	382
Northern Harrier	81.9	2	9.1 (Broad-winged Hawk)	22
Osprey	80.8	2	11.5 (Broad-winged Hawk)	26
American Kestrel	74.7	6	5.8 (Sharp-shinned Hawk)	86

Source: Kerlinger, Bingman, and Able 1985.

tion for these species, flock sizes for most species do not increase when they reach central Pennsylvania (S. Benz and B. Shelley, personal communication). Flocks of these species rarely exceed 10 individuals.

I have seen 50 Red-tailed Hawks together only once in central New York. Haugh (1972) reported occasional flocks of more than 100 for this species during spring migration at Derby Hill, on the southeastern shore of Lake Ontario. Generally, flocks of Red-tailed Hawks, Sharp-shinned Hawks, and American Kestrels are small and loosely organized. Elsewhere in New York during spring migration these hawks form small, ephemeral flocks (table 9.5). How long these flocks maintain integrity and cohesion is not known.

At all the sites where I have worked, mixed-species (heterospecific) flocks were observed regularly (tables 9.4, 9.5, 9.6). Species like Sharp-shinned Hawks and Ospreys join flocks of Broad-winged Hawks. Ospreys or Turkey Vultures are conspicuous when flying with Broad-winged Hawks, especially at low altitudes, but there are times when Sharp-shinned Hawks and even Peregrine Falcons fly with Broad-winged Hawks and are not readily discernible. When flocks are large, observers estimate numbers and cannot scrutinize flock members. For this reason some individuals of these and other species may be overlooked by counters. Observations by researchers using radio-telemetry also show that raptors like Peregrine Falcons join flocks of other species, flying with them for varying distances (Cochran 1975). Species like Bald Eagles, however, migrate alone (Harmata 1984). Because of the large number of individuals in some flocks and their similar size it is difficult to discern species like Sharp-shinned Hawks or

Table 9.5 Flocking Tendencies of Raptors Migrating during Spring 1976–1980 in Central New York

Species	Number Alone (%)	Maximum Flock Size	Number in Pairs (%)	Heterospecific Flocking Number (%)[a]	Sample Size
Sharp-shinned Hawk	201 (65.9)	4	62 (20.3)	22 (7.2) (BW)	305
Cooper's Hawk	10 (83.3)	2	2 (16.7)	0 (0.0) —	12
Goshawk	16 (93.8)	—	0 (0.0)	1 (6.2) —	16
Broad-winged Hawk	135 (45.9)	9	52 (17.7)	20 (6.8) (SS)	294
Red-shouldered Hawk	34 (77.3)	3	6 (13.6)	1 (2.3) (SS)	44
Red-tailed Hawk	148 (50.3)	5	78 (26.5)	6 (2.0) (SS)	294
Northern Harriers	30 (90.9)	2	2 (6.1)	1 (3.0) —	33
Osprey	50 (79.4)	3	6 (9.5)	4 (6.3) (BW)	63
American Kestrel	40 (65.6)	3	14 (23.0)	4 (6.6) (SS)	61

Note: Albany, Franklin Mountain, and Mohonk data sets used.

[a]Letters in parentheses are the species the listed raptor was associated with (BW = Broad-winged Hawk, SS = Sharp-shinned Hawk).

Table 9.6 Flocking Tendencies of Seven Species of Raptors during Spring Migration in South Texas

Species	Percentage in Mono-specific Flocks	Maximum Flock Size	Percentage with Broad-winged Hawks[a]	Sample Size
Turkey Vulture	77.6	200	19.9	619
Mississippi Kite	80.5	30	3.3	123
Sharp-shinned Hawk	9.7	3	20.9	134
Cooper's Hawk	0.0	—	26.3	23
Broad-winged Hawk	—[b]	3,000	—	89,346
Swainson's Hawk	95.3	1,500	2.3	3,751
American Kestrel	16.4	2	13.7	73

Source: Kerlinger and Gauthreaux 1985a, b.

[a]A small percentage of these individuals were observed with species other than Broad-winged Hawks.

[b]Flocking of this species is presented in detail in the text and in figure 9.6.

Cooper's Hawks among hundreds of Broad-winged Hawks when they fly at 800 m above the ground.

Choosing among the Explanations for Flocking

The evidence available supports the hypothesis that flocking during migration helps hawks find and use thermals. Both the local enhancement and random encounter models have merit. After takeoff at Santa Ana National Wildlife Refuge, individual Broad-winged Hawks that located strong thermals were joined quickly by other migrants. Most hawk watchers have observed similar events. Sometimes the birds fly off in the same direction and sometimes they do not. The latter behavior may indicate that the raptors had joined others in a thermal only after receiving information regarding its location and strength. That different species entered the same thermal, presumably as a result of local enhancement, tends to support the thermal location and utilization hypothesis rather than the foraging efficiency or orientation/navigation hypothesis.

Evidence for the random encounter model comes from the configuration of flocks and spacing of individuals within the flocks during interthermal glides. As shown above, species like Broad-winged Hawks, Swainson's Hawks, and Turkey Vultures space themselves perpendicular to the direction of migration so that they "sample" a large airspace while gliding between thermals. The ultimate flock shape is an extended cluster with the frontal area being 20 m to over 100 m across. Small flocks of 2–5 birds also space themselves perpendicular to the axis of migration. If other flocking species, such as those from the Old World, formed extended clusters, the random encounter model would be supported. This hypothesis could be tested by simulations in which the known size of flocks, glide ratios of the species involved, thermal sizes, thermal spacing, and thermal height were incorporated.

Four characteristics that are shared by most of the regularly flocking species promote flocking and may lend support to the thermal location and utilization hypothesis. First, most species migrate long distances, breeding primarily at temperate latitudes and moving to tropical latitudes during the non-breeding season. By finding thermals quickly and using them efficiently, these hawks will benefit more from flocking than species that migrate shorter distances. Second, a large number of these obligate flocking species are insectivorous, at least for part of the year. Although many depend on vertebrate prey during the breeding season, it seems that some, such as the Swainson's Hawk, switch to insects after breeding is completed. Even some of the species that do not form large migratory flocks occur in flocks during the non-breeding season (i.e., Sooty Falcon, Eleonora's Falcon), when they switch from vertebrate to insect prey. Species that spend the non-breeding season in flocks depend upon locally abundant and ephemeral insects such as locusts, grasshoppers, and emerging termites.

The third characteristic that the regularly flocking species have in common is that they have similar migration goals. For instance, nearly all Broad-winged Hawks, Swainson's Hawks, and Mississippi Kites, as well as a large number of the Turkey Vultures that breed in North America, leave that continent via Texas and Central America. Returning from South and Central America, these hawks also have similar migratory goals, at least for the first half of the journey. Similarly, large numbers of Common Buzzards, Black Kites, Honey Buzzards, and other species that migrate in large flocks from northern and central Europe move through the Strait of Gibraltar or through the Bosporus on their way to non-breeding sites in Africa and Madagascar. Another example of species having a common goal would be the hundreds of thousands of smaller falcons that move from Europe and Asia to restricted non-breeding sites in east-central Africa. Similar migration pathways and common goals may promote flocking among these migrants.

The final characteristic common to obligate flockers is that they are usually numerous species, at least over part of their breeding or non-breeding range. Kerlinger, Bingman, and Able (1985) found that flock size of a migrant species in central New York was positively correlated with the abundance of a species ($r > 0.8$, $N = 9$ species). Thus, if a species is rare or widely spaced in the breeding or non-breeding range, it will probably not often migrate in flocks of conspecifics, although it may join flocks comprising other species. Those species that are obligate flockers and are not numerous frequently breed in colonies (Eleonora's Falcon). Long distance migrants such as Ospreys and Peregrine Falcons (from northern populations) breed at low densities over large areas and may not encounter conspecifics often enough for flocking to occur. These species frequently rely on vertebrate prey that are more evenly distributed than insects and are more difficult to hunt and catch. Finally, common species form flocks that "hitchhiking" species use in a mutual or commensal manner that has nothing to do with foraging or orientation/navigation. Such individuals may derive energetic advantages by flying with the more numerous flocking species.

From the log-linear analysis and over evidence, it is likely that flocking serves more than one function. When flocking originated it may have had only one purpose—for example, the efficient location of patchily distributed prey. Today, however, it is likely that for some species flocking serves in prey location as well as for location and efficient use of thermals. An example of this may be Swainson's Hawks, which migrate very long distances and remain in foraging flocks that follow insects. Flocking during migration may help them locate thermals and may also set the stage for flocking during the remainder of the non-breeding season. In other species flocks may serve only one function. Broad-winged Hawks and many other species that form huge flocks during migration (e.g., Lesser Spotted Eagle, Honey Buzzard, Mississippi Kite) do not forage in flocks, so that flocking by these species may be a specific adaptation for locating and utilizing thermals. It is impossible at this time to say whether flocks of Broad-winged Hawks or other hawks orient or navigate better than individual migrants.

By advocating the thermal location and utilization hypothesis I have slighted the foraging efficiency and orientation/navigation hypotheses. Although, I believe the orientation/navigation hypothesis has little merit, the foraging efficiency hypothesis (or some variant of it) is probably a plausible explanation for the evolution of flocking, especially among the smaller falcons and kites. Unitary hypotheses concerning the evolution of behavior are sometimes naive. Hilborn and Stearns (1982) have made the cogent point that testing synergistic or nonindependent hypotheses in evolutionary studies is often more fruitful than testing alternative unitary hypotheses. Neither the thermal location nor the foraging efficiency hypothesis can be rejected using available data. The function of flocking in migrating hawks remains a wide open and fruitful area for study.

Factors That Inhibit Flock Formation and Integrity

Although it is clear that flocking promotes faster and more efficient migration, there are several factors that potentially inhibit flock formation and cohesion (Kerlinger 1985c). These factors may influence flocks of conspecifics or mixed-species flocks.

Intra- or interspecific aggression may preclude close association of migrants for any length of time. Agonistic behavior between conspecifics or mobbing of larger species by smaller ones (e.g., Red-tailed Hawks by Sharp-shinned Hawks) diverts energy and time from migration and may outweigh the benefits of flocking. Even with the possibility of predation and aggression, both homospecific and heterospecific flocks are common. It seems that aggression (territorial or otherwise) of some species is "turned off" during migration, suggesting that flocking is an important adaptation for migration. In addition, larger hawks sometimes prey on smaller hawks (Klem, Hillegass, and Peters 1985), and it may be dangerous for species like Sharp-shinned Hawks or American Kestrels to associate with larger species. Rudebeck (1950–1951) observed predation on three migrating European Sparrowhawks by a single male Peregrine Falcon hear Falsterbo, Sweden. He also mentions several other unsuccessful hunting attempts by Peregrine Falcons on this species and the successful capture of a Eurasian Kestrel. Although these events are rare, they occur regularly enough so that smaller birds of prey must be aware of the larger ones.

A second factor that potentially inhibits formation as well as integrity of flocks is disparate aerodynamic performance. If migrants do not climb in thermals at the same rate or if they cannot realize similar air speeds and sink rates when gliding between thermals, it is unlikely they will remain close to each other for long. Differences in morphology among species and between sexes (dimorphism in size) translates into differences in aerodynamic performance. Even a difference in fat load between conspecifics of the same sex or age class might result in a difference in aerodynamic performance that could influence flock integrity.

The negative influence of disparate morphology and aerodynamic performance on flock integrity may be outweighed by other factors. Consider the example of a Sharp-shinned Hawk migrating with a

Broad-winged Hawk. While thermaling, the smaller Sharp-shinned Hawk should, by virtue of its lighter wingload, be able to climb faster in a small thermal than the Broad-winged Hawk. Between thermals, however, it will realize a slower air speed at the same rate of sink as the Broad-winged Hawk (chap. 6). The differences in performance during thermal soaring and gliding may cancel each other out. The Sharp-shinned Hawk should leave a thermal before the Broad-winged Hawk, if they entered the thermal together. During the subsequent glide the Broad-winged Hawk would overtake the Sharp-shinned Hawk by realizing a faster air speed and greater glide ratio. By incorporating flapping, as the species often does during interthermal glides, the Sharp-shinned Hawk may be able to keep up with the Broad-winged Hawk. If there are many Broad-winged Hawks in the flock, the Sharp-shinned Hawk may be able to stay within the flock for a long time, thereby realizing the advantages of flying with a flock. This is speculation, but it can be examined in the field.

The third factor that inhibits flock formation and integrity is different migratory destinations among or within species. If two species or individuals are not trying to fly toward the same goal, then it is unlikely that they will remain together for long distances. Differences in mean headings among species were demonstrated in central New York and south Texas (chap. 7). Nearly all of the Broad-winged Hawks from northeastern North America have the same migratory goal (Texas and then Panama), whereas Sharp-shinned Hawks (Clark 1985a; Evans and Rosenfield 1985) or Ospreys (Lincoln 1952) do not migrate toward Texas. For this reason (and others) it is unlikely that Sharp-shinned Hawks or Ospreys stay with Broad-winged Hawk flocks for long distances. Another interesting case is that of Broad-winged Hawks in spring migration that move from Colombia or Panama through Central America into Texas and eventually northward into midwestern and eastern North America. These birds share a common goal until they reach central Texas. Birds returning to New England and the Maritime Provinces of Canada should take a different course from the midwestern birds they have been traveling with. Large flocks may disintegrate in mid- to north Texas. If Broad-winged Hawks from these areas migrate at different times of the spring, flocks could maintain their integrity for longer periods.

Summary and Conclusions

Many raptors migrate in flocks that vary in size and in frequency of occurrence. Some species rarely migrate alone, forming flocks that exceed 100 or 1,000 individuals. About half of all migratory species are known to form small flocks, composed of their own or mixed species. These latter flocks may be ephemeral in occurrence. Smaller hawks of the genus *Falco* are among the most social during migration and the non-breeding season. The species that flock most often and form the largest flocks are characterized by long distance migration, usually breeding in the north temperate zone and wintering near or south of the equator; dependence upon thermals; frequent dependency on ephemerally abundant (patchy) food resources such as some insects and, to a lesser degree, carrion; having a similar migratory goal or direction; and being abundant or breeding in colonies. Insectivorous and carrion-eating raptors often form flocks during autumn migration and remain in flocks through the non-breeding season. Three factors were suggested that may disrupt flock formation and integrity: agonistic or predatory behavior, dissimilar aerodynamic performance, and different migratory goals (flight direction). The adaptive function of flocking may be related to location and utilization of thermals, foraging, and orientation and navigation. Because preliminary evidence supports the thermal location and foraging efficiency hypotheses, flocking may serve more than one function. Furthermore, the function may vary from species to species or from time to time.

10

Water Crossing Behavior

It takes only a glance at a map of the world to see that a migrating bird cannot fly far without encountering water. The migratory pathways of most birds are influenced by lakes, oceans, seas, and even rivers. Water crossing behavior has been examined extensively among passerines, particularly in North America. Several excellent studies have demonstrated that many passerine species cross the Gulf of Mexico in spring and autumn, a flight of 1,000 km or more between the southeastern United States (Florida to Texas) and Yucatán, Mexico, in spring and autumn (Lowery 1945; Gauthreaux 1971, 1972; Able 1972; Buskirk 1980; Moore and Kerlinger 1987). After tedious hours of skywatching through a telescope Lowery found that warblers, tanagers, and other species arrived at the Louisiana coast at high altitudes during daylight hours. Later, Gauthreaux (1971, 1972) used weather radar to document the phenomenon.

One of the longest water crossings by birds occurs along the eastern coast of North and South America. During autumn migration, Blackpoll Warblers (*Dendroica striata*) and some shorebirds undertake nonstop flights from New England to northeastern South America (Nisbet 1970; Williams and Williams 1978). The trip takes 2–4 days of continuous flight. Other species change direction at the coast or return after flying out over the ocean. They sometimes are observed flying back to shore during daylight. These migrants eventually reach South America by crossing the Gulf of Mexico, a flight that requires more than 12 hr.

Raptors have been characterized as reluctant to cross water barriers (Rudebeck 1950; Lack 1954; Haugh 1972; Heintzelman 1975; MacRae 1985). As a result of this reluctance, many migrating hawks are counted at coastlines, isthmuses, and peninsulas. The isthmus of Panama (Smith 1973, 1980), the Strait of Gibraltar (Evans and Lathbury 1973), the Israel isthmus (Christensen et al. 1981), Cape May Point (Allen and Peterson 1936; Dunne and Clark 1977), along Lake Ontario (Haugh and Cade 1966), Lake Superior (Hofslund 1966),

Lake Erie (Field 1970) and similar sites are the best places to see large numbers of raptors. Few studies of the water crossing behavior of migrating raptors have been conducted. The goals of the current chapter are to outline the risks and benefits associated with flying over large bodies of water; review what is known about water crossing by migrating raptors; discuss empirical studies of water crossing behavior; and examine morphological correlates of water crossing behavior.

Risks and Benefits Associated with Crossing Water

Why are some migrating raptors reluctant to cross water? Why do some raptors fly hundreds of kilometers out of their way rather than make water crossings of only a few kilometers? To answer these questions we must examine the risks and benefits associated with crossing water. Factors that may play a role in the decision whether to cross a given body of water include weather (precipitation, wind, lift, visibility), length of the crossing, size of the body of water (in directions other than the preferred crossing direction), physiological state of the bird, time of day (daylight may be needed for orientation), aerodynamic performance (flight speed capabilities), experience, and interactions of these factors.

Crossing water can save time and energy, depending on distance of the crossing relative to the distance around the water barrier (ratio of distance around to distance across). As this ratio increases, benefits associated with crossing become greater. This is simplistic because it does not consider the difference in relative cost of flight over land and over water, nor does it consider weather at the time of crossing. For all but the shortest crossings hawks cannot use soaring flight, because thermals over water are usually weak and scarce. Without thermals raptors must resort to powered flight, and the cost of transport will increase by three to six times the cost of soaring flight. Assuming no wind, this means that a 10 km crossing using powered flight is equivalent in energy to a gliding flight of 30–60 km around the barrier. If the migrant glides across part or all of a barrier, the cost of crossing will be reduced. This is most important for crossings less than 20 km. A bird with an 8:1 glide ratio flying at a height of 800 m can glide 6–7 km before it must resort to powered flight. These relations change depending on strength and direction of the wind.

Most prominent among risks associated with water crossings are fatigue as a result of powered flight; becoming lost in fog, rain, or foul weather; being blown off course by adverse wind; and being preyed upon by gulls or larger hawks. Although these risks are listed separately, they act in concert. Hawks that are lost in the fog or are blown off course will become fatigued. After all, there are no places for a hawk to rest at sea. Any of these risks can result in death. Therefore, natural selection may act against individuals that do not attempt to minimize risks. The risk of mortality probably increases with the absolute distance of the crossing.

The model is too simplistic, of course. First, the experience of migrants arriving at a water crossing is not considered. Adult birds that have experienced a particular water crossing may have prior "knowledge" as to the distance around the water barrier. Immature birds do not have this "knowledge." A bad experience such as being blown off course over water may dissuade an individual from making future flights over water. Second, the physiological condition of an individual migrant may influence the decision to cross. A strong, fat bird may be more likely to undertake a crossing than one with depleted fat deposits or one that is starving. Similarly, a bird that had migrated for most of the day might be less likely to undertake a water crossing than if it encountered the barrier shortly after takeoff in the morning. Third, the wind, visibility, and precipitation at the time a water barrier is encountered may influence a raptor's decision. It is easy to model a system without considering these variables, and the cost-benefit functions may be naive for this reason. Only by conducting empirical studies at actual crossings can we determine whether the models are of value as they are or if other variables must be considered.

Are the risks outlined above real? To answer this we must look for evidence of mortality incurred during water crossings. The best evidence is dead birds. A second, though not conclusive, source of evidence is exhausted birds landing on ships at sea (MacRae 1985), with little hope of reaching shore. Although few publications document the mortality of raptors, other birds are known to die in large numbers during water crossings (personal observations). The best example of mass mortality of raptors during a water crossing was reported by Zu-Aretz and Leshem (1983). More than 1,300 hawks were found along one beach in Israel during April 1980. The authors state that "easterly winds swept the raptors over the Mediterranean Sea, where they died, being unable to glide over the sea." The birds involved were 826 Common Buzzards, 4 Long-legged Buzzards, 312 unidentified buzzards, 6 Black Kites, 3 Short-toed Eagles, 7 Lesser or Greater Spotted Eagles, 4 Booted Eagles, 4 Tawny Eagles, 124 eagles (*Aquila* spp.), 1 harrier (*Circus* sp.), 2 European Kestrels, 8 Sparrowhawks, 8 Griffon Vultures, 8 Egyptian Vultures, 8 unidentified raptors, and about 100 nonraptors. It is possible that these birds were blown out to sea or that they may have been undertaking a crossing. Raptors migrating along the Atlantic coast of eastern North America descend to low altitudes when offshore winds threaten to blow them off course (Kerlinger and Gauthreaux 1984). Zu-Aretz and Leshem did not provide weather conditions at the time of the event, so it is difficult to speculate on what happened.

Additional evidence comes from T. Erdman (personal communication), who found more than 30 dead hawks along the shore of Lake Michigan, in only a few years of casual observations. Sharp-shinned (> 20), Broad-winged, Red-shouldered, and Cooper's hawks were found on the beaches of Lake Michigan on the Door County peninsula. Birds were apparently crossing Lake Michigan between the Garden peninsula of Michigan and the Door County peninsula. D. Brinker (personal communication) informs me that many raptors make this crossing (<10 km from island to island) annually. Erdman hypothesizes that migrants become lost in fog, which develops within a few minutes with certain wind shifts.

A similar report comes from Project Recovery of the Long Point Bird Observatory, where systematic searches of beaches from all five Great Lakes have been undertaken (Lambert 1983; A. Lambert and M. McNicholl, personal communication). From 1977 to 1983 several hundred 2 mile transects of beach were walked, and 46 raptors were found including 14 Sharp-shinned Hawks, 6 Red-tailed Hawks, 3 Cooper's Hawks, 2 Broad-winged Hawks, 1 American Kestrel, 1 Northern Harrier, 1 Rough-legged Hawk, 1 Red-shouldered Hawk, and 17 unidentified raptors. These numbers may not seem large, but samples were obtained along a small proportion of the shorelines of the Great Lakes, with little coverage of Lake Superior. It is likely that hundreds of hawks perish every year in the Great Lakes.

Finally, there are hundreds of reports of hawks from offshore islands and ships (Moreau 1953; Lack 1954; Kuroda 1961b; Rogers and Leatherwood 1981) that document the occurrence of migrants at sea. Kuroda's report is especially interesting because he reports numerous migrants at Volcano Island and other islands over 1,000 km from the Asian mainland. All the species were from the Asian mainland and seem to have strayed during water crossings. Many of the migrant hawks reported from offshore islands, oil platforms, and ships are not in prime condition. Although they may survive long flights over water, their reproductive fitness is negatively affected.

Table 10.1 Selected List of Known or Suspected Sites Where Water Crossings
Are Made by Large Numbers of Migrating Hawks

Crossing	Body of Water	Distance (km)	Reference
Cape May Point–Delaware[a]	Delaware Bay/Atlantic Ocean	17–18	Allen and Peterson 1936; Kerlinger 1984
Whitefish Point–Ontario	Lake Superior	18–29	Kerlinger 1984
Spain–Morocco	Strait of Gibraltar, Mediterranean Sea	<25	Evans and Lathbury 1973
Falsterbo, Sweden–Denmark	Baltic Sea	20	Rudebeck 1950
Europe–Asia	Bosporus	<5	Porter and Willis 1968
Point Diablo–San Francisco, California	San Francisco Bay	<5	Binford 1977
Malaya–Sumatra	Strait of Malacca	20	Baker 1981
Yemen–Ethiopia	Bab el Mandeb (Red Sea)	<25	Probable

Note: Crossings are usually <25 km, although a few longer crossings are included.
[a]This is one of many sites where raptors make water crossings along the eastern coast of the United States and Canada.

Summary of the Water Crossing Tendencies of Falconiforms

Although there have been few direct studies of the water crossing behavior of falconiforms, there are numerous reports of crossings as well
as faunal lists and distributional maps from which crossings can be
inferred. A literature search reveals that a surprising number of raptors make water crossings as a regular part of migration. The information presented is tentative, and more documentation is needed for
many species.

At least 64 of the 133 (48.1%) species of migratory falconiforms
engage in water crossings (appendix 1). The distance of crossings is
related to the type of migration (e.g., complete, partial, or irruptive;
$r = 0.529$, $p < 0.01$, $N = 133$; or $\chi^2 = 44.54$, df $= 6$, $p < 0.001$).
Complete migrants tend to make longer crossings than partial migrants, and partial migrants tend to make longer crossings than irruptive migrants. About one-half ($N = 34$, 53.1%) of the species that
cross water do not make crossings exceeding 25 km. These include
crossings of larger lakes, bays, and straits. (More species undoubtedly
make crossings of less than 5–10 km, but there is no way of determining how many.) At these sites (table 10.1) it is common to see large
numbers of hawks soaring and "milling around" before crossing (or
not crossing). Their reluctance to cross is sometimes obvious. At these
crossings raptors can see their destinations except when fog, rain, or

haze makes visibility poor. Another common feature is the margin of error afforded a bird that does not make a direct crossing. For instance, a bird attempting to cross the Strait of Gibraltar, the Strait of Malacca, or Lake Superior at Whitefish Point (fig. 7.4) can deviate from the shortest route and make landfall without difficulty. This is not true of all crossings. At Cape May Point, birds that are blown to the east of their preferred crossing direction must fly long distances before making landfall (fig. 7.4). This is potentially dangerous.

At minimum, thirty species of raptors cross water barriers that are more than 25 km. These crossings are risky because they require prolonged powered flight and some orientation or navigation ability. Hawks flying at altitudes of several hundred meters may see their destinations when the distance is less than 50–100 km. For longer flights (>100 km), migrants rarely see their destinations and therefore must orient or navigate. The risk of becoming lost, especially with strong crosswinds or poor visibility, is greater for these flights than for shorter flights.

Species of the genus *Falco* account for fifteen of the thirty species (50%) that undertake flights of more than 25 km over water. Four harriers (*Circus*), three eagles (*Haliaeetus*), and two *Accipiter* species account for another 30% of the thirty species. The remaining six species are from six genera. The reasons for the high proportion of falcons and harriers in this group will be discussed below.

Crossings of over 100 km seem to be rare, and falcons account for seven of the fourteen species known to undertake such crossings. These crossings are little known even though they occur in many parts of the world. Table 10.2 includes more than the fourteen species listed as making crossings of over 100 km, since I have attempted to be conservative in my assessment of water crossing tendencies. For some of the species listed in table 10.2, few individuals are involved. Some northern species like the Rough-legged Hawk undertake longer crossings in very remote locales in the Arctic where they would not be noticed or where water barriers may be frozen. In Europe, evidence from banding recoveries suggests that this species is not terribly reluctant to make longer water crossings (Schuz 1972). Again, the scarcity of information on water crossing by raptors reflects how little we know about the phenomenon.

The most spectacular of crossings is that of the Amur Falcon from the west coast of India to the east coast of Africa, more than 3,000 km one way. Ali and Ripley (1978) note that this species stages on the coast of India before crossing and birds are eaten by locals, who maintain that they are quite fat. While staging these migrants may deposit fat, as do passerines and shorebirds. They may also be waiting for optimal weather for crossing. Prevailing northeast winds over the Indian Ocean during autumn promote faster and safer passage to Africa. The flight of Merlins from Iceland to the British Isles during autumn also depends on favorable winds (Williamson 1954). Merlins initiate autumn flights from Iceland to Scotland with strong northwest winds and passing cold fronts.

Table 10.2 Partial List of Known or Suspected Long Distance Flights (>100 km) over Water and Some of the Species Involved

Location	Distance (km)	Species
Florida to Cuba and South America via islands in the Caribbean Sea and Gulf of Mexico	>125	Osprey Northern Harrier Merlin Peregrine Falcon
Italy to Tunisia/Libya via Sicily and Malta–Mediterranean Sea	>200	Osprey Honey Buzzard Black Kite European Sparrowhawk (?) Red-footed Falcon Hobby
Southeast Asia to East Indian islands	>100	Marsh Harrier
Australia to East Indian islands (New Guinea)	<100	Marsh Harrier Australian Kestrel Little Falcon
Japan to Philippines via Ryukyus (and Borneo)	>600	Gray-faced Buzzard Eagle Japanese Lesser Sparrowhawk
China to Philippines via islands	>200	Gray Frog Hawk
Banks Island to Canadian Mainland–Beaufort Sea	>100	Rough-legged Hawk Gyrfalcon
Mediterranean islands to Africa	>200	Eleonora's Falcon
Greece/Turkey (via Crete and islands in Mediterranean Sea) to Cyprus and Libya	<100–>285	Montagu's Harrier Red-footed Falcon Saker Falcon
Red Sea at various places	>200	Lesser Kestrel (?) Amur Falcon Sooty Falcon
India to Africa–Indian Ocean	>2,000 (?)	Lesser Kestrel (?) Amur Falcon
Tasmania to Australia–Bass Strait	>75	Brown Hawk (?)
Iceland to Great Britain via the Faeroe Islands–North Sea	>450–800	Merlin
Greenland to Baffin Island and Labrador–Davis Strait	>350–700	Gyrfalcon (?) Peregrine Falcon
Newfoundland to Nova Scotia–Cabot Strait	>120	Osprey Merlin Peregrine Falcon

Note: Evidence for these crossings has been gleaned from the literature, and most are not well documented.

It is not known how many of the raptors listed in table 10.2 depend upon following winds or how selective of weather these migrants are when initiating flights over water. Blackpoll Warblers migrating from eastern North America to northeastern South America over the Atlantic Ocean in autumn do so only when winds are favorable (Nisbet 1970). They wait for northwest winds associated with a cold front and then fly downwind. This allows them to travel 2 to 5 times as fast (ground speed) as flight with no wind. By waiting for appropriate winds they save time, energy, and their lives. Raptors like Peregrine and Amur falcons may also choose favorable weather conditions.

Most of the water crossings listed in table 10.2 are less than 300 km and require only a few hours to complete. Some are more than 500 km and require 8 h or more, depending upon wind conditions. Unless these longer flights are initiated early in the morning, they cannot be completed during daylight. Nocturnal migration therefore may be necessary for species such as the Amur Falcon, Osprey, Gray-faced Buzzard Eagle, Japanese Sparrowhawk, Gray Frog Hawk, Lesser Kestrel, Gyrfalcon, and Peregrine Falcon. Cochran (1985) reported that Peregrines occasionally make nocturnal flights over water along the eastern coast of the United States. While following birds fitted with radio transmitters, he found that some initiated flights in the evening and returned to land before dawn. Although raptors are not considered night migrants, nocturnal flight is of interest because it suggests that some have evolved navigation and orientation systems that function during both day and night.

Island Hopping

Not all crossings of large bodies of water are nonstop. Shorebirds, some passerines, and hawks often use islands as resting and refueling stations. One of the longest flights over water involves resting and refueling on an island. Lesser Golden-Plovers (*Pluvialis dominica*) stop in the Hawaiian Islands during their autumn migration from Alaska to the South Pacific (Johnston and MacFarlane 1967). During their stay in Hawaii they forage and accumulate fat before continuing migration. Because of their great mass and the cost of powered flight, raptors also make stopovers on islands.

In the Mediterranean Sea raptors "hop" from Italy to Sicily to Malta to the African coast and perhaps some smaller islands on the way. Thousands of Honey Buzzards and other raptors make the trip annually (Beaman and Galea 1974; Thake 1980). Many species also make stops on Crete (Casement 1966) and Cyprus. From there they fly to the coast of Lebanon-Israel or directly to Africa in autumn and the reverse in spring. More than a dozen species are involved (Casement 1966; Beaman and Galea 1974). Recent studies by Galea and Massa (1985) and Vagliano (1985) summarize crossings and island hopping in the Mediterranean.

The Gulf of Mexico and the Caribbean Sea are obstacles to migration of many North American species. A few, such as the Peregrine Falcon, Merlin, Osprey, Northern Harrier, and certain others make crossings and frequent islands (gathered from faunal lists of several islands). Many of these islands have been colonized by species that are not known to cross large expanses of water. For instance, Broad-winged and Sharp-shinned hawks have established breeding populations on several islands. Some Broad-winged Hawks are suspected to cross the Gulf of Mexico/Caribbean (Smith 1980), although the evidence is not convincing. Nevertheless, large flocks of Broad-winged Hawks are known from Trinidad (flock of about 500, Rowlett 1980) and other islands in the Caribbean (ffrench 1980), and the Florida Keys (Edscorn 1974). This documents only successful crossings, but it does show that some individuals initiate crossings. It is possible that Broad-winged Hawks seen leaving Key West or Trinidad return to shore or eventually perish.

H. Darrow (personal communication) reports that Turkey Vultures, Sharp-shinned Hawks, Broad-winged Hawks, and other raptors migrate into the Florida Keys every autumn. Some, like Turkey Vultures and Broad-winged Hawks, use thermaling flight to hop from island to island. When winds are strong and unfavorable for migration these birds have difficulty making progress. Sharp-shinned Hawks also use updrafts from bridges or fly in the lee of bridges when passing from island to island. Some Broad-winged Hawks may attempt to cross to Cuba, whereas others remain in the Keys for the winter.

The appearance of hundreds of raptors does not go unnoticed by indigenous peoples. Raptors landing on and passing over islands have been persecuted for hundreds of years. In the Mediterranean Sea an undetermined number of hawks are shot on Sicily, Malta (Beaman and Galea 1974), Crete, Cyprus, and other islands. They are shot for food or for sport, and some become mounted displays.

In the Ryukyus, the island chain extending from Japan to the Philippines, the Gray-faced Buzzard Eagle has been persecuted unmercifully (R. Kennedy, personal communication). The species breeds in Japan, Korea, northeast China, and southeast Siberia and migrates to Southeast Asia, southern China, the Philippines, New Guinea, and Celebes. After landing on islands like Miyako Jima, midway between

Japan and the Philippines, or Batanes, in the northern Philippines, birds were captured by hand in roost trees or on specially made platforms. They were then eaten or kept as pets. The practice has ended in Japan (McClure 1974), but in the Philippines many are still killed. Because it can be captured easily, banding and recapture information

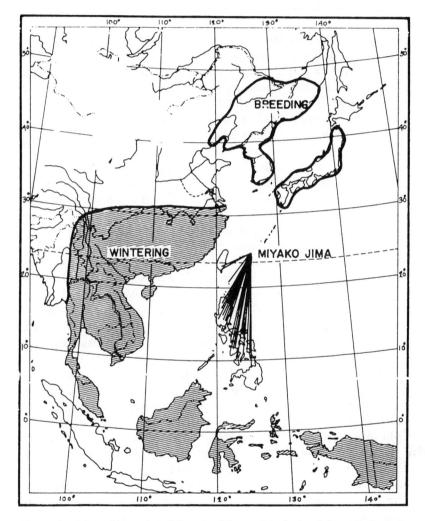

Figure 10.1 Map of recoveries of Gray-faced Buzzard Eagles banded on Miyako Jima in the Ryukyu Islands of Japan showing an island-hopping route from Japan to the Philippines (from McClure 1974; courtesy of H. Elliott McClure). Persecution of the species has ceased in the Ryukyus but continues unabated in the Philippines.

is extensive for the Gray-faced Buzzard Eagle. McClure (1974) reports that 106 of the more than 2,486 (about 4%) banded by his colleagues on Miyako Jima were recovered, primarily in the Philippines (fig. 10.1). McClure's data indicate that both adult and immature birds make the crossings and that many survive if they are not captured and eaten. Thus, raptors seeking refuge and rest on islands may find hostile and hungry human hosts.

One of the most visible and well-known cases of island hopping by migrating raptors occurs along the northeastern coast of the United States. Each autumn thousands of raptors fly from the mainland to islands and back to the mainland again. Islands such as Seal Island, Nova Scotia; Martha's Vineyard, Massachusetts; Block Island, Rhode Island; Fishers Island, Long Island, and Staten Island, New York (Ferguson and Ferguson 1922; Hawk Migration Association of North America newsletter—autumn reports 1985, 1986); and many more are excellent places to observe migrating raptors. The autumn passage of these migrants is presumably shortened by island hopping instead of following the mainland coast. Species such as Peregrine Falcons, Merlins, American Kestrels, and Sharp-shinned Hawks find birds to feed upon, while Ospreys forage in the ocean or bays. Some of these species also forage on the open sea (Rogers and Leatherwood 1981).

An ongoing study by Wormington, Finlayson, and Nisbet (A. Wormington, personal communication) on Caribou Island in east-central Lake Superior shows that hundreds of individuals of eleven species occur on the island during spring and autumn. Wormington believes some of these migrants are "lost/confused individuals that were fortunate enough to find the island." The nearest landfall is 40–50 km, and little food is available on the island. Hawks are harassed by gulls as they approach or leave. Nevertheless, the island serves as a resting and hunting site (for bird- and fish-eating species) until weather is suitable for completion of their journey. Although some individuals may have been lost, others were involved in a regular migration over water. Even those birds that were lost had to have been undertaking a flight over water when they became disoriented or fatigued.

In chapter 7 and above I suggested that the plethora of migrating hawks at coastal locations could be explained as a response to abundant avian prey. A similar explanation may pertain to island hopping. The behavior and taxonomic composition of migrant raptors observed on islands suggests that these places are excellent foraging sites.

During the spring of 1987 I spend fifty-one days on Horn Island (21 March-10 May) 15 km from the Mississippi coast. The island is one of a chain of barrier islands that parallel the shores of Alabama, Mississippi, and Louisiana along the northern coast of the Gulf of

Mexico. Although the focus of my research was the ecology of passerine migrants enroute, I noted raptors and their behavior. Bald Eagles and Ospreys were not included because they are resident. Seven species were seen, of which four were obligate bird eaters and two others were facultative avian predators. Three species were observed capturing or feeding on passerines that had recently migrated over the Gulf of Mexico. Many were caught easily over water, dunes, and marshes as they flew in from the Gulf. Merlins, Peregrine Falcons, and American Kestrels made stopovers of several days on the island. The barrier islands are excellent foraging and stopover sites for some migrant raptors, just as islands are excellent breeding places for Eleonora's Falcons that depend upon migrants (Walter 1979).

The origin of the raptor migrants seen on Horn Island during spring is not known, although some probably cross from Louisiana via Cat Island or the Chandeleur Islands (west and southwest of Horn Island). Others may have crossed from Texas or Mexico.

Island hopping by migrants is notable because it may explain the presence of island races (Heintzelman 1975) and species. Kestrels of the Seychelles and Mauritius could have originated from stocks of migrants crossing to Madagascar from Africa or to Africa from India. For a colonization to be successful, more than one bird must be involved. Flocking species such as Lesser Kestrels, Amur Falcons, and European Kestrels all make water crossings, thereby increasing the chance of successful colonization. Because raptors are long lived, flocks may not be necessary for successful colonizations.

Explaining how buteos such as the Io (*Buteo solitarius*) became established in the Hawaiian Islands or the Galápagos Hawk (*B. galapagoensis*) in the Galápagos is more difficult. These species may have originated from continental forms that migrate in flocks. Recent colonization of other islands like the Falklands off Argentina by several species, the Aleutians off Alaska by White-tailed Sea Eagles, the Canaries off Africa by Eleonora's Falcon, and Iceland by Merlins and Gyrfalcons is fascinating and shows that water crossing behavior of migrants may be important in colonization and speciation events. Heintzelman (1975) and Mindell (1985) discuss the importance of migration and the colonization by raptors of islands in the New World.

Comparative Behavior of North American Hawks at Water Crossings

Few studies have examined the behavior of raptors when they encounter water barriers (Rusling 1936; Binford 1977; Kerlinger 1984,

1985a), although some authors have speculated about crossings (Brown and Amadon 1968; Evans and Lathbury 1973; Heintzelman 1975; Smith 1980). Stone (1937) suggested that Sharp-shinned Hawks crossed Delaware Bay from Cape May Point regularly, but only at high altitudes. Farther down the Atlantic coast at Cape Charles, Virginia, Rusling (1936) observed that more hawks tended to cross Chesapeake Bay when winds were from the south during autumn migration than when winds were from the north or west.

To examine the behavior of migrating raptors at water barriers, I spent autumn 1980 at Cape May Point, New Jersey, and spring 1981 at Whitefish Point, Michigan (table 10.3; Kerlinger 1984, 1985a). Large numbers of raptors are counted at these locations. Because hawk counts have been conducted inland from the tips of these peninsulas, the behavior of migrants at the crossings was not observed. I monitored the behavior of thousands of migrating raptors at the ends of these peninsulas and determined what species made crossings, when crossings were initiated, how high migrants flew, and in what direction migrants flew while crossing.

Cape May Point and Whitefish Point are situated at the ends of long, narrow peninsulas (fig. 7.4). Birds arriving at the ends of these peninsulas are faced with a crossing over water of 17.7 km to more than 24 km depending upon flight direction. Migrants that opt not to cross must fly hundreds of kilometers around these barriers in directions that are inappropriate for migration. The opposite shoreline is usually visible, even to migrants flying at low altitude. At Cape May

Table 10.3 Details of Water Crossing Behavior Study of
Sharp-shinned Hawks and American Kestrels

	Cape May Point, New Jersey (4 September– 10 October 1980)		Whitefish Point, Michigan (27 April–6 May 1981)
	Sharp-shinned Hawks	American Kestrels	Sharp-shinned Hawks
Number of observation periods	28	20	25
Hours per observation period	1.5	1.9	1.4
Birds seen per observation period	192	30	83
Total number hawks seen	5,387	593	2,081
Number seen crossing (%)	4,343 (81)	407 (69)	970 (47)
Number seen going out and back (%)	347 (7)	59 (10)	251 (12)

Source: Sharp-shinned Hawk data from Kerlinger 1984.

Table 10.4 Summary of Water Crossing Tendencies of Ten Raptor Species during Spring 1981 Migration at Whitefish Point, Michigan, and Autumn 1980 Migration at Cape May Point, New Jersey

Species	Whitefish Point (N)	Cape May Point (N)	Percentage Crossing	Percentage Not Leaving Shore	Percentage out over Water and Back
Turkey Vulture	12	19	10	64	26
Osprey	16	67	93	5	2
Northern Harrier	37	75	93	4	3
Sharp-shinned Hawk	2,261	5,387	71	17	12
Broad-winged Hawk	1,364	759	26	33	41
Red-tailed Hawk	55	0	11	47	42
Rough-legged Hawk	125	0	69	17	14
American Kestrel	171	593	66	25	9
Merlin	5	41	76	15	9
Peregrine Falcon	1	13	100	0	0

Source: Kerlinger 1985a.

Note: Two Bald Eagles were noted to cross, whereas only one of two Golden Eagles crossed at Whitefish Point.

the possible risks of crossing to Delaware include predation by gulls (P. Dunne, personal communication), getting lost in storms or during times of poor visibility, and being blown (drifted) out over the Atlantic Ocean by north to west winds. Because land is not visible to the northwest of Cape May (in Delaware Bay), a migrant may "perceive" a risk in being blown out over Delaware Bay. The crossing at Whitefish is somewhat different in that there are several islands on which a bird may land and landfalls are visible in all directions to the east of the crossing direction (45°–75°). There is a risk of being blown out over the main body of Lake Superior by east to southeast winds. The crossings at Whitefish Point and Cape May Point are similar to those at many other sites.

The tendency to undertake crossings varied among the ten species observed (table 10.4). Species such as the Peregrine Falcon, Merlin, Osprey, and Northern Harrier rarely hesitated to initiate crossings. I have observed Merlins and Peregrine Falcons initiate crossings even in rain and when visibility was poor. Among the other species there was a range of behaviors. Red-tailed Hawks, Broad-winged Hawks, and Turkey Vultures made crossings least often, whereas Sharp-shinned Hawks, American Kestrels, and Rough-legged Hawks tended to be intermediate in their tendencies to cross Lake Superior and Delaware Bay.

Although Turkey Vultures, Broad-winged Hawks, and Red-tailed

Hawks were least likely to make crossings of Lake Superior, they sometimes left the shore flying in the direction appropriate for crossing. Those that did not cross turned back inland before crossing the coast or flew several hundred meters offshore before returning. Flocks of more than 30 Broad-winged Hawks and Turkey Vultures occasionally disappeared toward the opposite shoreline only to return after 5–10 min. A few Turkey Vultures returned to Cape May using continuous flapping flight at less than 20 m above the water. Their flight appeared labored, and they made slow progress into opposing winds. Some Red-tailed Hawks and Turkey Vultures did not return to Whitefish Point or Cape May Point and presumably made successful crossings.

Broad-winged Hawks crossed more often at Whitefish during spring than at Cape May Point during autumn, perhaps because most of the spring migrants were adults, while autumn migrants were both adults and immatures. An incident at Whitefish sheds light on the behavior of this species. Between 1000 h and 1100 h on 2 May 1981, two flocks of Broad-winged Hawks each numbering about 220–250 individuals (possibly the same flock seen two times) left the point and returned less than 10 min after leaving. The birds were flying at 300–400 m above the ground when they left the point and were below 100 m when they returned. At 1150 h about 600 birds (one flock of about 500 followed a few minutes later by a flock of 100 birds) flew out from the point and did not return. These migrants were so high they were barely visible without binoculars when directly overhead. These birds flew in the direction appropriate for crossing and used more flapping (partially powered glides) than is normal for the species. The official hawk counter did not see either flock depart toward Canada. These birds were not seen returning, leading me to speculate that more Broad-winged Hawks (and possibly other species) make this crossing than has been reported. If such crossings occur at high altitudes, they could pass unnoticed. Similarly, if hawks cross at high altitudes but abort crossings when flying at low altitudes, an observer might conclude that they did not make crossings. This behavior is similar to that of Sharp-shinned Hawks at water crossings.

The behavior of American Kestrels and Sharp-shinned Hawks at water barriers is more interesting than the behavior of other species, because it is variable yet predictable. Statistical analyses were possible because individuals of these species were numerous. Wind and visibility were robust predictors of water crossing behavior. Neither species made crossings in precipitation or fog. Indeed, these birds rarely migrate in fog or rain. A smaller percentage of Sharp-shinned Hawks and American Kestrels crossed at the sites when their destinations were not visible. Fog over Lake Superior halted kestrel migration at Whitefish Point even though there was little fog over land. On 4 May 1981, 17 kestrels ceased migrating at the tip of Whitefish Point, perching on bushes or driftwood. When the fog cleared enough so that Ontario was visible (to my eyes), the kestrels took off within a 5 min period and flew directly toward the opposite shoreline. This was a dramatic example of how ephemeral weather conditions influence water crossing.

When the opposite shoreline of Lake Superior and Delaware Bay were visible, wind strength was a strong predictor of crossing behavior of both Sharp-shinned Hawks and American Kestrels. Regressing the percentage of birds observed that made crossings on wind speed (table 10.5) revealed that both species crossed more often when winds were weak. Lateral wind—that is, the wind component perpendicular to the crossing direction—yielded stronger negative relations (fig. 10.2, table 10.5), and I concluded that birds preferred not to cross

Table 10.5 Regression Equations for Tests of Water Crossing Predictions for Sharp-shinned Hawks and American Kestrels at Cape May Point, New Jersey, and Whitefish Point, Michigan

Dependent Variable	Independent Variable	Regression Equation	r^2
SHARP-SHINNED HAWKS, CAPE MAY POINT ($N = 24$)			
Percentage crossing[a]	Lateral wind speed	$y = -7.83x + 82.35$	0.57^{**}
Percentage crossing	Following/opposing wind speed	—	0.02
Percentage crossing	Absolute wind speed	$y = -1.16x + 91.29$	0.28^*
Percentage crossing	Altitude	$y = 11.24x + 21.31$	0.32^*
Altitude	Absolute wind speed	$y = -0.28x + 4.61$	0.29^*
SHARP-SHINNED HAWKS, WHITEFISH POINT ($N = 19$)			
Percentage crossing	Lateral wind speed	$y = -4.05x + 62.59$	0.63^{**}
Percentage crossing	Following/opposing wind speed	$y = -3.53x + 42.43$	0.23^*
Percentage crossing	Absolute wind speed	$y = -4.81x + 67.40$	0.51^{**}
Percentage crossing	Altitude	$y = 12.74x + 7.08$	0.17^*
Altitude	Absolute wind speed	—	0.08
AMERICAN KESTRELS, CAPE MAY POINT ($N = 17$)			
Percentage crossing	Lateral wind speed	$y = -7.62x + 83.48$	0.55^{**}
Percentage crossing	Following/opposing wind speed	—	0.14
Percentage crossing	Absolute wind speed	$y = -4.85x + 81.24$	0.18^*
Percentage crossing	Altitude	$y = 9.04x + 33.60$	0.16^*
Altitude	Absolute wind speed	$y = -0.26x + 4.08$	0.26^*

Source: Sharp-shinned Hawk data from Kerlinger 1984.
[a]Percentages were arc-sine transformed before analysis.
$^*p < 0.05$; $^{**}p < 0.01$.

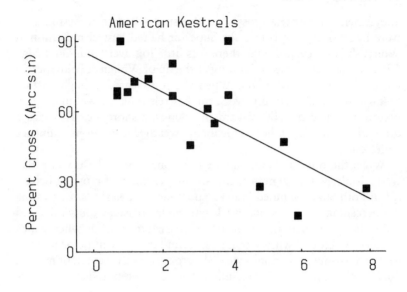

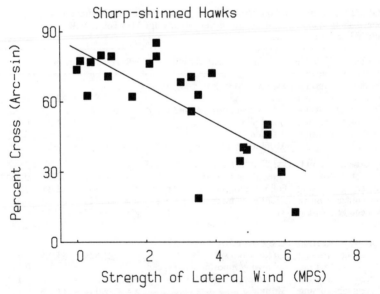

Figure 10.2 Regression of the percentage (arc-sin transformed) of Sharp-shinned Hawks and American Kestrels that made crossings during an observation period on wind lateral to the crossing direction. The crossing was from Cape May Point, New Jersey, to Cape Henlopen, Delaware (see fig. 7.4). See table 10.5 for regression equations.

when the chance of being blown off course over the inhospitable vastness of the Atlantic Ocean or Lake Superior was greatest. Relations for both American Kestrels and Sharp-shinned Hawks were curvilinear (fig. 10.2). The percentage of birds making crossings declined precipitously with lateral winds greater than 4 mps, possibly indicating the point at which migrants could no longer compensate for drift (a threshold). Although the exact regression coefficients may not be meaningful, the sign of the coefficient is important, as is the fact that the coefficients for the two species are similar.

An alternative hypothesis, that these hawks crossed only when tail winds would speed their journey, was tested by regressing the percentage making crossings on the following/opposing component of the wind. The result was a nonsignificant relation showing that they did not choose to cross with following winds (table 10.5). Although following winds promote faster and energetically less expensive passage, birds that initiate crossings with tail winds might experience difficulty returning to the place where they initiated the crossing. Thus, these species seem to have a conservative strategy for making water crossings; they "play it safe."

Simulating flight across Delaware Bay shows how costly crossings with strong lateral winds can be. Assuming an air speed of 11 mps for a Sharp-shinned Hawk or an American Kestrel (chap. 11) in level, powered flight, the time needed to cross Delaware Bay increases in a curvilinear fashion (fig. 10.3) as the strength of opposing wind increases. A second simulation using the same air speed, wind from the northwest (315°), and a heading of 270° shows that the maximum northwest wind these birds can compensate for and still make reasonable progress is about 7–8 mps. Thus, as opposing or lateral wind speed increases, crossings require more energy and ultimately cannot be completed. Crossing at the wrong time can result in death.

Although wind and visibility were important determinants of whether Sharp-shinned Hawks and American Kestrels would attempt crossings, altitude of flight was also correlated with crossing tendency (table 10.5). When migrants flew at high altitudes, they were more likely to cross. This relation is dependent secondarily on wind and updrafts. Better visibility at high altitudes may be important by permitting migrants to see their destination before crossing. This factor cannot be excluded from the overall analysis, but is difficult to control statistically (even using partial correlation/multiple regression).

Holthuizjen and Oosterhuis (1981) also studied the behavior of Sharp-shinned Hawks at Cape May Point. They made observations on the movements of 34 female migrants on which they placed radio transmitters. Only one migrant was tracked as it came into Cape Henlopen after crossing Delaware Bay. This finding suggests that Kerlinger (1984, 1985a, and above) may have overestimated the numbers of migrants that make crossings. On the other hand, it is possible that the transmitters inhibited water crossings or that birds were captured and equipped with transmitters at times when they were least likely to make crossings (i.e., strong winds). Capturing Sharp-shinned Hawks at Cape May is easiest when there are strong winds and migrants are at low altitudes, which is also the time when migrants do not initiate crossings. During ferry trips across Delaware Bay P. Dunne (personal observations) and Kerlinger and Gauthreaux (unpublished data) have recorded numerous Sharp-shinned Hawks near the Delaware side of the bay. Perhaps many birds cross the bay but few make landfalls at Cape Henlopen, where Holthuizjen and Oosterhuis placed their radio receiver.

The mechanism by which hawks decide whether to cross a water barrier is not known, but analyses show that "decisions" are made. Birds that fly back and forth at the shoreline or fly out over the water for short distances may be sampling wind conditions before embarking on a risky flight. By crossing only when winds are weak (especially

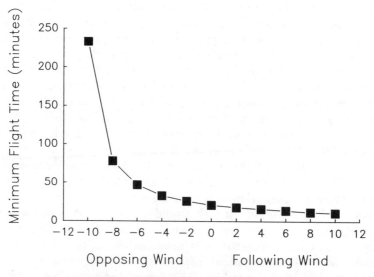

Figure 10.3 Time needed to cross from Cape May Point, New Jersey, to Cape Henlopen, Delaware, by a Sharp-shinned Hawk flying at 11 mps in opposing winds of various speeds.

crosswinds) migrants reduce the chances of being blown out to sea. Also, by not crossing with strong tail winds migrants can always return to the shore they started from. Decisions to cross are conservative in that birds avoid risk. It was interesting that some individuals did not cross even when winds were less than 2–4 mps, far below the calculated threshold for drift. Could these birds be in poor physiological condition? Whatever the mechanism, large bodies of water have undoubtedly imposed a strong selective force on the behavior of migrating raptors.

(My appraisal of water crossing tendency and behavior has been criticized by both European and American researchers. Some Europeans, including T. Alerstam [personal communication], believe that crossings of 25 km are made regularly by many species of raptors and that raptors are capable of flying much longer distances over the sea. An anonymous American reviewer suggested that very few raptors made crossings of more than 10–15 km. Thus, I have been criticized for being both too liberal and too conservative. It is obvious that more field research is necessary.)

Flight Morphology and Water Crossing Tendency

In several taxa, morphology is correlated with migration tendency or distance. Among insects, migratory species (or migratory populations in the case of locusts and milkweed bugs) have longer wings (higher aspect ratio) than nonmigrants (Dingle, 1980; Dingle, Blakeley, and Miller 1980). Similar findings have been noted among wood warblers in North America (Blem 1980) and migratory birds in Europe (Kipp 1958). The relation between morphology and aerodynamic performance was discussed in an earlier chapter, but mostly for soaring and gliding flight. During water crossings powered flight is of prime importance, since species capable of extended flights are more likely to make longer crossings.

Brown and Amadon (1968) were the first to attempt to correlate the water crossing tendency of a species with its morphology. They noted that raptors that predominantly use soaring flight do not make long flights over water. These statements are inexact; I have shown above that most raptors use soaring and gliding flight during migration. They also commented that species that engaged in long distance water crossings had light wing loading. A truly empirical test of their hypothesis is not possible because data on wing loading (mass/wing area) are extremely scarce (chap. 5). It is doubtful that wing loading is the primary or only morphological factor involved. Species such as

the large falcons have wing loadings that exceed those of many buteos, but the buteos are less likely to make crossings. In addition, species like harriers make long crossings, and their wing loading is much
less than that of the large falcons. A thorough analysis would include
mass of pectoral muscle, preflight fat reserves, wing shape, and aspect
ratio. A scaling procedure may also be necessary (Schmidt-Nielsen
1984), as is additional information on the water crossing tendencies
of more species of hawks.

The one quantitative treatment of morphology and water crossing
tendency is that of Kerlinger (1985a), in which behavioral and morphological data were used. Kerlinger (1985a) observed the behavior
of ten species of North American raptors at Whitefish Point and Cape
May Point. He proposed that species with high aspect ratios were
more adept at powered flight and were morphologically similar to
seabirds. The long, often pointed wings of some raptors are similar to
the wings of some seabirds (Savile 1957; Warham 1977). Species with
shorter, more rounded wings were reasoned to be less suited to powered flight but more adept at using thermals over land. Land birds
typically have disproportionately shorter wings than seabirds (Savile
1957). Although this relation is intuitively pleasing, a study using scaling of characters would be more complete. In figure 10.4 the proportion of each of the ten species that crossed Lake Superior and Delaware Bay is plotted against the aspect ratio of the species. Aspect
ratios were determined from wing span data on museum tags (Vertebrate Museum, University of California at Berkeley) and wing areas
published in Poole (1938). The relation is positive and significant, as
predicted, although not as strong as hoped. Species such as Peregrine
Falcons, Ospreys, and Merlins with the highest aspect ratios crossed
water more often than species such as the Red-tailed and Broad-
winged hawks with the lowest aspect ratios. The positive regression is
consistent with the aspect ratio hypothesis, but as pointed out by Kerlinger (1985a), the results are tentative and the relation should be
tested using a larger number of species from other water crossing situations. (Note that the aspect ratios of some species in fig. 10.4 differ
from those reported in chap. 6. These differences result from the different sources of data used and show the need for more thorough
consideration of morphological characteristics associated with flight.)

There are several reasons why birds having high aspect ratios are
capable of longer powered flights than birds with low aspect ratios.
First, longer wings reduce tip-vortex effects (Pennycuick 1972b,
1975). Induced drag is a negative function of wing span and therefore
is not proportionately as great in longer-winged species. A reduction

of induced drag allows powered flight at a lower cost per unit of distance traveled (Pennycuick 1975). In flight over water, where distance traveled is important, reduction of induced drag is crucial. A second advantage of reducing induced drag is faster air speed. In addition, high aspect ratio may facilitate soaring or gliding in weak updrafts. Seabirds such as gulls and tubenoses that have high aspect ratios make use of both of these sources of lift (Woodcock 1940, 1975).

There is additional evidence for the aspect ratio/water crossing hypothesis. An examination of the list of migrant species from the island of Malta (Beaman and Galea 1974) shows that long, narrow-winged species such as falcons, harriers, and kites are more numerous than wider-winged buteonine and vulturine species. The latter species fly around the Mediterranean Sea or make short water crossings (Porter and Willis 1968; Evans and Lathbury 1973). The crossing from Italy to Sicily is less than 20 km, from Sicily to Malta is over 90 km, and from Malta to Tunisia is more than 280 km (or from Malta to Libya is >310 km), a long and strenuous flight that may discourage all but

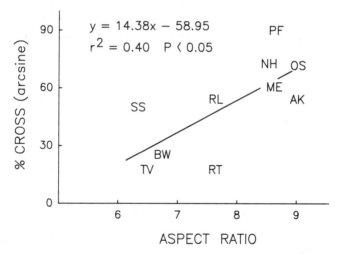

Figure 10.4 Regression of the percentage (arc-sin transformed) of a species that made crossings on the aspect ratio of that species. Crossings were of Lake Superior (redrawn from Kerlinger 1985a) from Whitefish Point, Michigan, to Ontario in spring and of Delaware Bay from Cape May Point, New Jersey, to Cape Henlopen, Delaware, in autumn (TV = Turkey Vulture, NH = Northern Harrier, OS = Osprey, PG = Peregrine Falcon, ME = Merlin, AK = American Kestrel, BW = Broad-winged Hawk, RL = Rough-legged Hawk, RT = Red-tailed Hawk, SS = Sharp-shinned Hawk; courtesy of the Wilson Ornithological Society).

the most proficient fliers. Lists of species from islands and ships also show that falcons, harriers, and other species with high aspect ratios tend to undertake water crossings more often than species with broad, rounded wings.

A list of migrants seen offshore in the North Atlantic Ocean reveals a paucity of short or broad-winged species (Kerlinger, Cherry, and Powers 1983; table 10.6). Peregrine Falcons, Merlins, and Ospreys accounted for over 60% of the migrants seen during five years of autumn and spring cruises in the waters off the United States north of 35° N. The only short-winged migrant was the Sharp-shinned Hawk, which occurred in a smaller proportion than in migration counts on shore. These few individuals may have been blown offshore while attempting crossings of Delaware Bay from Cape May Point to Delaware, from Rhode Island to Long Island, or elsewhere. An observer at Fire Island Inlet (Darrow 1963) on Long Island also reported a difference in behavior among species migrating along the south shore of Long Island. Darrow noted that Peregrine Falcons left Long Island on a course of 230°, which would have brought them to the central New Jersey coast. Other raptors, mostly Sharp-shinned Hawks and American Kestrels, continued west toward New York City and did not venture out over the ocean. Observers on Fire Island (Long Island) also report Peregrine Falcons and Merlins leaving the shoreline on south-southwest courses (W. Wegner, personal communication).

Table 10.6 Comparison of Hawks Seen Offshore in the North Atlantic Ocean and along the Coast of New Jersey during Autumn Migration

		Offshore		Coastal	
Species	Number of Hawks	Percentage of Autumn Total	Mean Distance from Land ± SD (km)	Mean Number of Hawks	Percentage of Total
Osprey	19	20.0	118 ± 53	1,170	2.0
Sharp-shinned Hawk	15	15.8	91 ± 81	41,876	72.6
American Kestrel	19	20.0	86 ± 66	13,643	23.6
Merlin	25	26.3	87 ± 56	862	1.5
Peregrine Falcon	17	17.9	84 ± 50	149	0.3

Source: Kerlinger, Cherry, and Powers 1983.

Note: Offshore data are from 26 autumn cruises (138 cruise-days). Coastal data are average numbers seen during the 1976–1980 migration seasons at Cape May Point, New Jersey.

Table 10.7 Raptors Seen on Offshore Oil Platforms during 1984 in the
North Sea between Norway, England, and the Netherlands

Species	Number of Individuals	Greatest Distance from Nearest Land (km)
Osprey	2	>250
Red Kite	1	170
European Sparrowhawk	37	>180
Common Buzzard	4	>150
Common Kestrel	42	>250
Merlin[a]	5	>200
Gyrfalcon[a]	1	>200
Peregrine Falcon	7	>200
Total individuals	99	

Source: Data extracted from Anderson 1985.
[a]Origin could have been Iceland.

Hawks are also sighted on offshore oil platforms in the North Sea. Within one migration season dozens, if not hundreds, of sparrowhawks, kestrels, Merlins, Peregrine Falcons, and Ospreys are reported on rigs several hundred kilometers from land (table 10.7; Anderson 1985). This list and a list of migrants seen from ships in the North Sea by Murray (1931) are not too different from the list in table 10.6. Whereas the Merlins, Peregrine Falcons, and Ospreys were undoubtedly making routine crossings, some sparrowhawks and kestrels probably were in dire straits.

Finally, the relation between aspect ratio and tendency to make long distance water crossings is supported by the list of species in table 10.2 that undertake crossings. The list includes eight falcons (53.3%) and two harriers (13.3%), which have long, narrow, and often pointed wings. Other harriers are known to make fairly long flights over water but were not included in the list (e.g., the Northern Harrier, which flies from Florida to Cuba, a distance of >140 km). The remaining species on the list include two "kites" and two accipiters. The flight morphology of these four species is not known, but they are predicted to have greater aspect ratios than congeners of similar size that do not make crossings.

Until more and better investigations of water crossing and flight morphology (or flight ability) are undertaken, the relation between aspect ratio and the tendency to undertake crossings is tentative. In addition, researchers should focus on the physiological and anatomical aspects of the flight capabilities of raptors.

Summary and Conclusions

Water barriers are an important selective force shaping the migration pathways of raptors in all parts of the world. Migrants encountering water must decide whether to cross or to fly around these barriers. By making crossings a migrant may save considerable time and energy, but it runs the risk of being lost at sea because of poor weather or fatigue. The mortality of raptors attempting even short water crossings is well documented. About half ($N = 64$) of migratory raptors make crossings, but only thirty species make crossings of more than 25 km. Species with long, narrow wings (high aspect ratio) such as falcons and harriers make water crossings more often, and make longer crossings, than species with short or rounded wings (low aspect ratio). High aspect ratio wings promote faster and less energy-expensive powered flight. Crossings occur most often when weather conditions promote safe passage. Studies of the water crossing behavior of Sharp-shinned Hawks and American Kestrels show that these species have conservative strategies (risk aversive), crossing most often when visibility is good and winds are weak (especially crosswinds). Longer water crossings by species such as Merlins, Amur Falcons, Ospreys, Gray-faced Buzzard Eagles and some others are likely to be timed to coincide with favorable (following) winds.

11

Selection of Flight Speed:
Maximizing Distance Traveled

Because raptors are capable of a wide range of air speeds, they must choose the appropriate speed during migration. If a migrant flies too fast it may deplete its energy reserves; if it flies too slowly it may not complete migration. Powered migrants pay for fast flight with energy, whereas gliding migrants pay with a rapid loss of altitude. The way a migrant chooses its speed has presumably been shaped by natural selection: individuals that fly at speeds that promote a safe yet fast migration are favored.

Studies of migratory and nonmigratory birds show that flight speed is adjusted to either minimize energy expended or maximize distance traveled. In this chapter I show that migrating raptors select speeds that promote a fast yet energy-efficient migration. In addition, I present air speeds of raptors gliding between thermals, gliding in ridge lift, and in powered flight over open terrain where lift is scarce or absent. These air speeds are used to test the hypothesis that migrating raptors select flight speeds to maximize distance traveled.

Flight Speed Selection during Powered Flight

Theory

A comprehensive theory has been developed by Pennycuick (1969, 1975, 1978) and others regarding how fast powered migrants should fly, based upon the curvilinear relation between air speed and power required for flight (power curve in chap. 5). The theory will be referred to as distance maximization theory in this chapter for both powered and gliding flight. The hypothetical power curve presented in figure 11.1 shows air speed for maximum range and minimum power. For a migrant to travel the maximum distance for a given amount of energy, it should set its air speed at the maximum range speed. In this case maximum range speed pertains only when wind is not present. Maximum range speed varies with wind component parallel to a bird's heading and is determined by drawing a line tangent to the power curve from that point on the *x*-axis corresponding to the wind speed at that time. With a following wind the maximum range speed is slower than with no wind or with an opposing wind (fig. 11.1; but see Pyke 1981). For a bird to travel the maximum distance per unit of energy expended, it must adjust its air speed to the component of wind parallel to its heading. Thus, migrants are predicted to use faster air speeds in opposing winds than in following winds.

There is at least one exception to the predictions outlined above. A bird that is disoriented over water or caught in rain or fog over water is predicted to fly at the minimum power speed. By doing so it can stay aloft longer than at the maximum range speed. This prediction applies only when the migrant cannot maintain a course to the nearest landfall.

Empirical Findings

There are several problems associated with testing hypotheses of flight speed selection by migrating raptors during powered flight. First, there has been only one study of the energetics of powered, level flight by a raptor (American Kestrel, Gessaman 1979). We know too little to make precise predictions about flight speed for raptors. A second problem is that few raptors use purely powered flight, including con-

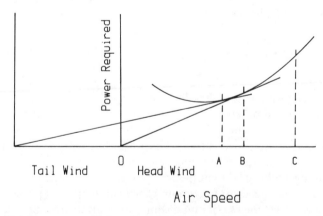

Figure 11.1 Power curve of a hypothetical bird showing strategy for optimal selection of flight speed based on wind component parallel to the bird's path.

tinuous flapping or intermittent gliding. At low altitudes over most terrain, wind-deflected or thermal updrafts are present. These updrafts are undependable, so that migrants must resort frequently to flapping. When updrafts are not available they should select speeds in the same manner as passerines and other birds that use powered flight. Flight over large expanses of water is ideal for testing whether raptors adjust flight speed to maximize distance in accordance with wind. The paucity of thermals over water (chap. 4) makes powered flight a necessity for those raptors that attempt water crossing.

Many hawks engage in low altitude powered flight for short distances, but the occurrence of these bouts is not predictable. They are often mixed with partially powered glides when updrafts are available. In eastern North America the most conspicuous migrants that do this are the Sharp-shinned Hawk and American Kestrel. To determine whether these migrants maximize distance traveled per unit of energy expended, I measured ground speed of migrants as they flew through a 65 m course (toward 280°) near Cape May Point, New Jersey. I used a stopwatch to time migrants as they flew over an open field where updrafts were limited and monitored wind direction and speed using a hand-held compass and a hot-wire anemometer. Air speed, heading, and wind components parallel to flight path were determined by vector analysis. Flight speeds of migrants flying into opposing winds (from 285° at 5.1 mps) were compared with those of migrants flying with following winds (from 35° at 4.4 mps).

Sharp-shinned Hawks used intermittent gliding and spent about half the time flapping. Flight varied from about 2 m to 20 m above the ground and was rarely level or straight. Birds that deviated more than 40° from straight flight were excluded from the analysis. As predicted by distance maximization theory, air speeds of migrants were significantly slower with following wind than with opposing wind (fig. 11.2; $F = 17.8$, df $= 1, 47$, $p < 0.01$). The difference between mean air speed with opposing and following winds averaged more than 2.0 mps. Ground speeds (fig. 11.2) of birds flying with following winds averaged 3.1 mps faster than with opposing winds ($F = 41.1$, df $= 1$, 47, $p < 0.01$). Seven migrants for which air speed was measured when no wind was evident averaged about 2 mps slower than migrants flying with tail winds. This may indicate that migrants flying in following and opposing winds used updrafts and generated more thrust by using power that would have been used for lift had there been no updrafts. If this is the case, the true air speeds of Sharp-shinned Hawks flying without lift may be slower than their speeds with opposing and following winds. Flight speeds of American Kestrels could be measured only with opposing winds, at which time they averaged about 2.2 mps faster than those of Sharp-shinned Hawks (fig. 11.2) flying in opposing winds. The long, pointed wings of kestrels promote faster flight than is achieved by Sharp-shinned Hawks as well as allowing powered flight for longer periods.

Speeds of migrating Merlins and Peregrine Falcons using level "flapping flight" were measured by Cochran and Applegate (1986; fig. 11.2). Air speeds of these species were slightly faster than those of Sharp-shinned Hawks but slower than those of American Kestrels, at least when the latter were flying into an opposing wind. Statistical comparisons among species are not possible at this time for three reasons. First, before a comparison of flight speeds of these migrants can be undertaken, larger samples are needed and wind conditions must be controlled, possibly by statistical means. By controlling for wind, flight speed selection can be tested for these species. Second, Cochran and Applegate did not test the hypothesis, nor did they publish wind conditions or flight directions for the birds they tracked. Finally, the distance over which these species were tracked varied. The Peregrines

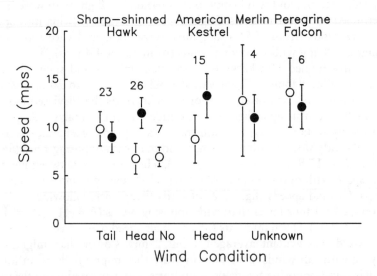

Figure 11.2 Mean (± SD) air (solid circles) and ground (open circles) speeds of four species of migrants during level, powered flight. Faster air speeds with opposing wind than with following wind for Sharp-shinned Hawks conform to predictions of distance maximization theory.

and Merlins tracked by Cochran and Applegate traveled farther than the Sharp-shinned Hawks and American Kestrels studied near Cape May. If the Peregrines or Merlins made slight turns during their flight they would have traveled farther than the measured (straight line) distance, and air speeds would have been faster. It is likely that air speeds reported for these two species are underestimates of their actual speeds. It is also possible that the speeds reported for Sharp-shinned Hawks and American Kestrels are not comparable because the birds may have used updrafts.

Flight Speed Selection during Gliding Flight

Theory

Because energy for flight is independent of air speed during gliding, the process of selecting a glide speed differs from that involved in powered flight. A gliding migrant must decide how fast it should fly based upon its sink rate at various air speeds and the lift available. In still air the bird will travel the greatest distance by gliding at its maximum glide ratio (maximum L/D or best glide) air speed. Maximum L/D is determined (as in chaps. 5 and 6) by drawing a line tangent to the glide polar from the origin. When there is no horizontal or vertical wind, a gliding bird should fly at or near its maximum glide ratio air speed. Still air is rare and is synonymous with poor soaring conditions, so migrants do not fly in these circumstances. To maximize distance during gliding flight a migrant must adjust air speed to prevailing updraft conditions. When updrafts are abundant and strong (the two are usually correlated) a migrant should fly faster than when updrafts are weak and scarce. Lift is paramount for selecting flight speed. Sailplane pilots use this strategy during cross-country flight. The reason for this should be obvious. When thermal or other updrafts are strong and abundant they are easy to locate. Furthermore, climb rates are faster so a migrant spends less time climbing and more time gliding toward its goal. A migrant should never fly slower than its best glide air speed except when soaring in thermals or other lift or when searching for lift, when sink rate must be minimized.

A graphic model shows how glide speeds should be selected to realize the maximum distance possible during cross-country flight. Lift and sink rates are given as the y-axis and air speed as the x-axis (fig. 11.3). Glide polars in figure 11.3 are for a soaring bird and a sailplane. To determine how fast a glide should be, a line is drawn tangential to the polar from the y-axis. This line is called a MacCready

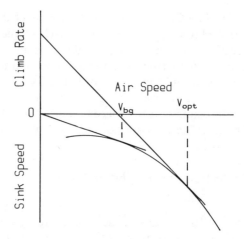

Figure 11.3 Glide polar of a hypothetical bird showing strategy for optimal selection of speed during interthermal and ridge gliding. As lift increases a migrant should increase its air speed. Tangents drawn from the *y*-axis (climb rate) that determine the optimal interthermal speed are called MacCready tangents.

tangent.[1] The location on the *y*-axis where the tangent originates is equal to the rate of climb within a thermal. Thus, as lift increases the correct air speed also increases. With "zero" lift (fig. 11.3) a bird or sailplane should fly at air speed V_{bg}, air speed at best glide. With some positive amount of lift V_{opt} is the best air speed at which to fly. By using these air speeds a bird realizes the maximum distance possible during migratory flight. V_{bg} and V_{opt} are best air speeds to use given the available lift. A near linear relation exists between lift and correct air speed.

Sailplane pilots usually glide slower than the MacCready speed ring indicates because the chances are small of finding a thermal as strong as the one they are in. By flying slower, they extend their glides to

1. Paul MacCready, the man for whom the MacCready tangent is named, is a pioneer in the field of light aircraft design. His achievements include the design of the Gossamer Condor, the first human-powered aircraft; the Gossamer Albatross, the first human-powered aircraft to cross the English Channel; a life-sized (5.5 m wing span) flying model of a pterosaur (called "QN" for *Quetzalcoatlus northropi*); and the MacCready speed ring, a "calculator" that enables sailplane pilots to determine MacCready tangents while in flight. He has received the coveted Kremer Prize (on two occasions) and the Engineer of the Century Award and was the 1956 International Soaring Champion. For more information on MacCready and his inventions see the July 1986 issue of *Science 86*.

increase the chance of locating a strong thermal (Reichmann 1978). Raptors that fly in flocks may not have to fly slower, because they can use smaller thermals and can locate thermals more readily than sailplane pilots.

It is easy to see that air speed selection based on lift varies as characteristics of the glide polar change. Glide polars of birds are steeper than those of sailplanes. That is, the rate of sink increases with air speed more rapidly for a bird than for a sailplane. A corollary is that at faster air speeds a sailplane sinks less slowly than a bird (see fig. 6.9 and table 6.4), permitting sailplanes to use much faster air speeds during interthermal glides. This is a result of different flight morphology. For a bird small changes in lift translate to small changes of air speed. Selection of air speed for a migrating bird is therefore different from that for a sailplane. Furthermore, as a bird loses weight (fat) during migration, its glide polar characteristics change, necessitating changes in behavior associated with air speed selection.

Empirical Findings: Ridge-Gliding Flight

Two tests of the hypothesis that hawks adjust air speed to maximize distance during ridge-gliding migration are described in this section. The tests involve migrants gliding along the Kittatinny Ridge at Raccoon Ridge, Blairstown, New Jersey, and at Hawk Mountain, Kempton, Pennsylvania, about 100 km to the southwest of Raccoon Ridge. The ridge at both sites is vegetated and nearly identical in physiognomy and orientation (toward 250–260°). Thousands of hawks migrate along the ridge during autumn. These migrants use continuous gliding flight, gliding with intermittent soaring flight, gliding with flapping bouts varying in duration (partially powered glides), and continuous flapping flight. Flight behavior changes frequently.

Flight speeds at Hawk Mountain were measured by Broun and Goodwin (1943) along a one-half mile course (about 800 m) using stopwatch and telephone. The study was one of the first to measure flight speeds of migrating birds. Broun and Goodwin included in their paper raw data for ground speed, wind speed and direction, height of flight, and type of flight used by most migrants. They did not analyze their data statistically.

With the aid of an assistant I measured flight speeds of migrants at Raccoon Ridge using a stopwatch and two-way radios. We noted height above the ground and type of flight used for migrants passing through a 330 m course. Migrants flying at more than 30 m above the ridge were eliminated from the analysis, as were migrants that made turns, changed altitude, or engaged in activities other than migration (e.g., landing, mobbing other hawks, chasing prey). High-flying migrants were excluded because parallax problems jeopardized the accuracy of the measurements. Wind speed and direction were measured at 3 m above the vegetation with a hot-wire anemometer (model W141, Weather Measure Corporation, Sacramento, CA) and compass.

Air speed was determined by vector analysis from measured ground speed, track direction of the bird (the same as ridge orientation), and wind speed and direction. Lift was determined by calculating the component of the wind perpendicular to the ridge. This is not to say that lift is strictly equal to the wind component perpendicular to the ridge. Furthermore, winds parallel to the ridge also produce updrafts because the ridge is not strictly linear and because of discontinuities in the ridge caused by topography and vegetation. I examined the relation between lift and air speed for Sharp-shinned Hawks, Red-tailed Hawks, Ospreys, and Broad-winged Hawks. Air and ground speeds of other species are also presented, although small samples preclude statistical analyses. To test the flight speed selection hypothesis, air speed was regressed on lift. If air speed is selected in a manner consistent with distance maximization theory, regression coefficients (slopes) will be positive.

Hawks migrating along the Kittatinny Ridge did not always use gliding flight. Broun and Goodwin (1943) measured ground speeds of 137 individuals of eight species and noted whether migrants used gliding, gliding and flapping, flapping and gliding (both power gliding), or flapping flight. Kerlinger (1982a and unpublished) repeated these procedures at Raccoon Ridge ($N = 166$ individuals, 8 species). The findings of both studies (table 11.1) revealed that most species incorporated some flapping in their gliding flight. Broad-winged Hawks, Ospreys, and Red-tailed Hawks used flapping flight of any type less than 45% of the time, whereas Sharp-shinned Hawks, Cooper's Hawks, Red-shouldered Hawks, Northern Harriers, and American Kestrels mixed flapping with gliding more than 45% of the time. This behavior was also noted by Haugh (1972). Sharp-shinned Hawks had the most varied flight along the Kittatinny Ridge, although Northern Harriers and Cooper's Hawks also varied their flight behavior. A graphic presentation of the Sharp-shinned Hawk data shows a significant relation between lift (wind speed perpendicular to the ridge) and the amount of gliding flight used by migrants (fig. 11.4). As lift increased Sharp-shinned Hawks used less powered flight, presumably reducing the energy cost of migration. In a similar study, Kerlinger

Table 11.1 Summary of Locomotory Behavior during Ridge-Gliding Flight at Hawk Mountain, Pennsylvania, and Raccoon Ridge, New Jersey

Species	N	Flight Mode[a]			
		Flapping	Flap and Glide	Glide and Flap	Glide
Sharp-shinned Hawk	127	21 (16.5%)	19 (15.0%)	17 (13.4%)	70 (55.1%)
Cooper's Hawk	14	6 (42.9%)	3 (21.4%)	2 (14.3%)	3 (21.4%)
Broad-winged Hawk	42	0 (0.0%)	1 (2.4%)	1 (2.4%)	40 (95.2%)
Red-tailed Hawk	76	0 (0.0%)	3 (3.9%)	13 (17.1%)	60 (78.9%)
Red-shouldered Hawk	8	5 (62.5%)	0 (0.0%)	2 (25.0%)	1 (12.5%)
Osprey	31	1 (3.2%)	3 (9.7%)	7 (22.6%)	20 (64.5%)
Northern Harrier	8	1 (12.5%)	2 (25.0%)	2 (25.0%)	3 (37.5%)
American Kestrel	17	4 (23.5%)	5 (29.4%)	2 (11.8%)	6 (35.0%)

[a]Pooled data from Hawk Mountain (Broun and Goodwin 1943) and Raccoon Ridge (unpublished data by author). Flap and glide flight is characterized by more flapping than gliding, whereas glide and flap flight is characterized by more gliding than flapping.

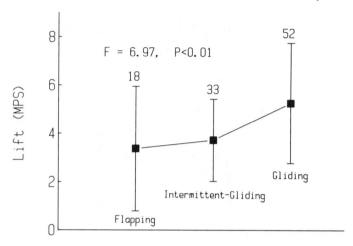

Type of Flight

Figure 11.4 Type of flight used by Sharp-shinned Hawks during ridge gliding migration plotted against lift (mean ± wind component perpendicular to the ridge).

and Gauthreaux (1984) noted that this species used less flapping when flying at high altitudes than at lower altitudes. The reduced flapping by Sharp-shinned Hawks in stronger updrafts promotes greater distances per unit of energy expended.

Average ground speeds of hawks flying along both ridges ranged from about 11 mps for Sharp-shinned Hawks, American Kestrels, and Northern Harriers to over 15 mps for Red-shouldered Hawks, Red-tailed Hawks, and Ospreys (table 11.2). Air speeds for Red-shouldered Hawks were the slowest, averaging 12.2 mps, whereas those of Ospreys and Cooper's Hawks averaged over 16 mps. A statistical comparison of air speeds among species is not possible because wind and lift differed at the time of flight; amount of flapping flight varied among species; and samples were small. Hereafter, only the four most numerous species will be discussed.

Mean air speeds for Broad-winged Hawks, Red-tailed Hawks, Sharp-shinned Hawks, and Ospreys were faster than air speed at best glide by over 1–2 mps (one-sample t-tests, ts = 3.93–9.25, df = 29–109, p < 0.01). (Air speeds at best glide for these species are given in table 6.4.) These results are consistent with predictions of distance maximization theory: when lift is available air speed should be faster than at best glide. By flying faster than air speed at best glide when

Table 11.2 Summary of Air Speeds and Ground Speeds of Eight Species of Hawks during Ridge-Gliding Flight at Hawk Mountain, Pennsylvania, and Raccoon Ridge, New Jersey

Species	Air Speed (mps)				Ground Speed (mps)		
	N	Mean	±SE	Range	Mean	±SE	Range
Sharp-shinned Hawk							
Hawk Mountain	37	16.4	1.0	8.9–32.9	13.5	0.9	7.2–26.9
Raccoon Ridge	74	13.2	0.2	8.0–19.6	9.5	0.3	5.5–14.7
Combined	111	14.2	0.4	8.0–32.9	10.8	0.4	5.5–26.9
Cooper's Hawk[a]	15	16.6	1.0	10.7–25.8	13.1	1.0	9.4–24.7
Broad-winged Hawk							
Hawk Mountain	8	13.7	0.6	12.1–16.2	14.2	1.1	8.9–17.9
Raccoon Ridge	30	13.7	0.5	10.5–24.5	10.4	0.7	6.1–22.8
Combined	38	13.7	0.5	10.5–24.5	11.1	0.7	6.1–22.8
Red-tailed Hawk							
Hawk Mountain	54	16.0	0.3	11.7–20.8	13.0	0.3	9.0–17.9
Raccoon Ridge	19	14.7	0.6	10.8–21.1	13.2	0.8	8.9–20.9
Combined	73	16.6	0.3	10.8–21.1	13.1	0.3	8.9–20.9
Red-shouldered Hawk[a]	8	12.2	1.0	13.2–17.1	15.5	0.5	8.0–15.2
Osprey							
Hawk Mountain	15	18.3	1.7	9.4–33.3	16.8	1.3	9.1–25.5
Raccoon Ridge	15	14.2	1.0	10.4–24.4	10.8	1.1	7.9–22.8
Combined	30	16.2	1.0	9.4–33.4	13.8	1.0	7.9–25.5
Northern Harrier[a]	7	12.9	0.8	10.3–16.6	10.8	1.5	6.5–17.0
American Kestrel[a]	15	14.4	0.8	11.2–21.5	10.8	0.7	7.0–16.1

Source: Hawk Mountain data from Broun and Goodwin 1943; see text for method of determining air speeds. Raccoon Ridge data from Kerlinger 1982a and unpublished material.

[a]Data from Hawk Mountain and Raccoon Ridge were pooled because of small sample sizes.

updrafts are present, migrants travel farther than if they had flown at slower air speeds.

Linear regression of air speed on lift revealed that Broad-winged Hawks, Red-tailed Hawks, Sharp-shinned Hawks, and Ospreys increased air speed as lift increased. Regression coefficients (slopes) for species at both locations were positive, although not always significant (table 11.3). At Raccoon Ridge lift explained between 5% and 70% of the variance of air speed, whereas at Hawk Mountain this variable explained between 1% and 60% of the variance. Regression coefficients varied from 0.20 for Sharp-shinned Hawks to 0.99 for Ospreys and Broad-winged Hawks at Raccoon Ridge and from 0.51 for Red-tailed Hawks to 2.38 for Ospreys. Combining data from the two sites resulted in positive, significant regressions for all species with coefficients of determination ranging from 0.07 to 0.51 (7% to 51% of the variance). Results were consistent with the hypothesis that these migrants select air speeds that allow them to travel a greater distance than if flight speeds were random or independent of lift. The large variance that was unexplained is interesting and merits further study.

Although wind gusts and variation in lift may explain part of the variance in the relations above, some of the large variation of air speed reflects variance in behavior among individuals. As demonstrated earlier in this section, hawks flying in ridge lift do not use gliding flight solely. That is, some individuals incorporate varying amounts of flapping, which for Sharp-shinned Hawks was correlated with the strength of ridge lift. To maximize distance, raptors that use partially powered glides must select speeds based upon both lift and wind component parallel to the flight path (following/opposing winds). By using multiple regression analysis with lift and following/opposing wind component as independent variables, and air speed as the dependent variable, Kerlinger (1982a) showed that air speed was related positively to updrafts (essentially the same as reported above) and negatively to following/opposing wind component (table 11.4). Lift was the more important predictor, explaining about two times as much variance as following/opposing wind. Together the variables explained 27% to 64% of the variance for Broad-winged Hawks, Ospreys, and Sharp-shinned Hawks. The addition of the head/tail wind increased the explained variance by 12% for Sharp-shinned Hawks, 13% for Ospreys, and 8% for Red-tailed Hawks.

Only the Red-tailed Hawk did not select air speeds as predicted by distance maximization theory. Why? To examine this problem, the migrant's height above the ground was included in a multiple regression analysis as a categorical variable (1 = low, 2 = moderate, 3 = high).

Table 11.3 Linear Regression Analyses for Air Speed Selection during Ridge Gliding

	Sharp-shinned Hawk	Broad-winged Hawk	Red-tailed Hawk	Osprey
Raccoon Ridge, NJ				
N	74	30	19	15
r^2 (r)	0.05 (0.22)[a]	0.50 (0.70)**	0.19 (0.44)[a]	0.23 (0.48)[a]
Equation	y = 0.20x + 12.22	y = 0.98x + 10.22	y = 0.71x + 9.96	y = 0.99x + 10.68
Hawk Mountain PA				
N	37	7	54	15
r^2 (r)	0.40 (0.63)**	0.01 (0.10)[b]	0.17 (0.41)**	0.60 (0.78)**
Equation	y = 1.45x + 10.22	—	y = 0.51x + 13.77	y = 2.38x + 5.41
Combined				
N	111	37	73	30
r^2 (r)	0.15 (0.39)**	0.45 (0.67)**	0.07 (0.27)*	0.51 (0.71)**
Equation	y = 0.70x + 11.07	y = 0.94x + 10.36	—	y = 1.86x + 7.94

[a] p = 0.06.
[b] p > 0.10, two-tailed test.
*p < 0.05; **p < 0.01.

Table 11.4 Summary of Multiple Regression Analyses for Tests of the Air Speed Selection/Distance Maximization Hypothesis for Ridge-Gliding Migrants

Species	N	Independent Variables			R^2_1	R^2_2	R^2_3	Total R^2	R
		Lift[a]	Wind[b]	Altitude[c]					
AIR SPEED									
Sharp-shinned Hawk	111	0.46*	−0.36*	—	0.15	0.12	—	0.27	0.52
Broad-winged Hawk	37	0.77**	−0.13	—	0.45	ns	—	0.45	0.67
Red-tailed Hawk[d]	73	0.27*	−0.33*	—	0.08	0.07	—	0.15	0.39
Red-tailed Hawk[e]	43	0.58*	−0.31*	0.41*	0.27	0.05	0.23	0.41	0.56
Osprey	30	0.74**	−0.36*	—	0.51	0.13	—	0.64	0.80
GROUND SPEED									
Sharp-shinned Hawk		0.29*	0.14	—	0.11	ns	—	0.11	0.32
Broad-winged Hawk		0.71**	0.02	—	0.54	ns	—	0.54	0.73
Red-tailed Hawk[d]		ns	0.40*	—	ns	0.15	—	0.15	0.39
Red-tailed Hawk[e]		ns	0.34*	0.56*	ns	0.06	0.38	0.44	0.66
Osprey		0.70**	0.24*	—	0.51	0.06	—	0.57	0.75

[a,b,c]Variables are explained in text—statistics given are standardized regression coefficients. Variance accounted for by the independent variables is given in columns marked by subscripts, 1, 2, and 3, which correspond to lift, wind, and altitude respectively.

[d]Analysis conducted with entire data set.

[e]Analysis conducted with part of data set for which altitude was known.

* $p < 0.05$; ** $p < 0.01$.

Some of the Red-tailed Hawks for which ground speed was measured by Broun and Goodwin (1943) were flying at "very high" altitudes, undoubtedly higher than the permissible 30 m maximum at Raccoon Ridge. Including altitude as a third variable changed the relations among variables. The three variables explained 56% of the variance of air speed (table 11.4). Height and lift were related positively to air speed, whereas the head/tail wind component was related negatively. Height of a migrant above the ridge is probably a result of the strength of lift. These results suggest that Red-tailed Hawks, like other migrants, adjust air speeds during ridge flying in a manner that approximates a distance maximization strategy.

In summary, lift was the most important predictor of air speed among ridge-flying migrants, whereas the following/opposing component of the wind is less important. By increasing air speed as lift and opposing wind increase, hawks seem to be maximizing the distance they travel, thereby reducing the time and energy necessary for migration.

Glide ratios of ridge-gliding migrants varied with air speed. Although the glide ratios (with respect to the air) of the migrants listed above ranged from about 8:1 for Sharp-shinned Hawks (for partially powered glides) to about 10:1 for Ospreys, these birds probably realized greater ratios (with respect to the ground) by gliding continuously in lift. In theory, these migrants could realize glide ratios of more than 100:1 (ground speed to sink rate), which is one reason they use ridge lift.

Thermal-Gliding Flight

Air and ground speeds of hawks gliding between thermals were measured with a tracking radar by Kerlinger (1982a) near Albany, New York. During three spring and two autumn migrations, air speeds (tracks >30 sec) of individuals of nine species were measured. Average speeds were faster (table 11.5) than those of ridge-flying migrants by over 4 mps for most species. Similar interthermal air speeds have been reported by observers using motor gliders to follow migrants in Massachusetts and Connecticut during autumn (Broad-winged Hawks, $N = 11$, $\bar{x} = 19.5$ mps, range $= 17.0–26.0$ mps; Osprey, $N = 5$, $\bar{x} = 21.6$ mps, range $= 18.4–26.0$ mps; Sharp-shinned Hawks, Red-tailed Hawks, and Turkey Vultures when pooled, $N = 6$, $\bar{x} = 22.2$, range $18.4–26.0$ mps; Hopkins et al. 1979). Elsewhere these researchers report similar interthermal speeds (Hopkins 1975; 22.4 mps for Broad-winged Hawks), although Welch (1975) reports interthermal glide speeds of only 18–20 mps for Broad-winged Hawks. Leshem (1987,

Table 11.5 Summary of Interthermal Air and Ground Speeds of Nine Species of Hawks Migrating in Autumn and Spring Measured with a Tracking Radar in central New York

Species	N	Air Speed (mps)		Ground Speed (mps)	
		Mean	SE	Mean	SE
Sharp-shinned Hawk	103	22.5	0.5	22.7	0.5
Cooper's Hawk	8	21.3	0.8	20.6	1.9
Goshawk	7	22.5	2.1	20.7	2.7
Broad-winged Hawk	112	24.2	0.5	23.7	0.5
Red-tailed Hawk	109	23.9	0.5	24.4	0.6
Red-shouldered Hawk	18	21.8	1.9	21.6	2.1
Northern Harrier	11	18.7	1.2	19.4	2.5
Osprey	24	24.9	0.9	23.7	0.9

personal communication) reported interthermal glide speeds of 16.8–22.4 mps for Honey Buzzards, Levant Sparrowhawks, and Lesser Spotted Eagles. These birds were migrating during autumn in Israel.

When air speed was regressed on the rate of climb within a thermal (before or after a glide) no significant relation emerged for Sharp-shinned Hawks, Broad-winged Hawks, Red-tailed Hawks, or Ospreys. Because a positive relation was predicted from distance maximization theory, it seems that air speeds were not selected based on the strength of thermal lift at the time of flight.

To determine if air speeds were selected by these species in a non-random manner, the speeds used by migrants were compared with the ranges of speeds at which migrants can fly (chap. 6). If a migrant does not select an air speed, a uniform (possibly random) frequency distribution of speeds is expected. The distributions were significantly non-uniform (table 11.6), with speeds clustered at the fast end of the distribution of air speeds. Migrants avoided flying at speeds near the air speed at minimum sink and best glide (L/D_{max}). Mean air speeds for interthermal glides averaged about 10 mps faster than the air speed at minimum sink.

That interthermal air speeds were faster than speed at best glide or minimum sink was not surprising and was consistent with predictions of distance maximization theory. It was surprising that interthermal air speeds were not related to thermal strength. How may these discrepancies be resolved? Perhaps hawks use an alternative strategy when gliding between thermals that allows them to migrate distances that are not much shorter than if they adjusted air speeds to lift conditions. A post hoc analysis revealed that mean air speeds of Broad-

Table 11.6 Tests of the Hypothesis That Migrants Select Interthermal Glide Speeds in a Nonrandom Fashion

Species	N	df	Chi-square
Sharp-shinned Hawk	103	21	70.08**
Broad-winged Hawk	112	23	83.03**
Red-tailed Hawk	109	20	43.96**
Osprey[a]	24	2	6.63*

Note: Counts in each category (an air speed, e.g., 16 mps) in the chi-square analysis represent the measured flight speed of one migrant.

[a]Sample sizes for Ospreys were too small to conduct a chi-square analysis using 1 mps as the category size. Three cells were constructed as follows: 15–20 mps, 21–26 mps, >27 mps.

*$p < 0.05$; **$p < 0.01$.

winged Hawks, Red-tailed Hawks, Sharp-shinned Hawks, and Ospreys were within 1–2 mps (table 11.7) of the speeds predicted by distance maximization theory (a MacCready tangent) if migrants were selecting interthermal speeds based on average lift conditions. Thus, it appears that hawks migrating at high altitudes from thermal to thermal do not adjust speed to lift conditions at the time of flight but fly at speeds (and glide angles) expected if thermal conditions were averaged. By doing so they may not travel as far as if they had fine-tuned their behavior, but they fly farther than if there were no selection of speed or if they flew at speeds that were close to air speed at best glide.

Glide ratios (air speed to sink rate, determined from glide polars) of these species were lower than during ridge-gliding migration.

Table 11.7 Summary of Realized and Predicted Best Air Speeds for Migrants to Use Based on MacCready Tangent Derived from Average Lift Conditions (Climb Rate) from Times When Hawks Were Migrating

Species	N	Mean Climb Rate ± SD (mps)	Predicted Air Speed (mps)[a]	Realized Air Speed Mean ± SD (mps)	Mean Length of Glide ± SD (sec)
Sharp-shinned Hawk	103	2.7 ± 0.8	23–24	22.5 ± 4.7	49 ± 24
Broad-winged Hawk	112	3.2 ± 1.2	24–25	24.2 ± 4.8	48 ± 29
Red-tailed Hawk	109	2.9 ± 1.1	26–27	24.9 ± 5.0	52 ± 29
Osprey	24	3.1 ± 1.1	26–27	24.9 ± 4.4	55 ± 36

[a]Derived from glide polars for the species presented in Kerlinger 1982a.

Ospreys performed the best with an average glide ratio of about 8.5:1, whereas Sharp-shinned Hawks flew at about 7:1 using some partially powered glides. Glide ratios during interthermal glides for Broad-winged Hawks reported by Hopkins et al. (1979) were 9–9.5:1, which is close to the 8.1:1 ratio determined from tracking radar studies (Kerlinger 1982a). Welch (1975) and Hopkins (1975) reported glide ratios between 16:1 and 25:1 for Broad-winged Hawks, which probably were glide ratios over the ground (not through the air). These are not unreasonable when considering Pennycuick's (1972a) observation of a vulture realizing a glide ratio of 60:1 (ground speed to sink rate) using "dolphin flight" over the Serengeti.

Schmid, Steuri, and Bruderer (1986) also tested the hypothesis that migrating hawks approximate a distance maximization strategy. Data gathered with tracking radar in a valley in the Swiss Alps showed that European Sparrowhawks and Common Buzzards adjusted air speeds to thermal strength at the time of flight. Migrants flying later in the season realized slower climb rates in thermals and adjusted their air speed accordingly by slowing air speeds and reducing glide angle. Glide ratios for these species were slightly greater than for North American congeners, ranging roughly between 12 and 22. Many of these birds were probably using partially powered glides or were gliding in updrafts. Most birds flew between 10 and 12 mps (air speed) during glides. One reason for the slower air speeds and higher glide ratios for Schmid's birds is that they were flying at lower altitudes than many of the hawk migrants studied in North America. At less than 400 m above the ground, these migrants were not much higher and realized similar air speeds (table 11.2) to ridge-gliding raptors in northeastern North America.

Pennycuick, Alerstam, and Larsson (1979) demonstrated that migrating Common Cranes flew faster between thermals when lift was strong than when it was weak. It is likely that this behavior is widespread among soaring migrants. They reported air speeds of between 15 and 26 mps for glides and partially powered glides by Common Cranes.

Studies of foraging African vultures and hawks by Pennycuick (1971a, 1972a) are of importance to studies of flight speed selection of migrating hawks. Speeds for the Martial Eagle were only 14.0–16.5 mps, but this species can undoubtedly fly faster during interthermal glides. One individual observed by Pennycuick (1972a) had its feet extended downward and its head down as if looking for prey. This behavior slows glide speed by several mps in addition to increasing sink rate. Tawny Eagles usually flew at less than 18 mps, whereas vultures (African White-backed Vulture and Ruppell's Griffon) flew between 18 and 23 mps. These speeds are slightly slower than inter-thermal air speeds of the migrating raptors reported above. Foraging raptors are not attempting to maximize distance. Instead, they must reduce sink rate to search a large area and stay aloft as long as pos-sible. Therefore they should avoid fast interthermal glides. Pennycuick stated that during some (presumably longer) cross-country flights vul-tures adjusted air speed to lift conditions, a finding that is consistent with distance maximization theory and the behavior of some migrat-ing raptors.

Discussion

Flight speeds of migrating hawks were adjusted in a manner consistent with a distance maximization theory in three different flight situa-tions. Flapping migrants flew faster with opposing winds than with following winds; ridge-gliding migrants flew faster with strong lift than with weak lift and faster with opposing winds than with follow-ing winds; and migrants gliding between thermals flew faster than migrants gliding in ridge lift and flew close to the air speed predicted by distance maximization theory if lift conditions were averaged. Nat-ural selection has probably shaped the behavior of these migrants such that they fly at speeds that promote a quick and energy-efficient journey.

Several questions must be addressed before we leave the topic of flight speed selection. These questions pertain to flight speed selection mechanisms of raptors versus other groups of migrants; the large amount of variance of air speed not explained by the predictor vari-ables; and the differences between the flight speeds of raptors and non-raptorial migrants.

The Mechanism of Selecting a Flight Speed

By what mechanism(s) do raptors determine and adjust air speed? I do not mean to imply cognitive processes, although some "thought" must occur (Griffin 1984). Instead, I refer to the cues migrants use to assess their rate of travel. The theory of maximizing distance does not consider mechanisms by which organisms determine their own speed and make adjustments. Instead, the theory predicts only how birds should adjust flight speeds to fly the greatest distance with a given set of atmospheric conditions.

The primary mechanism by which a bird can judge its rate of travel is its position in relation to landmarks (except during long flights over water). By maintaining a constant speed over the ground with follow-ing and opposing winds, a powered migrant will change its air speed in a manner that approximates predictions of distance maximization theory. Avoidance of powered flight in strong opposing wind alleviates the necessity to fly at fast, energy expensive speeds. A reduction of air speed with following wind reduces the energy expended by a migrant while maintaining the same or a faster ground speed. When using powered flight raptors usually fly at low altitudes where it is possible to use landmarks to assess progress and adjust air speeds.

Gliding migrants may rely on other mechanisms to select flight speeds. In addition to movement relative to the ground, a migrant can use information such as height above the ground. Remember, at faster speeds a bird sinks faster. If a migrant sinks too quickly it will be forced to resort to powered flight, which is energy expensive. This sort of mechanism may explain the discrepancy between the flight speed selection behavior of ridge-gliding and thermal-gliding migrants and between the birds tracked by Kerlinger and by Schmid.

Migrants gliding in ridge lift need only maintain a level course in relation to a ridge to behave in a manner consistent with predictions of distance maximization theory. By flying faster in strong lift the mi-grant avoids flying too high and, perhaps, above the region of strong-est lift. Conversely, by flying slow in weak lift the migrant avoids sink-ing too quickly and flying too close to the ridge, thereby leaving the area of strongest lift or entering surface turbulence. If ridge-gliding migrants simply maintain level flight over the ridge, does it mean that they are not selecting air speeds correctly? Because these migrants se-lect flight speeds nonrandomly and because they travel farther than if speed was not adjusted correctly, I believe the answer is yes. Whatever the mechanism, the behavior is consistent with distance maximization theory. Before alternative explanations of the mechanism for selection of a flight speed can be tested, we need a knowledge of lift and tur-bulence patterns along ridges.

A similar mechanism for flight speed selection by migrants gliding between thermals may be invoked. By adjusting air speed in relation to altitude, a migrant selected flight speed as predicted. As migrants glide lower they should fly at slower air speeds, realizing slower sink rate and forestalling powered flight. Flight at lower altitude occurs when thermals are weak and scarce (more time spent at low altitudes) and when migrants have been gliding a long time after leaving a ther-mal. At either time the migrant should slow its air speed to reduce sink rate.

The discrepancy between the results of Kerlinger and those of Schmid may be reconciled by the difference in the migrants' altitude between the studies. Migrants tracked by Kerlinger in central New York did not adjust flight speed according to lift conditions, whereas those tracked by Schmid did. Schmid's migrants rarely flew as high as those Kerlinger tracked, however. Because Kerlinger could not track raptors below about 300–400 m, air speeds were not available for these migrants. Perhaps migrants flying at high altitudes have difficulty judging altitude and therefore fly within a narrow range of air speeds. At lower altitudes it is easier to assess altitude changes, and flight speed can be adjusted accordingly. This difference may explain the discrepancy between findings from New York and Switzerland. If this is the case, birds flying at low altitudes in New York are predicted to fly slower.

My observations of migrants using thermaling flight between 50 m and about 300 m above the ground in New York, New Jersey, Texas, and Michigan suggest that low flying migrants fly slower than higher flying migrants. Although I have not measured the speeds of these migrants, their wing and tail planform reveals relative air speed. At low altitudes hawks do not sweep their wings back as far as during high altitude flight, so they must by flying slower. They appear to decelerate as they descend by gradually opening their wings, resulting in an increasing glide ratio. At still lower altitudes they sometimes reduced sink rate even more by incorporating flapping (powered glides), as is seen among Common Cranes (Pennycuick, Alerstam, and Larsson, 1979), Sharp-shinned Hawks (fig. 11.4; Kerlinger and Gauthreaux 1984), and other gliding birds.

Thus, migrating raptors adjust air speed in a manner that allows them to approximate a distance maximization strategy. The observed behavior differs slightly from the predicted behavior and may be related to the aerodynamics, energetics, and perceptual (cognitive?) abilities of these migrants. Kerlinger calculated that the reduction in distance traveled between the observed behavior and a strict distance maximization strategy amounted to a difference of less than 10%. By averaging the strength of thermal lift over the array of conditions during which migration occurs, raptors may be demonstrating long-term optimization (sensu Katz 1974). Selecting the optimal flight speed and glide ratio for particular atmospheric conditions is a complex task. Sailplane pilots use programmed calculators (or a MacCready speed ring) to assist them. Migrants do not have this luxury. They must make many decisions simultaneously, only one of which is selecting an air speed. It may be that a simple, long-term optimization strategy is easier and more dependable than a strategy in which speed is fine-tuned to lift conditions.

Because coefficients of determination were low in some ridge-gliding analyses, the selection of flight speeds by migrants merits further inquiry and explanation. Potential sources of unexplained variance in the air speed data include fluctuations of wind speed or direction; alternative sources of lift (or downdrafts) such as abatic and anabatic air currents; error in measurement of either flight speed or wind; and of course true (inherent) variability of behavior among individuals. The first three explanations undoubtedly account for a small yet constant portion of the unexplained variance, whereas individual behavior is probably an important source of variance. Because many of the migrants Kerlinger tracked were within their first days of autumn migration, it is likely that many birds, especially young of the year, had little migratory experience. Without such experience variance is expected to be large. Finally, some individuals may have been short distance migrants whose ancestors had not been subjected to strong selective pressure for optimizing flight speed. For this reason, long distance migrants such as Broad-winged Hawks, Lesser Spotted Eagles, Honey Buzzards, and Ospreys should be more likely to optimize flight speeds than shorter distance migrants.

Comparison of Flight Speed and Speed Selection of Raptors and Other Avian Groups

So far in this discussion I have examined flight speed selection of migrating hawks only in relation to theory. How do speeds and speed selection behaviors of migrating hawks compare with those of other groups of migrants? By comparing the flight speed and behavior of various groups we may gain insight into mechanisms—that is, how selection shapes flight speed selection and how strongly it has acted.

A graphic comparison of the flight speeds of the major groups of migrants is given in figure 11.5. For convenience I categorized all migrants into five groups: nocturnally migrating passerines, diurnally migrating passerines, shorebirds, waterfowl, and hawks. The hawks were partitioned by flight type: level, powered, ridge-gliding, and interthermal-gliding flight. Note that the speeds given for hawks are not average cross-country speeds, which are slower when soaring flight is included. Chapter 12 deals with cross-country speed of raptor migrants using thermaling, ridge-gliding, and powered flight.

For powered flight, hawks compare favorably with passerines, exceeding the speeds of most nocturnal migrants and overlapping with those of many diurnal fliers. Air speeds of some hawks using powered flight are slower than some shorebirds and waterfowl, although the speeds of falcons are likely to overlap with those of shorebirds and waterfowl. The speed of gliding hawks overlaps with the speeds of shorebirds and waterfowl and exceeds those of most passerines. The large variance within each of these groups makes generalizations tenuous.

From the perspective of energy cost, powered flight of raptors is more expensive than that of passerines, shorebirds, and waterfowl. Because they are larger than most passerines, the cost of powered flight is greater for raptors (chap. 5; Pennycuick 1975) even though they fly at about the same speeds. By switching to gliding flight these birds realize the same speeds as most migrating shorebirds and waterfowl at a fraction of the cost (ca. 0.20–0.25 using standard metabolic rate as the currency to avoid scaling difficulties). This may be why these species avoid powered flight and why flight over water is limited.

Schnell and Hellack (1979) and Tucker and Schmidt-Nielsen (1971) demonstrated that nonmigrating gulls, terns, and other birds adjusted air speed during powered flight in a manner consistent with distance maximization theory, as other researchers have done for passerines and waterfowl (Bellrose 1967; Bruderer and Steidinger 1972; Able 1977; Emlen 1974; Demong and Emlen 1978). These birds fly

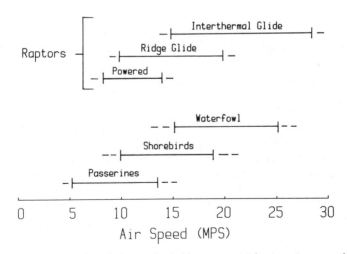

Figure 11.5 Synopsis of the flight speeds used by nonraptorial avian migrants and raptors. Dashed lines indicate speeds near the upper and lower extremes reported in the literature; solid lines represent speeds cited most often.

faster with opposing wind and slower with following wind. Kerlinger (1982b) did not find such a relation for loons migrating at high altitude over land, presumably because the loons were flying downwind. Thus, the flight speed selection behavior of many nonraptorial birds during both migratory and nonmigratory powered flight is similar to that of raptors using powered flight.

In a more detailed examination of flight speed selection Schnell and Hellack (1979) reported that the air speeds of larids during foraging flights near a breeding colony were not consistent with predicted speeds. Instead, speeds were "a compromise between those predicted for minimum metabolic rate and for minimum cost of transport over the ground." In addition, they reported a large variation in speed. These behaviors may be a result of their proximity to a nest colony. These birds are not flying long distances like migrants, so fine-tuning of behavior is not as important.

A more quantitative comparison of the flight speed selection of raptors with that of other birds is not possible, since we lack studies in which a power curve was either calculated or determined empirically. Because of the convergent behavior among unrelated taxa, I conclude that behaviors associated with flight speed selection have resulted from strong selection pressure.

Summary and Conclusions

The hypothesis that raptors adjust flight speeds to maximize distance traveled was tested among migrating Sharp-shinned Hawks, Broad-winged Hawks, Red-tailed Hawks, and Ospreys by measuring their flight speeds with different wind and lift conditions. As predicted by distance maximization theory, migrating hawks used faster air speeds when flying into the wind than when flying with the wind during level, powered flight (often mixed with intermittent gliding). Ridge- and thermal-gliding migrants also performed as predicted by distance maximization theory by flying faster than air speed at best glide and by flying faster with strong lift than with weak lift. Mean air speeds of migrants flying at high altitudes between thermals was within 1–2 mps of that predicted from a MacReady tangent generated from average lift conditions (thermal strength) encountered by migrants. These behaviors permit migrating raptors to travel farther than if flight speeds were random or independent of lift.

Mean air speeds of Sharp-shinned Hawks, American Kestrels, Merlins, and Peregrine Falcons ranged from 9 mps to nearly 13 mps during level powered flight. Eight species of migrants gliding in ridge lift averaged between 11 and 16 mps; lower than air speeds of the same migrants gliding between thermals, which averaged 17–24 mps. The air speed of raptor migrants using powered flight was faster than that of migrating passerines but slower than that of migrating waterfowl or shorebirds. Air speeds of raptors gliding between thermals compared favorably with the speeds of waterfowl and shorebirds, whereas during ridge gliding raptors flew slightly slower. Raptors rarely resort to powered flight because it is more energetically expensive and slower than gliding flight.

12

Daily Flight Distance: Simulations and Data

Selecting an appropriate flight speed allows birds to complete migration safely and swiftly using a minimum of energy. By selecting the correct speed a migrant can maximize distance flown, thereby reducing the number of days it must fly. But how far do raptors travel in a given day? To date, little is known about the daily flight distances of migrating birds. Radiotelemetry has permitted a few researchers to follow migrating birds for several days or even weeks. Much of the information is unpublished, although data are available for geese (Wege and Raveling 1984), thrushes (*Hylocichla*; Cochran, Montgomery, and Graber 1967), and a few other migrants. Empirical determinations of daily flight distance for hawks are based on a small number of same- or next-day recoveries of banded migrants, migrants equipped with radiotelemetry devices, or migrants followed for long periods with aircraft. Banding data are less reliable than those from telemetry or aircraft because flight time (days or hours) and flight path are unknown.

The simplest means of determining the daily flight distance of migrants is to divide the total distance traveled by the number of days in the migration season. This gives the minimum distance a migrant must fly on each day to complete migration. Both seasonal timing of migration and the location of breeding and nonbreeding sites of a species must be known to make this computation. For example, a Broad-winged Hawk that migrates from Colombia to New York (>6,000 km) in spring spends about 30 days migrating (15 March-15 April). During this time it must fly a minimum of 150 km per day. A few minimum daily flight distances are presented in table 12.1. The middle-distance migrants (Sharp-shinned Hawks and Bald Eagles) must fly more than 100 km per day, whereas the longer-distance migrants must fly more than 140 km per day. Because migrants do not migrate on all days of the migration season, daily distances given in table 12.1 are underestimates.

In the section that follows, three simulations of daily flight distance

Table 12.1 Estimates of the Minimum Daily Flight Distance of Selected Migrants

Species	Migration Distance (km)	Days	Minimum Daily Distance (km)	Days Needed If Daily Distance Equals 220 km
Sharp-shinned Hawk	2,000	30	67	9.1
Broad-winged Hawk	6,000	40	150	27.3
Swainson's Hawk	8,000	50	160	36.4
Lesser Spotted Eagle	6,000	40	150	27.3
Bald Eagle	2,500	35	74	11.4
Eleonora's Falcon	7,000	50	140	31.8

are presented: powered flight over relatively flat terrain; gliding in declivity currents along ridges; and thermal to thermal gliding at higher altitudes. Data used in the simulations are measured flight speed, altitude, climb rate in thermals, and direction of flight presented in this volume. For simplicity I have not incorporated the time migrants spend foraging or resting. Because some species take on more fat than others before or near the beginning of migration, the time needed by different species (and different individuals of the same species) undoubtedly varies. The researcher interested in simulations of the time needed for a seasonal migration of an individual would have to include these factors.

The Broad-winged Hawk was chosen for simulations of migrants gliding in ridge lift and gliding and soaring in thermal lift, whereas the Peregrine Falcon was chosen for the powered flight simulation. Flight speeds are known for these species. Daily flight distance for other species can be simulated using the data on flight parameters presented in this book. To validate the simulations, I will present empirical data from banding, aircraft, and radiotelemetry studies.

Powered Flight Simulation

Although level, powered flight is not commonly used by raptors, some species use it when updrafts are unavailable and to fly over water. Because of their wing morphology, Peregrine Falcons, Merlins, Ospreys, and Northern Harriers are more capable of powered flight than buteos, vultures, and accipiters. The speed of the Peregrine Falcon in extended powered flight (Cochran and Applegate 1986; fig. 11.2) was held constant at 12.5 mps. Flight speed selection based upon wind will not be considered in the present simulation.

Daily flight distance (D) was calculated as the product of ground speed (V_g) and time in flight (T):

$$D = (T)(V_g). \qquad (12.1)$$

Table 12.2 Values of Flight Parameters Used in Daily Migration Distance Simulations

	Type of Migration		
Parameters	Powered	Ridge-Gliding	Thermal-Gliding
BEHAVIORAL			
Air speed (mps)	12.5	Varied with lift	24.0
Heading	212°	Varied with wind	212°
Track	Varied with wind	260°	Varied with wind
Sink rate in glide (mps)	NA[a]	NA	3.0
Flight time (h)	6.0	6.0	6.0
Efficiency	1.0	1.0	1.0
ENVIRONMENTAL			
Wind speed (mps)	0,3,6,9	0,3,6,9	0,3,6,9
Wind direction	45–360°, 45° intervals	45–360°, 45° intervals	45–360°, 45° intervals
Climb rate in thermals (mps)	NA	NA	2,3,4

[a]NA = not applicable.

Ground speed and track direction of the migrant were determined by vector analysis from wind strength and direction, air speed (12.5 mps) and heading (held constant at 212°). This heading was chosen to simulate a migrant flying in eastern North America. Time in flight was 6 h, while wind speed was varied from 0 to 9 mps at 3 mps intervals and wind direction was varied according to the eight major compass points (parameters summarized in table 12.2). In a second simulation, following and opposing winds were varied between +9 mps (following) and −9 mps (opposing) at 3 mps intervals.

Results of the Powered Flight Simulation

The simulations show profound differences in distance flown with differing wind conditions (fig. 12.1). For example, flight distance was approximately 20% greater with calm conditions than with opposing winds of 3 mps and 185% greater with a 6 mps following wind than with an opposing wind of the same magnitude. Thus, flying with a following wind expedites migration, making it less expensive per unit of distance traveled. A flight speed selection strategy would alter the distance flown and reduce the energy cost of travel.

When wind from different directions is factored into the simulation, the track must be considered in addition to distance traveled (fig. 12.2). Distance flown was greatest with north and northeast winds, followed closely by winds from the east, whereas the least progress

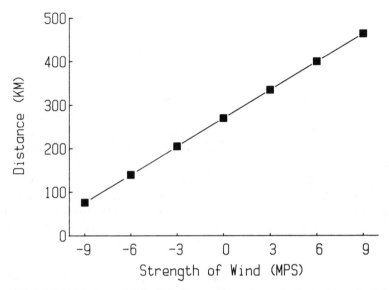

Figure 12.1 Simulations of daily flight distance for a powered migrant (Peregrine Falcon) flying in opposing and following winds for 6 h at a speed of 12.5 mps.

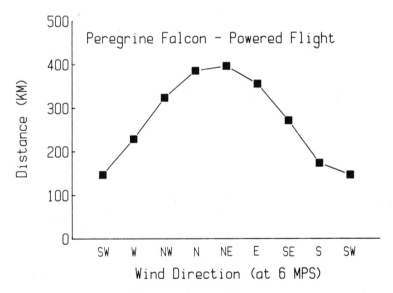

Figure 12.2 Simulation of daily flight distance for a powered migrant (Peregrine Falcon) flying in winds of 6 mps from varying directions. Flight time is 6 h at an air speed of 12.5 mps and a heading of 212°. Numbers above the symbols are realized compass directions (tracks).

was made with winds from the south and southwest, followed by west winds (fig. 12.2). Winds perpendicular to a migrant's flight path will cause drift if it does not compensate by heading into the wind. For example, a 6 mps wind perpendicular to a migrant's heading results in a 30° deviation from its heading.

The Peregrine Falcon was chosen, in part, because of its propensity for flying long distances over water. The simulations permit accurate calculation of the time needed for water crossings. A falcon can fly from Florida to Cuba, about 200 kilometers, in 4.4 h with calm conditions. With a headwind of 3 mps the flight would take only 5.9 h. Flights from Malta to Tripoli, Libya (340 km), necessitate continuous powered flight for 7.6 h, assuming there is no wind. These simulations show that bodies of water such as Delaware Bay, Chesapeake Bay, and the Great Lakes are not major barriers for raptors except with strong winds. Opposing winds of more than 6 mps make some water crossings dangerous (fig. 12.3).

When coupled with bioenergetic data on the cost of powered flight for a range of speeds, the simulations will serve for generating predictions as to when (weather and time of day) and where migrants should

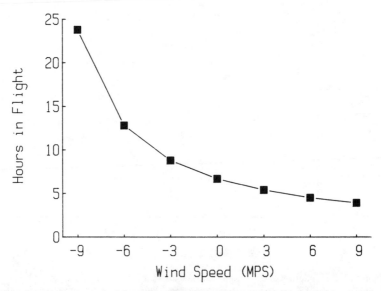

Figure 12.3 Time required for a Peregrine Falcon flying at an air speed of 12.5 mps to cross a 300 km water barrier with following and opposing winds of varying strength. Opposing winds of over 6 mps make crossings risky and energetically expensive.

cross water. In addition, predictions can also be generated as to when migrants should resort to powered flight over land.

Ridge-Gliding Simulation

To calculate the distance a bird travels using ridge lift, the bird's aerodynamic performance, its behavior, and the wind at flight time must be considered. The ultimate constraint of the simulation is the bird's aerodynamic performance, which determines the range of air speeds with associated sink rates. In the previous chapter migrants were shown to vary air speed during ridge gliding, approximating predictions of distance maximization theory. By increasing air speed as wind lift increases, these birds fly farther than if they flew at speeds not selected in this manner (i.e., random or fixed air speeds). Because lift is created by wind deflected by a ridge, the wind component perpendicular to the ridge is correlated with lift and was used here as the lift variable. The wind component parallel to the ridge (following and opposing winds) also affects ground speed. Behavioral variables include the flight speed selected (from chap. 11), the length of time a bird is in migratory flight, and the efficiency of flight (E). Time in flight includes the time from initiation of migratory flight until the bird finishes for the day (assuming uninterrupted flight). Efficiency is defined as the proportion of time the migrant spends in goal directed flight and does not include time spent foraging or resting. When a bird selects air speed or uses lift inappropriately, efficiency decreases. Activities other than migratory flight, such as pursuit of prey, perching, and mobbing a predator, detract from migration efficiency. Such behaviors are often observed among migrants but have not been quantified.

In the ridge-gliding simulations wind direction was varied from 45° to 360° at 45° intervals, and wind speed was varied from 3 to 9 mps at 3 mps intervals. Lift (L) was calculated from wind speed and direction as the wind component perpendicular to the ridge. Direction of flight (track) was toward 260°, as is the case with migrants flying along the Kittatinny Ridge in Pennsylvania. Because air speed of migrants is adjusted to lift conditions (chap. 11), air speed for Broadwinged Hawks was determined by the following relation:

$$V_a = 0.94\ L + 10.36. \tag{12.2}$$

From air speed determination, ground speed was calculated by vector analysis. (The slope of this equation must be interpreted carefully because lift is available with parallel winds, and L would be zero.)

Finally, distance traveled is the product of ground speed, time in flight, and efficiency of flight:

$$D = (V_g) (T) (E). \tag{12.3}$$

Time in flight was set at 6 h, and efficiency was assumed to be unity (an unreal assumption that birds are perfectly efficient). The ranges of values used for each parameter are summarized in table 12.2.

Results of the Ridge-Gliding Simulations

The simulations of hawks gliding along a ridge varied greatly (fig. 12.4), with wind speed and direction interacting to influence daily flight distance. A migrant flying with strong favorable winds could travel about 500 km in 6 h, whereas with unfavorable winds that same migrant might not fly more than 100 km. At slower wind speeds, wind direction seems to be less important for influencing distance flown (fig. 12.4). That is, the difference between distance flown with favorable and unfavorable winds is less with weak winds than with stronger winds. The curve in figure 12.4 is flatter for winds of 3 mps than for winds of 9 mps. West and southwest winds were least favor-

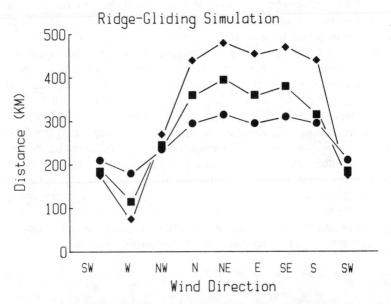

Figure 12.4 Simulation of distance flown by a Broad-winged Hawk during 6 h of ridge gliding (ridge oriented toward 255°) with winds from different directions at 3 (circles), 6 (squares), and 9 (diamonds) mps. Air speed was varied with lift (see text).

able for migrants, whereas northeast winds were most favorable (fig. 12.4). It was surprising that east and southeast winds resulted in flights that were nearly as long as with northeast winds. Although east and southeast winds are favorable for migrants, they are often accompanied by rain in the northeastern United States. By flying with a 6 mps northeast wind a migrant realizes more than three times the distance of a migrant flying with a west wind of the same magnitude. The difference between distance traveled with 6 mps and 3 mps winds from the northeast amounts to about 23%. When wind is absent, declivity currents are unavailable to migrants and ridge gliding does not occur.

The simulation is based on measurements of flight speeds of hawks flying a few hundred meters along a particular ridge. Ridges vary in slope, orientation, continuity, and vegetation and thus in the lift they provide. These factors would be reflected in the flight speeds measured at other ridges and are probably specific to a ridge. Thermals and abatic lift are often present along ridges and may augment lift generated by horizontal wind, allowing faster air speeds than predicted by equation 12.2 and greater distances than predicted by the simulation. Finally, I have assumed that the relation between wind and lift is linear, which may not be the case over a wider range of wind speeds.

Thermal-Gliding Simulation

Simulating daily flight distance for migrants using thermaling flight is more complex than for those using powered or ridge-gliding flight. The parameters needed to simulate flight for thermal-gliding migrants include climb rate within a thermal, air speed between thermals, wind speed, wind direction, efficiency of flight, time spent in flight, and heading of the bird. Not only are there more parameters, but the relations among the parameters are more complex than in the other simulations. Because a large portion of a migrant's flight is spent climbing in thermals, flight must be partitioned into time spent climbing in thermals and gliding between thermals. The time for one glide and one climb is referred to hereafter as a bout (B), comprising time spent gliding (T_g) and climbing (T_c) as follows:

$$B = T_c + T_g. \tag{12.4}$$

The time spent in these activities is determined by the altitude to which a bird climbs, its climb rate, and the rate at which it glides to the next thermal. Many migrating hawks fly at 600–800 m above ground level, with changes of about 400 m between the top of a climb and the bottom of a glide. This air space was termed the height band (H_b) in chapter 8. Thus, no gain or loss of altitude occurs from bout to bout. Time spent gliding and climbing are:

$$T_g = H_b/V_z;$$ (12.5)

$$T_c = H_b/V_v.$$ (12.6)

Values for climb rate were 2, 3, and 4 mps, corresponding approximately to the mean climb rate ($\pm$ 1 SD) achieved by Broad-winged and other hawks. Sink rate was constant at 3 mps, corresponding to the rate at an air speed of 24 mps.

Following determination of B, the horizontal distance and direction of the climb and glide were determined by addition of vectors. The horizontal distance of a climb (D_c) was the product of time spent climbing and wind speed (W_s):

$$D_c = (T_c) (W_s).$$ (12.7)

The direction of the climb is the same as wind direction, because thermals move with the wind, as do soaring migrants (chap. 7). The distance of a glide (D_g) is the product of time in a glide and ground speed:

$$D_g = (T_g) (V_g).$$ (12.8)

Ground speed and direction of the glide were determined from air speed, heading, and wind speed and direction. A heading of 212° was used, approximating those used during autumn in central New York by this species, and the same as used in the simulation of powered flight. No attempt was made to incorporate correction for lateral winds in the simulations. By doing this the amount of drift that can potentially occur has been illustrated. Distance traveled during one bout (D_b) and the resulting direction was determined by adding vectors D_c (in same direction as the wind) and D_g (in the direction of the glide; fig. 12.5).

To arrive at the final distance, the number of bouts in a day's flight is determined (by dividing the time in flight by the time of one bout) and multiplied by the distance traveled in each bout and an efficiency factor.

$$D = (D_b) (B) (E).$$ (12.9)

Flight time was 6 h, within the 4–10 h range reported for migrating hawks (Mueller and Berger 1973; Cochran 1972, 1975; Smith 1980, 1985a; Harmata 1984; Kerlinger and Gauthreaux 1985a). The efficiency factor reflects errors in behavior such as incorrect air speed, time spent searching for thermals, flight in an inappropriate direction, and time in nonmigratory activities. As in the ridge-gliding simulation, efficiency was assumed to be 1.0.

Results of the Thermal-Gliding Simulation

Summarizing the outcome of the thermal-gliding simulations entails examining distance flown (figs. 12.6 and 12.7) and the migrant's direction (tables 12.3 and 12.4). With no wind and average strength thermals the cross-country speed of a Broad-winged Hawk is only about 12 mps, or one-half the air speed during interthermal glides. This rate is equal to about 43 kph or 260 km per 6 h flight. The least favorable winds for thermal to thermal migration are west, south, and southwest, whereas the most favorable are north and northeast. Wind speed and direction interact, as can be seen by comparing the change in direction from west to northwest with 6 and 9 mps winds. With west wind increasing from 6 to 9 mps, the change in distance is minimal (2% increase), although the resultant direction changed from 182° to 165° (table 12.3). The corresponding change in distance with northwest winds is 49 km (312–361 km), equal to an increase of 16%

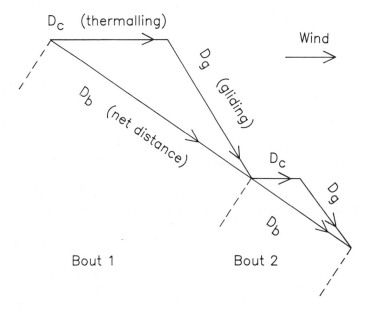

Figure 12.5 Schematic diagram of the distance and direction traveled during one bout (D_b) of gliding(D_g) and climbing in a thermal (D_c).

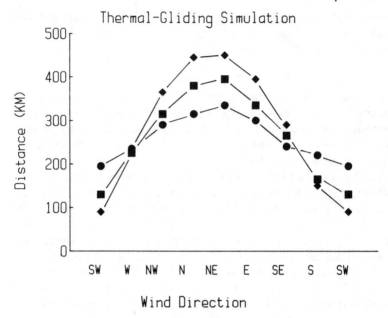

Figure 12.6 Simulation of daily flight distance of a Broad-winged Hawk during 6 h of thermal-gliding flight (see table 12.2 for parameter values) in winds from different directions. Wind speeds were 3 (circles), 6 (squares), and 9 mps (diamonds).

but a change of only 8° in the resulting vector. Clearly, as wind speed increases wind direction is more important for determining distance traveled.

Climb rate affects distance traveled by requiring that a greater portion of flight time be spent climbing when thermals are weak and less when they are strong. The influence of climb rate on distance traveled is shown in figure 12.7. The distance realized when climb rate increases from 2 to 4 mps is about 70 km for all wind directions with a 6 mps wind, equivalent to an increase of 86% of distance in a southwest wind and a 27% increase in a northeast wind. Differences in the direction of a resultant bout ranged from 2° with a northeast wind (225°) to 15° with a 6 mps south wind (table 12.3). This means that as climb rate decreases the direction deviates from the glide heading more than when climb rate is faster. From these differences a premium on aerodynamic and behavioral performance is evident, with better performance yielding greater distance toward a migratory goal.

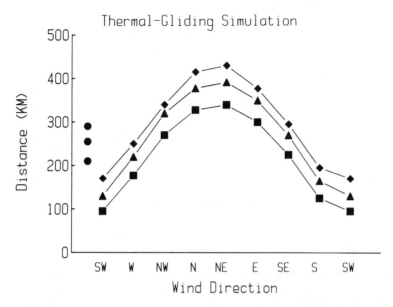

Figure 12.7 Simulation of daily flight distance of a Broad-winged Hawk during 6 h of thermal-gliding flight (see table 12.2 for parameter values) in winds of 6 mps from different directions. Climb rate of migrants was 2 mps (squares), 3 mps (triangles), and 4 mps (diamonds). The closed circles arranged vertically nearest the y-axis represent the distances for simulations in which no wind was present and climb rate was 2, 3, and 4 mps (bottom to top).

Table 12.3 Directions Realized in Simulations of Thermal-Gliding Migrants with Climb Rate Held Constant at 3 mps and Wind Varying in Speed

Direction of Wind in Simulation (from)	Direction Realized during Thermal-Gliding Flight with Wind Speed		
	3 mps	6 mps	9 mps
Southwest	208°	200°	180°
West	198°	182°	165°
Northwest	199°	188°	180°
North	206°	201°	198°
Northeast	215°	216°	218°
East	223°	231°	236°
Southeast	226°	241°	253°
South	222°	237°	260°
Southwest	208°	200°	180°

Table 12.4 Directions Realized in Simulations of Thermal-Gliding Migrants with Climb Rate Varying and Wind Speed Held Constant at 6 mps

Direction of Wind in Simulation (from)	Direction Realized during Thermal-Gliding Flight with Thermal Strength		
	2 mps	3 mps	4 mps
Southwest	192°	200°	202°
West	174°	182°	186°
Northwest	184°	188°	191°
North	200°	201°	202°
Northeast	217°	216°	216°
East	234°	231°	229°
Southeast	247°	241°	237°
South	226°	237°	232°
Southwest	192°	200°	202°

Comparison of the Flight Simulations

In the three simulations wind direction and speed profoundly influenced the distance a migrant travels in a given day. Distance also varied with wind direction and speed, with southwest and west winds being the least favorable for migration in eastern North America and north and northeast winds being most favorable for all types of migration. Although air speed for ridge-gliding migrants was sometimes one-half that of thermal-gliding migrants, the distances realized were similar because thermaling hawks lose time when climbing in opposing or side winds. Some differences were noted; with south winds, ridge migration becomes more favorable as wind speed increases. Most differences are less than with this condition.

The major difference between ridge-gliding and thermal-gliding simulations was the direction migrants realized. The direction depended on wind, heading, and climb rate. Ridge-migrating hawks cannot vary tracks, nor can they be blown off course easily. Thermal migrating hawks can vary their heading but do not have the ridge to act as a leading line. At such times they run the risk of being blown off course.

Because flight direction and distance differ between thermal- and ridge-gliding migrants with some weather conditions, a migrant must make a choice as to which type of lift to use. Without wind or thermal updrafts at ridge sites, birds should use thermal-gliding flight if convection is available. The greatest numbers of birds counted on the ridges of Pennsylvania are with west, northwest, and north winds (Haugh 1972). Results of the simulations show that hawks should use the ridge in preference to thermals with south winds. However, with west and northwest winds the advantage of flying on the ridge is flight earlier and later in the day, when thermals are not available. In addition, with strong winds from the west and northwest migrants with preferred directions to the west of south may be displaced from their migration pathways if they use thermal lift. The distance they are displaced from headings increases as climb rate decreases (time spent soaring increases) and as wind speed increases. Thus with west and northwest winds and poor climb rates, it may be to a migrant's advantage to use the ridges of the Appalachian Mountains.

Empirical Studies of Daily Flight Distance

Few researchers have attempted to quantify the distance migrants fly in one day. The paucity of measurements is related to technical difficulties and the expense of following birds for long distances. Researchers have been thwarted by topography, geopolitical borders, weather, equipment failure, and expenses. For example, an Israeli research team using a motorized sailplane to follow migrants has been forced to abandon its pursuit at the borders of Lebanon, Egypt, and Syria (Leshem 1987). Similarly, Cochran (1975) was unable to follow a radio-tagged Peregrine Falcon when it passed into Mexican airspace after being tracked from Wisconsin. Cochran (1985) has also found it difficult to follow radio-tagged Peregrine Falcons during flights over the Atlantic Ocean, and more than one bird was "lost" because the plane lacked the fuel required to continue pursuit. Simpler insurmountable problems have plagued other researchers following radio-tagged migrants. Harmata's (1984) studies of Bald Eagle migration were sometimes foiled by "logistical problems," transmitters falling off (see Cochran 1972, 1975), weather, and "a paucity of funds." Even with these obstacles, researchers have provided significant information about the daily distances migrants travel.

In the sections that follow information relating to daily flight distance is presented for studies involving aircraft, radiotelemetry, and banding recoveries. After the empirical findings are summarized, I will discuss the findings as they pertain to the time needed to complete migration and compare the findings with distances calculated in the simulation models.

Aircraft Studies

Of the three research groups that have used motor gliders or powered aircraft to follow migrating hawks, all have chosen to follow species that fly in large flocks. Before these studies Pennycuick (1971a, 1972a, 1975) developed the technique to study soaring birds in Africa. (Donald Griffin [1964] and Charles Walcott [Walcott and Green 1974] used aircraft to follow individual birds during homing experiments.) Shortly after Pennycuick's efforts the New England Hawk Watch began to use light aircraft and motor gliders to follow Broad-winged Hawks and other migrants over Massachusetts and Connecticut (Hopkins 1975; Welch 1975; Hopkins et al. 1979). Cross-country speeds of flocks of Broad-winged Hawks flying 40 to 150 km during autumn migration ranged from 37.5 to 44 kph ($N = 6$). It took 4 h for a flock to travel 150 km (37.5 kph) with winds perpendicular to the flight path at about 9 mps and slow climb rates (<1.0 mps). Similarly, it took 2.5 h to fly some 105 km (41.5 km). In an 8 h day of flight these birds would travel about 330 km if they maintained a cross-country speed of 41.5 kph. With good lift conditions (climb rates >3 mps) these migrants might average 50 kph and travel 400 km in an 8 h flight.

Estimates of cross-country speeds of Broad-winged Hawks flying through Panama are slightly slower but overlap with those reported by the New England researchers. Smith (1985a) and Smith, Goldstein, and Bartholomew (1986) give estimates of cross-country speeds ranging from 30 to 40 kph with migrants flying from about 0800 to 1800 h, or 10 h per day. Because Smith does not present his data, it is difficult to know if these estimates are for short or long flights or if they represent estimates, means, or individual measurements. Assuming these speeds are accurate, Broad-winged Hawks migrate 300 to 400 km per 10 h day in Central America.

Studies by Leshem and his colleagues from the Israel Raptor Information Center (Society for the Protection of Nature in Israel) followed raptors for long distances on 44 days of autumn migration in Israel. By equipping a motor glider with a large gas tank, they were able to stay aloft for entire days, following flocks of Honey Buzzards, Levant Sparrowhawks, Lesser Spotted Eagles, White Storks, and other soaring migrants for distances of over 200 km. Leshem (1987, n.d.) states that these species cross Israel in 5–7 h, a distance of about 220 km. Cross-country speeds ranged between 31.5 and 44.0 kph depending upon the length of time taken. Leshem also reports "average speeds" of 45–50 kph. Wind conditions were not reported by Leshem, so there is no way of knowing how winds or thermal conditions influence daily flight distance. He concluded that flocks averaged 350–500 km per 10 h day of flight.

Pennycuick (1971a, 1972a) reported cross-country speeds of foraging White-backed, Lappet-faced, Hooded, and Ruppell's Griffon vultures over the Serengeti Plain that range from 36 to 75 kph. A speed of "47 km per hour is representative of a Ruppell's Griffon's cross-country capabilities in the dry season" (Pennycuick 1972a). Because these birds are known to fly for more than 5 h in a given day in search of carrion, they realize distances of over 200 km per day. During 2 h flights Pennycuick followed some individuals for nearly 100 km and reported foraging flights exceeding 100 km one way.

Radiotelemetry Studies

Only four research groups have studied the migratory flight of raptors using radiotelemetry (not including habitat use studies). Two of the four studies have focused on endangered species and two on more common species. The success of these groups varies greatly, but together they yield an interesting picture of the ecology and behavior of migrants en route. The results of Hunt, Ward, and Johnson (1981) remain unpublished and will not be reported.

Cochran's (1972, 1975, 1985) are the best known radiotelemetry studies. He has placed radio transmitters on numerous migrants that were captured during migration. Studies of Peregrine Falcons in the midwestern United States and along the Atlantic coast show that one-day flights often exceed 200 km. Of three birds captured in Wisconsin during autumn migration, all flew more than 200 km on their first day following release (222, 238, 302 km) and on subsequent days flew up to 330 km. The average daily distance for all birds was 185 km (Cochran 1975). They used a "circle-soaring migratory flight" during most of the time they were tracked, indicating that these migrants rely on thermal lift. This finding contrasts with Heintzelman's (1975) statement that this species "does relatively little thermal soaring." Cochran reports that with strong thermals only 37% of the 5.3 h per day (average of all birds) in flight was spent circling, whereas with weak lift 62% of the time was spent circling. With weak lift migrants also commenced flapping flight until another thermal was located. Seldom did migrants fly below 150 m when thermals were strong. Cross-country speeds varied with thermal and wind conditions. With following winds migrants averaged 47 kph (282 km per 6 h day), whereas with crosswinds a speed of only 16 kph (96 km per day) was realized.

The autumn flight of an immature male Peregrine Falcon from Wisconsin to Mexico is most informative. Cochran (1975) followed this bird 2,620 km in 15 days. The average cross-country speed was 33.0 kph and the daily distance averaged nearly 175 km. Most of this was accomplished using thermal-gliding flight. On days when winds were opposed to the migratory direction, migration ceased by 1200 h, showing that there is a large variation in flight time with different weather conditions. During early morning and late afternoon the bird made low hunting flights using powered flight. These flights were oriented in the migratory direction. By flying in the appropriate migratory direction when hunting, the bird was able to save energy and time.

The migration of an adult female Sharp-shinned Hawk from Wisconsin to Alabama was also documented by Cochran (1972). This bird flew between 0 and 6 h on the 11 autumn days it was followed: 0 h on 2 days, 3–5 h on 4 days, and 6 h on 4 days (one day was incomplete, and exact flight time on all days was not given). The average distance on the 8 days of actual migration was 150 km (about 33 kph), although on some days the bird averaged less than 100 km and on others more than 200 km.

Like the Peregrine Falcons discussed above, the Sharp-shinned Hawk used thermals for lift. On one day it did nothing but circle in thermals, averaging less than 25 kph. Short "precursor" flights, including soaring and flapping flights at altitudes up to 100 m, were made on many days. These flights occurred 30 min to 1.5 h before the regular migratory flights. It is not known whether these flights were for hunting or testing updraft and wind conditions.

Unlike Cochran, Harmata and his colleagues (1984; Harmata, Toepfer, and Gerrard 1985) placed radio transmitters on migrants before they initiated spring and autumn migration. By doing this they knew where the Bald Eagles they tracked originated and in at least one instance followed an individual through its spring migration. This bird, an adult male, was equipped with a solar-powered radio transmitter at its winter range in the San Luis Valley of south-central Colorado. Harmata (1984) tracked this migrant 1968 km to its nest site in north-central Saskatchewan. During this 15 day migration the eagle spent 6 days "sitting out bad weather" and averaged about 220 km on the days it migrated. This bird and other males initiated spring migration before females, although the dates varied greatly among individuals and between years. By monitoring the breeding activities of some of his birds, Harmata found that telemetry devices did not adversely influence reproductive activity.

Although Harmata was unable to follow other eagles for a complete migration, several were followed for long distances. One flew from Fort Peck, Montana, to Reindeer Lake in northeastern Saskatchewan, more than 1,000 km in 5 days, again averaging more than 200 km per day. For about 100 km this bird flew along the Missouri Coteau, a ridge oriented north-northwest to south-southeast in south-central Saskatchewan. To use the updrafts provided by this ridge, the eagle had to deviate from its intended migratory direction to the northeast. It reoriented after leaving the ridge. (M. T. Myres of the University of Calgary has shown me 16 mm film of a surveillance radar in Saskatchewan that shows echoes of soaring birds, probably cranes, Red-tailed Hawks, and Rough-legged Hawks following the Missouri Coteau during spring migration. Perhaps some of the radar echoes were Bald Eagles.)

Harmata (1984) noted reverse migration by at least one migrant. After one day of northward migration, an eagle returned to the San Luis Valley after encountering a snowstorm. On both days the eagle flew a distance of 145 km. Several other eagles were noted to fly about 150 km per day during spring and autumn migration (A. Harmata 1984, personal communication).

The studies of the initiation of autumn migration by first-year eagles is particularly interesting (Harmata, Toepfer, and Gerrard 1985). The movements of two first-year eagles in central Saskatchewan averaged only 33 km on the first 6 days of migration. Flights were initiated between 1100 and 1300 h and ended between 1400 and 1550 h. All flights were southward. Cross-country speed in nonsoaring (powered?) flight was only 26 kph at altitudes of 54–180 m. Clearly, these migrants must have had to fly farther during ensuing days, or they would not complete their migration of more than 1,700 km (Gerrard et al. 1978).

Beske's (1982) studies of the initiation of autumn migration by first-year Northern Harriers show that other migrants besides Bald Eagles make short "exploratory" flights before initiating longer flights. Radio tracks of three Northern Harriers from nest sites in central Wisconsin were short and slow. One bird made a southeast flight of 27 km and them returned to the site where it was hatched. Another traveled only 71 km in 10 days after leaving the site where it was fledged. The longest flight Beske tracked was 171 km southeast, after which the migrant established a temporary home range for 46 days. These flights, and the short flights reported by Harmata, Toepfer, and Gerrard (1985), suggest dispersal movements (sensu Gauthreaux 1982a), in which an immature bird seeks a site where it may spend subsequent summers. Alternatively, these short, slow migratory flights may be "exploratory" movements (Baker 1978). Exploration may familiarize a naive bird with a large geographical area that can be used for navigation during ensuing migrations or for establishing future nest or foraging sites.

Recoveries of Banded Migrants

Recoveries of banded migrants also offer insight as to the distance hawks travel during one day or more of migration. Estimates of daily flight distance from band recoveries are less accurate than those from aircraft and radiotelemetry studies, however, because they do not tell us the number of hours or days when migration occurred between banding and recovery (or recapture). Stopovers of several hours or days cannot be detected from banding data, nor can reverse migrations. If Harmata had relied on banding data to determine flight distance of the adult male Bald Eagle he followed from Colorado to Saskatchewan (1968 km in 15 days), he might have concluded the bird traveled 131 km per day rather than 220 km per day. Because he used radiotelemetry, Harmata found that 6 of the 15 days of migration were devoted to stopovers.

The data set reported by McClure (1974) from recoveries of banded Gray-faced Buzzard Eagles is one of the few published banding studies that provides information about daily flight distance. Of more than 100 band recoveries that McClure reported, 3 were recaptured or recovered 5 or fewer days after banding. The daily average flight distance of these birds was 280 km (2 days), 213 km (3 days), and 192 km (5 days). The distance traveled per day declines as the number of days since banding increases. Daily distances averaged only 33 to 85 km per day for 5 individuals recovered 15 to 40 days after banding. All of these birds were tagged on Miyako Jima in the Ryukyu Islands of southern Japan and were recovered in the Philippines. This flight represents a long water crossing during which powered flight undoubtedly was required. Because we do not know how many hours these birds flew, or even if they flew every day, precise determinations of cross-country speed cannot be made. Some of the water barriers these migrants crossed were wider than 400 km (chap. 10). Suffice it to say that Gray-faced Buzzard Eagles (and probably many other species) can realize over 200 km on consecutive days using powered flight.

Another interesting banding recovery is that of a Sharp-shinned Hawk captured in the morning at Cape May Point, New Jersey, and recaptured 6 h later at Kiptopeke, Virginia (Foy 1983). The bird, an immature male, flew 223 km southwest with a 9 mps northwest wind. If it flew continuously it realized a cross-country speed of 37.5 kph. This cross-country speed is comparable to those reported by Cochran (1972) for this species. Because the sky was overcast, the migrant undoubtedly used a combination of powered (over Delaware Bay) and gliding flight in declivity currents created by vegetation and dunes. Air speed probably was greater than ground speed because northwest winds necessitated headings to the west of southwest.

There are many banding recoveries in the literature (or unpublished) that can be used to determine daily flight distance of migrants. Instead of accurate estimates of flight distance on a given day, banding data may also yield information about the geographic progress of a population—that is, the rate at which a population moves toward or away from its breeding range. Gerrard et al. (1978) reported the southward progression of immature Bald Eagles from nesting sites in Saskatchewan. By plotting the latitude and direction of recoveries by month, they showed a distinct pattern for birds of different ages. In a slightly different treatment of banding data, Paevskii (1973) showed that European Sparrowhawks from the northwestern Soviet Union banded during autumn migration along the Baltic Sea at the Courland Spit averaged more than 30 km per day. These studies show that original, creative analyses of banding data yield important information about the movements of entire populations. When these data are combined with radiotelemetry information, a precise picture of the ecology and behavior of migrants en route will result.

Comparison of Measured Daily Flight Distance with Simulations

Ideally, a researcher would use the simulations to make predictions about the distance birds *should* fly on a day with known atmospheric conditions. A comparison of this sort allows one to determine the efficiency (definition given) of a migrant and to compare the efficiencies of different species or individuals. For example, the efficiency of long distance migrants could be compared with that of short distance migrants, or individuals migrating alone could be compared with those migrating in flocks. These sorts of comparisons might yield insight about the advantages of flocking and the intensity of selection on migratory behavior. Individuals migrating in flocks and long distance migrants are predicted to have higher efficiencies than single and short distance migrants.

Few researchers who have followed migrants with radiotelemetry devices or aircraft have included information about the weather (atmospheric structure) at the time they followed migrants, so quantitative comparisons cannot be made. The data presented above from aircraft and telemetry studies show that migrants do not fly as far as the simulations predict. The data demonstrate that longer-distance migrants such as Honey Buzzards, Broad-winged Hawks, and lesser Spotted Eagles fly farther and for longer periods of time in one day than middle-distance migrants like Sharp-shinned Hawks and Bald Eagles. The former group usually flew more than 6 h and often up to 10 h per day, whereas the latter species flew less than 6 h on most days. The Peregrine Falcon flew for periods comparable to those flown by the middle-distance migrant, even though it migrates farther.

By virtue of longer daily flight time, longer distance migrants also flew farther during one-day flights than middle distance migrants. Daily flights of over 300 km seem to be normal for the longer distance migrants, whereas other species rarely flew more than 300 km per day and often less than 200 km per day.

When hourly cross-country speeds of the longer distance migrants are compared with those of the middle distance migrants there may not be a difference. Hourly rates for both seem to be somewhat greater than 40 kph, although the Sharp-shinned Hawk and possibly the Bald Eagle and Peregrine Falcon did not realize 40 kph as often as the long distance migrants. If this is correct, it suggests lower migration efficiency. The efficiency coefficients of Sharp-shinned Hawks for a few days of migration from which rough atmospheric data were available (data from Cochran 1972) ranged between 0.6 and 0.85, whereas for Broad-winged Hawks (data from Hopkins 1975; Hopkins et al. 1979) the values exceeded 0.8 (Kerlinger 1982a).

The difference in flight time, distance, and efficiency may be attributable to flocking behavior by the long distance migrants. By flying in flocks migrants are hypothesized to locate and use thermals more efficiently, resulting in greater flight time (earlier and later in the day) and greater distance traveled per hour and per day. Before better comparisons can be made, radiotelemetry or aircraft studies must either make the comparison themselves or report detailed atmospheric data for such comparisons.

Daily Flight Distance of Nonraptorial Migrants

After determining the approximate distances raptors fly during one-day migratory flights, it is of interest to compare these distances with those of other migratory birds. As with raptors, accurate data for single migratory (day or night) flights are scarce. I will focus on radiotelemetry studies by Cochran, Montgomery, and Graber (1967) in which thrushes (*Hylocichla* spp.) were examined during spring migration in Illinois and by Wege and Raveling (1983) in which Canada Geese (*Branta canadensis*) were examined during autumn and spring migration between Lake Manitoba, Manitoba, and Rochester, Minnesota. In the former study the six longest flights (other flights were adversely influenced by unfavorable flying weather) reveal similar distances flown by raptors and by thrushes during single flights. The six thrush flights averaged about 200 km in 8 h, with a range from less than 100 km to more than 200 km. One Swainson's Thrush (*Catharus ustulatus*) was followed for 1,512 km during six nights of migration. During that time six flights averaged 7 h and varied from 105 to 375 km (Cochran 1987).

The geese tracked by Wege and Raveling (1983) also varied in flight time and distance. Nonstop flights of over 855 km were noted during autumn for 4 of 7 radio-tagged birds. Minimum flights during autumn were not less than about 275 km. Flights in spring were shorter, although some exceeded the distances reported above for raptors.

It must be concluded that the large variation of flight distance within migrant groups obscures major differences between groups. However, most birds that undertake long distance migrations using powered flight probably fly farther during single flights than raptors. This is most obvious among nonraptors that fly over barriers such as the Gulf of Mexico, western Atlantic Ocean, and Mediterranean Sea. Raptors may not be able to migrate as far in a single flight as some powered migrants because of time restrictions imposed on raptors by the convective currents they need for lift. Migrants that depend upon thermal convection usually cannot fly for more than 10 h during the day.

Use of the Flight Distance Simulations

The simulations of flight distance are applicable to other species of raptors that use these types of flight and for which flight parameters are known. At present values for flight, parameters are available for few species, although estimations can be made for other species using data presented in this volume. The models can be used to determine the distances hawks can be expected to travel with a given set of environmental conditions, or they can be amended to make quantitative predictions regarding other questions. The simulations can be used to:

1. Predict the type of flight a migrant should use (powered, ridge, or thermaling) to realize the greatest distance given particular weather conditions.

2. Compare the efficiency factors of raptors flying by themselves as opposed to those of individuals migrating in flocks. The results of such comparisons can be used to determine whether flocking promotes a faster and more efficient migration.

3. Predict optimal headings for migrants with different wind conditions (a quantitative means of dealing with wind drift).

4. Predict when hawks should migrate and when they should not.

5. Compare efficiency factors of long distance and short distance migrants.

Summary and Conclusions

Simulations of daily flight distance of migrating raptors were presented for level powered (flapping), ridge-gliding, and thermal-gliding flight. Values of parameters (air speed, glide ratio, climb rate, lift, hours in flight, altitude, flight direction) for the simulations were from chapters 6–11. Wind direction and speed were varied to simulate winds following, opposing, and lateral to a migrant's flight path. The simulations demonstrate that the distances varied with type of flight used and weather. Migrants do no fly as far using level powered flight as when ridge or thermal gliding. Although ridge and thermal gliding yielded approximately the same distances, a disadvantage of ridge gliding is that migrants cannot choose flight direction. Thermal gliding permits a migrant to fly in any direction (except during some weather conditions). Advantages of ridge gliding become obvious when weather is not conducive to thermaling flight.

Measurements of daily migration distance are scarce and not always reliable or comparable. Banding data provide insight into how far individuals travel, although radiotelemetry and aircraft studies are more suited to gathering these data. Empirical studies show that migrants vary greatly in daily flight distance and do not travel as far as predicted by the simulations. Long distance migrants and species that migrate in flocks fly longer during the day and farther distances than species that are shorter distance migrants or that do not fly in large flocks. By dividing the distance predicted by the simulations by measured flight distance, one can estimate the efficiency of a migrant using ridge or thermal updrafts. Migrants such as Broad-winged Hawks seem to travel 200 to more than 400 km per day, Lesser Spotted Eagles and Honey Buzzards 350–500 km per day, Sharp-shinned Hawks about 100 to more than 200 km per day, Bald Eagles 150–220 km per day, and Peregrine Falcons about 100–300 km per day. Simulations will prove useful for formulating predictions about behavioral strategies of migrants that can be tested using radar, aircraft, or radiotelemetry.

13

Flight Strategies of Migrating Hawks and Future Research

The ultimate goal of science is to develop theory regarding underlying mechanisms of natural phenomena. That is, scientists seek to understand why natural patterns exist. For biologists, mechanistic questions pertain to how a phenomenon evolved or, more specifically, what selective pressures shaped it. Before we address questions about mechanisms, a phenomenon must be described, preferably in a numerical or quantitative manner. The description of natural phenomena, termed natural history, is no longer a fashionable or fundable research endeavor. Those who conduct modern studies of natural history will unfortunately be classed with earlier natural historians who knew little of research methodology (quantification and hypothesis testing) or evolutionary theory. Modern ecology (evolutionary ecology) and behavior (behavioral ecology) seem to be dominated by a belief that important research can be conducted without a firm knowledge of what animals actually do. Furthermore, the current attitude among those who dispense funds in this field is that only experimental studies be funded (but see Ricklefs 1987). It is not easy (or possible in most cases) to conduct controlled experiments on migration, so studies of migration as well as studies of natural history are difficult to fund. If these attitudes prevail, migration research will suffer. This will be unfortunate, because the field is nearing maturity and we are now in a position to learn much about the evolution of migratory behavior and the behavioral strategies of migrants.

In the previous twelve chapters I have attempted to describe quantitatively the migratory flight of hawks using an explanatory reductionist approach (chap. 3). Radar, direct visual observations, radiotelemetry, banding, and other methods have provided a large data base on how hawks migrate. Such data can be used to test hypotheses regarding migratory flight behavior, as was shown in the last seven chapters. Thus, the data presented in this volume represent a first step toward developing theory as to how natural selection has shaped the migratory strategies of hawks.

Before about 1980 little was known about the flight behavior and morphology of migrants. Because flight morphology and aerodynamics are now known thoroughly, at least for a few species, students of migration may begin to ask questions regarding emergent strategies as well as how and why such strategies evolved. In other words, the components of migratory flight have been characterized, and we must now consider the process of migration as a whole or as an emergent property that is greater than the sum of its behavioral components. The purpose of this chapter is to discuss the findings presented in this volume and use them to develop a theory of migration strategies of hawks; enumerate and discuss the shortcomings of the approaches used to study hawk migration (mostly my own), suggesting how future research should be conducted; and present specific questions that need to be addressed.

Flight Strategies of Migrating Hawks

The theme that the migratory flight of hawks is a diverse, complex, and variable phenomenon has dominated this volume. Migratory flight behavior of soaring birds is more complex and variable than that of powered migrants such as waterfowl, passerines, and shorebirds. A comparison of powered and soaring migrants (fig. 13.1) reveals that both must make similar sets of decisions but that the number of levels within the hierarchy of decisions and the number of decisions within each level are greater for soaring migrants. Powered migrants seldom resort to soaring because they are not aerodynamically designed to exploit thermal or ridge lift. Figure 13.1 shows that there are at least two more hierarchical levels of decision making for soaring migrants. The hierarchy in figure 13.1 is reminiscent of the one presented by Tinbergen (1951). In addition, learning may be a major factor influencing migrants' decision making.

Diversity and variability of behavior are evident both among and within species. Faced with migrating from location A to location B, different species and perhaps different individuals of the same species migrate successfully even though their flight behaviors differ. Throughout this volume differences of migratory behavior among species have been noted. Major differences among species were evident with respect to flocking, distance of migration, water crossings, flight speed, direction of flight, type of flight, and altitude of flight. A specific example is the difference between Broad-winged and Sharp-shinned hawks: Sharp-shinned Hawks fly lower, fly slower, use more powered flight, use different headings (i.e., in central New York), have steeper glide polars, and fly alone or in smaller flocks. A similar, though not as pronounced, variability was evident among individuals of the same species. Sources of variability within a species may be sex (sexual size dimorphism), age (experience and learning), population differences (e.g., leapfrog migrations), individual variability (developmental, genetic, or other constraints), and error variance. Variation among individuals of the same species has been noted repeatedly. In cage experiments Emlen (1967) and Moore (1984) demonstrated differences in the preferred directions of individual passerines as well as differences correlated with age. Differences in behavior among age and sex classes of raptor migrants was noted in chapters 1, 6, 7, 10, and 12. Thus, selection has operated differently in different species or even within a species.

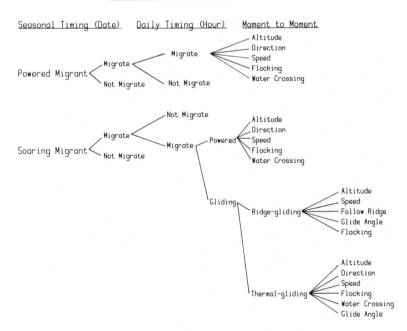

Figure 13.1 Comparison of the hierarchy of decisions made by soaring and powered migrants.

Why is there so much variability among and within species? Is this variability all adaptive? Some variability can be attributed to measurement error and some to error on the part of the migrants being observed, the latter being the material on which selection operates. The problem is to distinguish between adaptive changes and other sources of variability (Lewontin 1978).

The theme of variability may be a strategy in itself. An alternative way to conceptualize variance is to consider it behavioral plasticity. Because scientists are trained to describe and explain variance, we may confuse behavioral variability with plasticity. If variability, as noted in this volume, is thought of as behavioral plasticity, perhaps this is a clue to how selection has shaped the flight strategies of migrants. Viewed this way, the ability to change migratory flight behavior is an adaptation. If this is the case we have been too rigid in our interpretation of how selection has operated on migrants. Instead of asking Why so much variability? perhaps we should also ask Why is there so little stereotyped flight behavior among migrating raptors?

Observing migrants flying along a ridge or from thermal to thermal provides insight as to how often soaring migrants change behavior and make decisions. Migrants continually change wing and tail planform as well as incorporating an occasional flap or series of flaps that seemingly correct for movement of air around the bird. The flight behavior of these migrants is different from stereotyped courtship movements. Viewing migrating raptors is interesting and fun, because they must constantly change their behavior to extract maximum energy from convective motion in the atmosphere to stay aloft and continue to migrate. Ask a sailplane or hang glider pilot; constant vigilance is necessary even when lift is strong and abundant. Stereotyped behaviors in migrants surely would be selected against.

There are several reasons why selection should select against rigid or stereotyped migration behaviors. First, and most important, is weather. At both macro- and meso-local scales, weather changes constantly. Horizontal wind, updrafts, cloud cover, precipitation, and other weather variables rarely remain constant for more than a few minutes. These conditions vary more in the daytime, when hawks fly, than at night, a factor that has been suggested to be a prime selective force for the evolution of nocturnal migration (Kerlinger and Moore 1989). Atmospheric variability may also explain the greater number of decisions made by soaring as opposed to powered migrants (fig. 13.1). During autumn at temperate latitudes, weather conditions change from hour to hour. Cold fronts with associated airflow, lift, cloud cover, and precipitation move through North America throughout the year. A migrant that initiates its autumn migration after the passage of a cold front, as some researchers hypothesize (Broun 1935, 1949; Haugh 1972; Heintzelman 1975), will experience different weather conditions every day for the first week or two of migration. Blustery winds from the north to west, nearly 100% cloud cover, and precipitation characterize a passing cold front. At this time thermals are unavailable and migrants must utilize powered flight, declivity currents, or a combination of the two. Several hours to a day after the front passes, the weather clears and winds remain strong from the north to west. Thermal soaring becomes possible, although thermals may be difficult to use or may not extend high above the ground. A migrant in northeastern North America with a preferred southwesterly flight direction will experience potentially drifting winds, especially if it attempts to fly high. Low altitude flight along ridges is more profitable to migrants at this time.

From one to three days after a cold front, winds abate, skies remain clear, and temperatures rise. At this time migration and soaring conditions can be excellent, deteriorating when winds shift to the south-southeast. With the change in wind direction, cloud cover sometimes increases, warm air predominates, and precipitation is not unusual. This occurs from one to several days before the passage of the next cold front.

A migrating raptor experiences a variety of weather in a short time. It may choose not to migrate with some conditions, but it cannot wait too long for good weather. The best strategy for dealing with these changes is to fly as much as possible during the best weather, fly for varying amounts of time during marginal weather, and avoid flying during the worst weather. Thus, contingency plans are necessary along with adaptations (physiological, morphological, and behavioral) to exploit varying lift and wind conditions.

The pattern of cold front passage I have described is typical of temperate regions such as northeastern North America. Other parts of the world have endemic weather patterns, so descriptions must be modified for other locations. The important point is that flight conditions are not constant.

Autumn seasons in which few cold fronts pass through the northeastern United States are rated by counters as "poor" for hawk migration. During early September through mid-October 1982 few strong cold fronts passed through the eastern United States. The newsletter of the Hawk Migration Association of North America (August 1983) attributed the small numbers of migrants observed to a paucity of strong cold fronts. Of course this does not mean migration did not occur. In these "poor" years most migrants fly at higher altitudes (chap. 8) and are more dispersed, making them difficult to count. These "poor" years may be better in some ways for soaring migrants, because they promote soaring flight, which is less costly than powered flight. The second reason selection should work against rigid or stereotyped migration behaviors is related to the differing topographic and geographic situations migrants fly over. Migrants pass over temperate forest, marsh, tropical forest, brushland, desert, prairie, farmland, city, and other "ecosystems." In addition, the distance a bird has flown and has yet to fly changes during migration. Consider a Sharpshinned Hawk migrating from Quebec to Florida. During its flight it encounters forests, farmland, ridges, mountain ranges, large cities, marshes, beaches, bays, rivers, and possibly the Atlantic Ocean. Several models of migratory flight behavior predict that an animal's behavior should vary depending upon how far it is from its goal (Alerstam 1979b) or how long it has been flying (Gwinner 1986). With weather, topography, and geography changing constantly, rigid or stereotyped flight behaviors are maladaptive. An individual that cannot change its responses to changing situations will not complete migration or be as successful as one that can.

An animal with contingency "plans" is more plastic in its behavior than one without such plans. Do raptors have contingency plans, and if so what sort? The answer to the first question is yes. Throughout this volume the idea of variability (ability to change wing and tail planform and make decisions related to flight) was documented repeatedly. All of the species examined in this book demonstrated plasticity or contingency plans. The ability to change behavior presumably is a product of natural selection. The set of contingencies a migrant uses is its flight strategy.

Contingency plans are exemplified by the behavior of any of the species examined in this book. The migration of an Osprey to the West Indies or Venezuela from its breeding site in Quebec is characterized by changes of weather as well as the topography and geography it flies over. In completing its flight, the Osprey must use powered flight during takeoff, hunting, and flight over the Gulf of Mexico and the Caribbean Sea, as well as using ridge-gliding flight and thermaling flight for varying periods. The same individual also adjusts its speed during each type of flight, changes altitude, joins flocks of Broad-winged Hawks, changes flight direction, and modifies other behaviors as conditions change. The ability to modify flight behavior for different situations has certainly been selected during tens of thousands of years of evolution.

Readers may criticize my suggestion that selection has favored plasticity of migratory flight behavior. The hypothesis that migrants have open-ended or plastic flight strategies may lack predictive power—that is, it may explain everything and thus nothing. Perhaps it is better to view it from the perspective of selection working against stereotyped flight behavior. Consider, for example, a hawk arriving at a water barrier such as Delaware Bay where it meets the Atlantic Ocean (chap. 10). The behavior of species like Peregrine Falcons, Ospreys, and Merlins seems to be rigid. They do not hesitate to cross, presumably because they are capable of extended bouts of powered flight at relatively fast air speeds. If weaker fliers such as Sharp-shinned Hawks or American Kestrels did not hesitate to cross some large bodies of water, they would sometimes perish. Indeed, they have a mechanism by which they decide when crossing is safe. But when confronted by crossings of several hundred or a thousand kilometers, species like Merlins may behave just like Sharp-shinned Hawks at 15 km crossings. These examples show how selection could work against rigid or fixed behaviors. Clearly, individual Sharp-shinned Hawks possess contingency plans for crossing water barriers as well as other aspects of migratory flight.

Shortcomings of Radar and Visual Techniques: Questions for Future Research

Before ending this volume it is important to evaluate the progress made toward understanding the migration of hawks. Hawk migration research has progressed from count studies, in which the phenomenon was recognized and described qualitatively, to radar and visual studies that described flight behavior quantitatively and tested hypotheses regarding flight behavior. From these studies the flight strategy of migrating hawks has been examined briefly, but these data cannot elucidate flight strategy directly. What are the shortcomings of count, radar, and visual techniques that make them inappropriate for testing hypotheses concerning flight strategies? How might we test the hypothesis that behavioral plasticity is an important component of the strategies of migrating hawks?

Researchers attempting to test hypotheses about flight strategies of migrating hawks will encounter several problems. Most important is the absence of data relating to the physiology of migration. Second, data presented in this book either are incomplete or cannot be used to test hypotheses about flight strategies. Before such strategies can be elucidated, we need more quantitative description and hypothesis testing from more species at more locations. Furthermore, a new data set is needed from birds traveling through an entire migration. Such data will not be available in the near future. The remainder of this chapter outlines and discusses questions pertaining to flight behavior, physiology, morphology, and ecology.

Behavioral Questions

I will not repeat criticisms of the hawk migration count method detailed earlier. Suffice it to say that counts cannot be used to study flight behavior. Instead, I will limit my criticism and comments to radar and visual techniques. The primary criticism of data generated from radar and visual techniques is that such methods yield information from a small segment of a migratory journey. Consider visual and radar studies in which migrants were observed for a few minutes over a few kilometers. For a Sharp-shinned Hawk that migrates from Quebec to Florida, a 4 min radar track with simultaneous visual observations is a small portion of its migratory journey.

Accumulation of hundreds of radar and visual observations of flight behavior at several locations permits quantitative description of samples chosen at random from a population of migrants. Measurement of flight behavior with different weather conditions allows some of the variance of the behavior of migrants to be explained through statistical analysis. Lacking from radar and visual studies is information as to how individuals deal with changing flight conditions. What does a migrant do when weather changes and when it flies over differing topography and geography? These changes occur from moment to moment and location to location. Virtually nothing is known about this aspect of flight behavior. If we knew more about how *individuals* deal with changing conditions through a migratory journey, the behavioral plasticity hypothesis could be tested. For each situation quantitative predictions can be made based on weather, geographic location, topographic situations, and past behavior of the migrant. Data could be used to determine whether migrants make correct choices, as well as how long it takes migrants to respond to changing weather or topography (a potential measure of efficiency as in chap. 12).

It seems clear that radiotelemetry is the best method for testing hypotheses about behavioral plasticity and questions related to the behavior of individuals through their migration. To illustrate how radiotelemetry data can be used, hypothetical migration pathways of two Sharp-shinned Hawks flying from Quebec City, Quebec, to Gainesville, Florida (2,120 km, a rhumb-line course), and Asheville, North Carolina (1,590 km), are presented in figure 13.2. Because

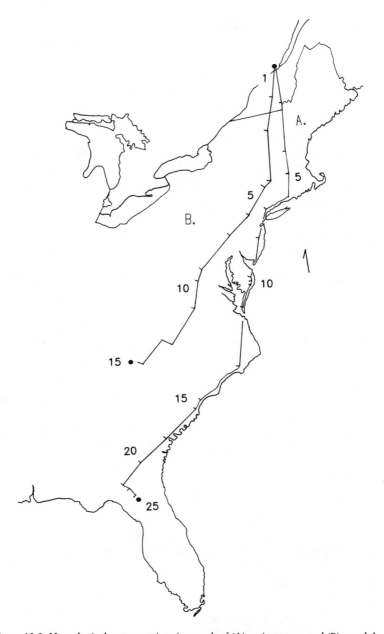

Figure 13.2 Hypothetical autumn migration track of (A) an immature and (B) an adult Sharp-shinned Hawk. Numbers on the tracks represent the number of days that the migrant has been in flight. For Sharp-shinned Hawk A the numbers correspond to the daily flight summaries in table 13.1.

migrants do not fly rhumb-line courses, actual distances will be farther. These flights were chosen because much is known about the migration of Sharp-shinned Hawks in eastern North America from banding (Clark 1985a), count (Broun 1949; Dunne and Clark 1977), and radar studies (Kerlinger 1982a, 1984; Kerlinger and Gauthreaux 1984; Kerlinger, Bingman, and Able 1985). The distance of the flight is also similar to the migration of an adult female Sharp-shinned Hawk radio tracked by Cochran (1972).

The hypothetical migration of Sharp-shinned Hawk A, as shown in figure 13.2 and detailed in table 13.1 deviates from a straight-line migration pathway. Actual distance flown was over 2,319 km. Nonlinear flights such as on day 6 and day 13 occurred when the migrant encountered water barriers. The difference between a nonlinear migration track and a rhumb-line course was slightly more than 200 km (9.5% of the rhumb-line distance), a typical one-day flight for this species. The "data" in table 13.1 show that 95% of the migration occurred on 12 days during which migrants flew 5.96 ± 1.56 (SD) h and traveled 31.6 ± 5.1 kph, or 183 km per day. Rain inhibited migration on 9 days, and on 2 days short flights (<20 km, <2.5 h) were observed. These short flights were in search of prey or suitable habitat. After the 24th day flights were short and irregular in direction, as though the migrant was looking for an acceptable home range for winter.

The migration of Sharp-shinned Hawk B was included in figure 13.2 to show how two or more individuals might be compared. Sharp-shinned Hawk B behaved differently at water crossings, oriented differently at the outset of migration, traveled a shorter distance, and did not take as long to complete migration as migrant A. This individual was an adult, whereas A was an immature bird.

Radiotelemetry data from fewer than 20 individuals of a species, if detailed as in figure 13.2 and table 13.1, would provide a data base relevant to the behavioral plasticity hypothesis.

Radiotelemetry studies such as the one outlined above may never be undertaken. Current funding for research and negative attitudes toward nonexperimental research preclude such projects. In addition to monetary difficulties, fieldwork is exhausting, and failures reduce data sets to small samples. Because few radiotelemetry studies have been published in the peer-reviewed literature, funding of such projects will be difficult in the future.

Data from radiotelemetry studies can be used to answer the following questions:

1. Do hawks utilize headings that exploit prevailing winds and re-

Table 13.1 Summary of Hypothetical Daily Flights of an Immature Sharp-shinned Hawk Migrating from Quebec City, Quebec, to Gainesville, Florida, during Autumn

Day	Distance (km)	Type of Flight	Resultant Direction	Wind (mps)	Duration (h)	Cross-Country Speed (km/day)	Comments
1	180	Thermal[a]	185°	290° (5)	6.0	30	Clear skies
2	235	Thermal[a]	190°	255° (2)	6.5	36.1	Clear
3	110	Thermal[b]	180°	215° (4)	4.0	27.5	Mostly cloudy
4	<2	—	—	160° (3)	—	—	Rain
5	<2	—	—	135° (1)	—	—	Rain in A.M.
6	165[d]	Thermal (low)	215[c]	315° (6)	7.0	23.6	Avoid New Haven harbor
7	140[d]	Flap and glide	203°	300° (2)	6.0	23.0	Cross New York harbor
8	190	Thermal[a]	206°	160° (3)	8.0	33.8	Cross Delaware Bay
9	18	Powered (low)	210[c]	130° (2)	<1.0	<20.0	Rain and fog
10	<2	—	—	0	—	—	Stratus clouds
11	10	Powered (low)	210°	145° (6)	<1.0	<10.0	Rain, hunting
12	95	Powered (low)	214°	300° (6)	3.0	31.7	Stop at Chesapeake Bay, cold front in P.M.
13	275	Thermal[a,c]	205°	20° (3)	8.0	34.4	Cross bay
14	235	Thermal[a]	232°	65° (5)	6.5	36.1	Clear
15	37	Powered (low)	230°	110° (2)	2.5	14.8	Hunting, clear
16	<2	—	—	135° (6)	—	—	Rain
17	<2	—	—	165° (7)	—	—	Rain
18	265	Thermal[a]	228°	270° (2)	7.0	37.9	Clear
19	165	Thermal[b]	230°	190° (3)	4.5	36.7	Clear
20	<2	—	—	200° (3)	—	—	Clear, hunting
21	<2	—	—	0	—	—	Rain
22	142	Thermal[a]	225°	245° (3)	5.0	28.4	Clear
23	<2	—	—	115° (4)	—	—	Showers
24	<2	—	—	230° (5)	—	—	Clear
25	25	Low, powered	345°	225° (1)	1.5	16.6	Clear
26	14	Low, thermal	15°	170° (3)	1.0	14.0	Clear

Note: This is hawk A in figure 13.2. Data from radiotelemetry studies can be like these.

[a]Less than 15% of flight at low altitudes (<100 m).

[b]More than 20% of flight at <100 m.

[c]Change of direction >45° when migrant encountered a water barrier.

[d]Distance was a direct line from initiation point to terminus of day's flight, not the actual distance flown, which was always greater.

sult in curvilinear migration tracks? By examining day-to-day migration of migrants coming from and going to known locations, a researcher can determine if hawks modify their orientation in accordance with past behavior and with changing wind. Furthermore, one can determine if hawks change headings when they encounter leading or diversion lines, as well as with what weather and topographic situations they follow these lines and for how far. Harmata's (1984) study of migrating Bald Eagles shows changes of flight behavior. Other researchers should follow his lead.

2. How far do hawks fly in a given day, and how efficient are they during migration? Quantitative measurements of daily flight distance from radiotelemetry studies can be used to validate (or invalidate) the simulation models presented in chapter 12 and determine how efficiently raptors find and use lift. To test the hypothesis that these animals make decisions about what type of flight to use for distance maximization, data from thermaling, ridge, and powered flight can be compared. Studies by Cochran (1975) and Harmata (1984) suggest that some raptors cease migration when wind or other conditions are adverse for rapid migration.

3. Do hawks feed during migration? What is the duration of stopovers? Whether all species make stopovers to feed during migration has been debated, especially for soaring species such as Broad-winged and Swainson's hawks. There is no doubt that Red-tailed Hawks, Sharp-shinned Hawks, Merlins, Peregrine Falcons, Northern Harriers, American Kestrels and many others feed, as is evident from the large numbers of these species that are trapped during migration. In addition to the normal radiotelemetry device used for following the movements of migrants, a second telemetry device might be implanted to monitor gut motility (M. Fuller, personal communication). Such a device would tell the researcher when migrants eat, how often they eat, and possibly the types of things they eat.

4. Do raptors select habitat in which to stop over and forage, and how is habitat utilized by migrants during stopovers? Researchers have begun to answer this question, although only two or three species have been studied. Holthuizjen, Oosterhuis, and Fuller (1985) radio tagged Sharp-shinned Hawks near Cape May Point, New Jersey, and found that they roosted and hunted in certain types of vegetation. Studies by Hunt, Ward, and Johnson (1981) of migrant Peregrine Falcons along the south Texas barrier islands and by Harmata (1984) of migrant Bald Eagles along the eastern edge of the Rocky Mountains also detail habitat use and are important for management purposes. For species such as American Kestrels (Enderson 1965; Koplin 1973;

Mills 1976) and European Sparrowhawks (Marquiss and Newton 1982) that segregate sexually by habitat during winter, the onset of differential habitat use could be monitored.

5. Do migrants cross water, and with what weather conditions are they most likely to cross? Also, what is the fate of migrants that fly over large bodies of water? Although water crossing by a few species has been studied, several questions remain. Radiotelemetry allows researchers to monitor the behavior of individuals when they encounter barriers such as water, mountain ranges, deserts, prairies, and cities. For example, how large does a body of water have to be before it is a barrier? It would be interesting to know the fate of Broad-winged Hawks migrating on islands in the Caribbean or on the Florida Keys.

6. What are the weather requirements for migratory flights? The hawk migration literature has been dominated by the hypothesis that raptors initiate daily or annual migration with particular weather conditions (Broun 1949; Mueller and Berger 1961; Haugh 1972; Heintzelman 1975; Titus and Mosher 1982). The data researchers used to arrive at this conclusion comes from hawk migration counts. Telemetry will permit us to determine what weather conditions promote migration, as well as what time of day migration is initiated and terminated. We may find that behaviors change during a migration season. That is, migrants may be more prone to take risks later in the season and fly with less regard for weather. Or instead of not flying when weather is unfavorable, we may find that some migrants change the type of flight they use. The ability to change from soaring flight to powered flight may allow migrants to "flee" from harsh weather at northerly latitudes, thereby avoiding starvation. Answers to these questions would shed light on the behavioral plasticity hypothesis.

7. Do the hawks that arrive earliest at breeding ranges in spring acquire the best quality territories (or nest sites and mates)? And do these individuals realize greater reproductive fitness? These fundamental questions are of particular importance among species that nest in cavities or cliffs and species that are polygynous. Comparing arrival time patterns of species that use a wide range of nesting situations with those that have limited nesting sites such as cavities or cliffs might provide insight about the evolution of seasonal timing of migration as well as competition for nest sites or mates by a single sex (sexual selection).

Problems associated with radiotelemetry were outlined in chapters 2 and 12. Some will be overcome by technological innovation. By the time this book is published numerous animals will have been radio tracked throughout their entire migrations using satellite telemetry as suggested by Craighead and Dunstan (1976). In addition, transmitters are becoming smaller, more powerful, and less expensive. Many of the questions listed above, however, will be answered only by using telemetry by standard techniques in which migrants are followed by automobile and light aircraft.

The data acquired from radiotelemetry studies present special problems of analysis and interpretation. Although data can be collected quickly once an individual is fitted with a tracking device, special analytical techniques are necessary because data points are not independent. Despite this drawback, radiotelemetry is a powerful tool. For analyzing data from single tracks, an individual may become its own control because its behavior is examined at different times. If a sufficient number of individuals are followed the normal statistical techniques (randomized designs) can be used to describe behavior and test hypotheses about populations. Awareness of these problems allows the researcher to design radiotelemetry research with particular statistical tests in mind, thereby avoiding inefficient methods of data collection and analysis.

Physiological Questions

Physiology is the least known aspect of hawk migration. More than fifty years have passed since the studies of Broun (1935) and Allen and Peterson (1936), yet few quantitative or controlled studies of migratory physiology have been conducted. Several authors suggest that raptors deposit migratory fat (Gessaman 1979, American Kestrel; Smith, Goldstein, and Bartholomew 1986, Broad-winged and Swainson's hawks; W. S. Clark, personal communication, Sharp-shinned Hawk; Geller and Temple 1983, Red-tailed Hawk), but few data substantiate these claims. Similarly, only one study has quantified the metabolic cost of powered flight for a falconiform (Gessaman 1979, American Kestrel).

The study of fat deposition by migrants was a dominant topic in the migration literature on passerines and other birds from the 1950s through the 1970s (Berthold 1975; Blem 1980). Many of the studies sought to determine the amount of fat (absolute and as a proportion of dry weight) that migrants accumulate. Others have examined the rate of fat accumulation during stopovers (Cherry 1982; Moore and Kerlinger 1987). Acquiring data pertaining to the rate of fat deposition is difficult and time consuming for passerines; for raptors it is nearly impossible among free-ranging individuals.

When studies of fat deposition are undertaken, the appropriate species must be selected. The best candidates are common long-distance migrants such as Broad-winged and Swainson's hawks (Smith 1980) from the New World, and the Honey Buzzard, Lesser Spotted Eagle, and Gray-faced Buzzard Eagle from the Old World. The Western Red-footed Falcon, Amur Falcon, Merlin, and Lesser Kestrel are also good choices because they use powered flight to complete their long flights over water.

Researchers may adopt methods used in earlier studies of fat deposition. These include collecting and comparing individuals at different times of the year or at different locations on the migration path (a randomized design); manipulating photoperiod among captive birds while monitoring weight and visual fat deposits (a repeated-measures design); and comparing blood chemistry of migrants with that of non-migrants (randomized or repeated-measures design) to determine whether the former group is metabolizing stored fat. Some birds may have to be sacrificed to determine anatomical placement and relative amount of fat in different parts of the body, but the resourceful researcher will devise a means of studying fat deposition without killing animals.

The species listed above are also ideal for determining the cost of different types of flight by migrants, especially in studies of fat deposition. Methods for examining the cost of flight include the use of wind tunnels in which the oxygen consumption of a migrant is measured at different air speeds (Tucker 1968; Baudinette and Schmidt-Nielsen 1974; Gessaman 1980) and during different types of flight and also the doubly labeled water (a radioisotope) technique. The latter method is used commonly in studies of free-ranging animals (Nagy 1980). In both types of studies raptors will have to be trained to fly with a variety of conditions, using different types of flight, and for varying periods of time. These methods are tedious and difficult, but there is no other way to acquire the information.

Why is a knowledge of the energetics and physiology of migration important? Implicit in many of the hypotheses discussed in this volume are differences of energy cost associated with different behaviors. Models of optimal behavior show that when given a choice, animals should choose the least costly alternative (as long as the risks are constant; Pyke, Pulliam, and Charnov 1977; Caraco 1981). In this volume migrants were shown to make choices with regard to water crossing, flight speed selection, altitude, flocking, flight direction, timing of daily initiation of flight, and type of flight. A knowledge of energy costs of flight would permit researchers to gain information about the relative costs for different migratory strategies. Armed with this information, they could test critical hypotheses relating to flight behavior and morphology as well as the relation between behavior, morphology, and physiology. For example, a researcher could determine how long a Sharp-shinned Hawk, Merlin, or Broad-winged Hawk can stay aloft using powered flight. Are long flights over water possible for all species? What sort of air speeds can be sustained and for how long? With this knowledge the data presented in this volume and drawn from studies suggested in this chapter will provide a more complete and meaningful picture regarding the evolution of flight strategies.

Morphological and Aerodynamic Questions

Chapter 6 presents a brief introduction to the morphological and aerodynamic adaptations of migrants. Although the data describe the morphology and aerodynamic performance of more than eight species, they cannot be used to answer questions pertaining to morphological adaptations for migration. Surprisingly, there are few if any studies of birds' morphological adaptations for migration. The absence of information reflects the fact that the topic has not been studied as well as the difficulties of addressing such questions.

The eight species for which morphology and aerodynamic performance were summarized in chapter 6 have diverse morphologies. Future research should focus on "morphotypes" that are unrepresented, such as the kites and harriers. Species that could be investigated are the Mississippi Kite, an abundant, long-distance migrant with typical kite characteristics, and the Northern Harrier, for similar reasons. These migrants use slower speeds during migration than other raptors (tables 11.3 and 11.6; personal observations), and some engage in long flights over water. In addition, information about the performance and morphology of the smaller (e.g., Amur Falcon or Lesser Kestrel) and larger (Peregrine Falcon) falcons would round out our knowledge of these aspects of migration and allow researchers to estimate the performance of any raptor. Such information would also be useful for making predictions pertaining to foraging, territorial defense, and other aspects of nonmigratory behavioral ecology.

Ecological Questions

With the exception of a brief summary in chapter 1, I have ignored the ecology of raptor migration. Recent reviews by Gauthreaux (1982a) and Ketterson and Nolan (1983), as well as an entire volume by Baker (1978), have focused on ecological aspects of migration in birds. The focus of many of these reviews has been differential migration by age-sex class or by population. Also of importance are the duration of stopovers and habitat selection and usage by migrants. I have omitted material related to the ecology of raptor migration because of the dearth of information about the subject (appendix 2).

Is differential migration by population or age-sex class of raptors common or unusual? Newton (1979) suggests patterns of seasonal timing of migration by different age-sex classes, such that among partial migrants immatures precede adults in autumn and among complete migrants adults precede immatures. Although there is evidence for differential seasonal timing, there are too few rigorous studies to draw conclusions regarding patterns among species. With respect to differential distance of migration or differential habitat use by migrating hawks there are even fewer studies.

Differential migration is of importance to studies of flight behavior and strategies because variation of migration ecology within a species may contribute to the variation of flight behavior as reported here. By measuring flight behavior of unmarked migrants at single locations, a researcher has no means of determining whether the migrants represent a homogeneous, heterogeneous, or random sample of a species or population. If raptors, like other birds, have differential migration by age-sex class, or leapfrog migration by population, it is likely that variance of flight behavior within species as reported throughout this book is in part a result of ecological differences among individuals. Samples taken at given times and locations comprise individuals from different populations as well as mixtures of age-sex classes that do not behave identically. This may give a false impression of behavioral variance among individuals, masking true variance within homogeneous populations. For example, when measuring the directions of migrating Sharp-shinned Hawks in central New York (chap. 7) Kerlinger, Bingman, and Able (1985) were tracking a heterogeneous or "mixed" sample, comprising individuals from the northern terminus of the range as well as from 100 km north of their study site. Also, they tracked adults, immatures, males, and females with different destinations.

A knowledge of age-sex class and place of origin of migrants might have allowed a more thorough partitioning of variance in flight direction, altitude, flocking, or flight speed by Kerlinger, Bingman, and Able (1985) and Kerlinger (1982a, this volume), thereby permitting more precise tests of hypotheses about flight behavior. Partitioning of variance by age-sex class and population is especially important for testing hypotheses regarding migrants' behavioral strategies. Again, the importance of radiotelemetry studies of individuals of known origin and age-sex class is evident.

Summary and Conclusions

A theme throughout this volume is the diversity and variability of flight behavior of migrating hawks both within and among species. Whereas variability is attributed to errors of measurement and errors by migrants, some variance is adaptive. An alternative interpretation of this variance is behavioral plasticity of individuals, with migrants changing their behavior as conditions dictate. Atmospheric, geographic, and topographic conditions vary constantly during migration, and birds that do not adjust their behavior will not migrate successfully. Rigid, fixed, or stereotyped flight behaviors should be selected against. Behavioral plasticity is widespread among raptor migrants and can be interpreted as an emergent property or strategy.

Although this volume presents data detailing behavioral components of migratory flight for several raptors, it should not be viewed as an end point. Instead, the information and approach presented here should serve as a starting point for new research. Questions for future research focus on physiological, ecological, morphological, and behavioral questions. The physiology of migration among raptors is virtually unstudied. Data on the energetics of flight and fat deposition are needed before some behavioral hypotheses can be tested. Radiotelemetry provides the best information for examining behavioral plasticity or other theories about strategies of migrants. The advantage of radiotelemetry over techniques like radar is the ability to observe how individuals respond to changing atmospheric, topographic, and geographic conditions. Radar and visual observations yield a limited picture of migration because they track individuals for only short distances. With data from radiotelemetry a researcher can examine the behavior of individuals of known age and sex, natal location, "wintering" location, and migration history (days or years before) and test hypotheses outlined here.

Appendixes

Appendix 1 Summary of Migratory Tendencies of 133 Species of Falconiformes

Species	Migration Distance[a]	Water Crossing[b]	Flocking Tendency[c]	Insect Follower[d]	Comments[e]
COMPLETE MIGRANTS					
Osprey *Pandion haliaetus*	2	3	1 (H)	0	
Honey Buzzard *Pernis apivorus*	2	3	2 (100s)	0	
Mississippi Kite *Ictinia mississippiensis*	2	0	2 (100s)	0?	
Short-toed Eagle *Circaetus gallicus*	2	1	1 (100s)	0	
Montagu's Harrier *Circus pygargus*	2	3	2 (<5)	0	
Gray Frog Hawk *Accipiter soloensis*	2	3	0?	0	
Gray-faced Buzzard Eagle *Butastur indicus*	2	3	1 (<10)	0	
Broad-winged Hawk *Buteo platypterus*	2	1	2 (1,000s)	0	
Swainson's Hawk *B. swainsoni*	2	1	2 (1,000s)	1	
Rough-legged Hawk *B. lagopus*	2	2?	1 (2–5 +)	0	
Lesser Spotted Eagle *Aquila pomarina*	2	2	1 (100s)	1	
Greater Spotted Eagle *A. clanga*	2	1	1 (5–6 +)	1	
Lesser Kestrel *Falco naumanni*	2	3	2 (1,000s)	1	
Red-footed Falcon *F. vespertinus*	2	3	2 (1,000s)	1	
Amur Falcon *F. amurensis*	2	3	2 (1,000s)	1	
European Hobby *F. subbuteo*	2	2	2 (100s)	1	
Eleonora's Falcon *F. eleonorae*	2	3	2 (12 +)	1	
Sooty Falcon *F. concolor*	2	3	1 (6 +)	1	

Appendix 1 *(continued)*

Species	Migration Distance[a]	Water Crossing[b]	Flocking Tendency[c]	Insect Follower[d]	Comments[e]
PARTIAL MIGRANTS					
Turkey Vulture	1	1	2 (1,000s)	0	SHN
Cathartes aura					
Black Vulture	1	0	1 (10s)	0	
Coragyps atratus					
African Cuckoo Falcon	0	0	0	0	
Aviceda cuculoides					
Black Baza	0	0	1 (<20)	0	
A. leuphotes					
Jerdon's Baza	0	0	0	0	ALT
A. jerdoni					
American Swallow-tailed Kite	1	1	1 (10s)	0	
Elanoides forficatus					
Black-shouldered Kite	0	0	1?	0?	WD
Elanus caeruleus					
Black-shouldered Kite	1	0	1?	0?	SHN
E. leucurus					
African Swallow-tailed Kite	1	0	1 (5–20)	1	WD, ITM
Chelictinia ricourii					
Snail Kite	1	0	1 (10s)	0	SHN
Rostrhamus sociabilis					
Plumbeous Kite	1	0	2 (<40)	0?	SHN, ITM
Ictinia plumbea					
Black Kite	1–2	2	2 (1,000 +)	1	ITM
Milvus migrans					
Red Kite	1	1	1 (3–5)	0	
M. milvus					
Whistling Hawk	0	0	2 (<100)	0	
Haliastur sphenurus					
Brahminy Kite	0	1	1 (10s)	0	WD, ALT, ITM, SHN
H. indus					
Pallas' Sea-Eagle	1	0	1 (5–7)	0	
Haliaeetus leucoryphus					
Bald Eagle	1 +	2	1 (2 +)	0	
H. leucocephalus					
White-tailed Eagle	1	2	0	0	
H. albicilla					
Steller's Sea-Eagle	1	2	0?	0	
H. pelagicus					
Lesser Fishing Eagle	0	0	0	0	ALT
Ichthyophaga nana					
Vulturine Fish Eagle	0	0	0	0	WD, ITM
Gypohierax angolensis					
Egyptian Vulture	1	1	1 (<5)	0	
Neophron percnopterus					
Lammergeir	0	0	1 (<5)	0	ALT
Gypaetus barbatus					
Indian White-backed Vulture	1	0	2 (10s)	0	
Gyps bengalensis					

Appendix 1 *(continued)*

Species	Migration Distance[a]	Water Crossing[b]	Flocking Tendency[c]	Insect Follower[d]	Comments[e]
Griffon Vulture *G. fulvus*	1	1	1 (<10)	0	ALT
Ruppell's Griffon *G. ruppelli*	1	0	1?	0	WD
African White-Backed Vulture *G. africanus*	1	0	1?	0	WD
Lappet-faced Vulture *Torgos tracheliotus*	1	0	0?	1	
European Black Vulture *Aegypius monachus*	2	1	0?	0	ALT
African Harrier Hawk *Polyborus typhys*	0	0	0?	0	ITM
Madagascar Harrier Hawk *P. radiatus*	0	0	0?	1?	WD
Spotted Harrier *Circus assimilis*	0	2	0?	1	SHN
Marsh Harrier *C. aeruginosus*	2	3	1 (3+)	0	SHN
Northern Harrier *C. cyaneus*	2	3	1 (<10)	0	
Cinereous Harrier *C. cinereus*	1	1	0?	0	SHN
Pallid Harrier *C. macrourus*	2	1	1 (<5)	1	
Pied Harrier *C. melanoleucus*	1	1?	1(<10)	0	
Long-winged Harrier *C. buffoni*	1	0	0	0	SHN
Northern Goshawk *Accipiter gentilis*	1	1	0	0	ALT?
Ovampo Sparrow-hawk *A. ovampensis*	1	0	0?	0	WD
Japanese Lesser Sparrowhawk *A. gularis*	2	2+	0?	0	
Besra Sparrowhawk *A. virgatus*	2	0	0	0	
European Sparrowhawk *A. nisus*	1	1	1 (<5)	0	ALT
Sharp-shinned Hawk *A. striatus*	1	1	1 (<5)	0	ALT?
Levant Sparrowhawk *A. brevipes*	2	1+	2 (100s)	1	
Shikra *A. badius*	1	0	0?	0	WD
Cooper's Hawk *A. cooperii*	1	1	1 (<3)	0	ALT?
Dark Chanting Goshawk *Melierax metabates*	0	0	0	0	WD
Gabar Goshawk *M. gabar*	0	0	0	0	WD

Appendix 1 *(continued)*

Species	Migration Distance[a]	Water Crossing[b]	Flocking Tendency[c]	Insect Follower[d]	Comments[e]
Grasshopper Buzzard Eagle *Butastur rufipennis*	1	0	0?	1	WD
White-eyed Buzzard *B. teesa*	0	0	0	0	
Common Black Hawk *Buteogallus anthracinus*	1	0	0	0	
Bay-winged Hawk *Parabuteo uncinctus*	1	0	0?	0	SHN?
Gray Hawk *Buteo nitidus*	1	0	0	0	
Red-shouldered Hawk *B. lineatus*	1	1	1 (3+)	0	
White-tailed Hawk *B. albicaudatus*	1	0	1 (10–30)	1	SHN
Red-backed Buzzard *B. polysoma*	1	1+	0?	0	SHN?
Zone-tailed Hawk *B. albonotatus*	1	0	0	0	
Red-tailed Hawk *B. jamaicensis*	1	1	1 (<50)	0	
Common Buzzard *B. buteo*	1	1	1 (50+)	0	
Long-legged Buzzard *B. rufinus*	2	0	1 (<10)	0	
Upland Buzzard *B. hemilasius*	1	0	0	0	
Ferruginous Hawk *B. regalis*	1+	0	0	1	
African Red-tailed Buzzard *B. auguralis*	1	0	1 (3–4)	1	WD, ITM
Steppe Eagle *Aquila rapax*	2	1	1 (<100)	1	
Imperial Eagle *A. heliaca*	1	1?	0	0	
Wahlberg's Eagle *A. wahlbergi*	0	0	1 (<5)	1	WD, ITM, SHN
Golden Eagle *A. chrysaetos*	2	1	0	0	ALT
Black Eagle *A. verreauxi*	1	0	0?	0	
Bonelli's Eagle *Hieraaetus fasciatus*	1	1	1 (<5)	0	
Booted Eagle *H. pennatus*	2	1?	1 (<5)	0	
Martial Eagle *Polemaetus bellicosus*	0	0	0	0	WD, ITM
Chimango *Milvago chimango*	1	1	1?	1	SHN
Fox Kestrel *Falco alopex*	1	0	0	1	WD, ITM
American Kestrel *F. sparverius*	1+	1	1 (<7)	0	

Appendix 1 *(continued)*

Species	Migration Distance[a]	Water Crossing[b]	Flocking Tendency[c]	Insect Follower[d]	Comments[e]
Eurasian Kestrel *F. tinnunculus*	1	2	1 (100s)	1	
Australian Kestrel *F. cenchroides*	1	2+	2 (<6)	0	SHN
Gray Kestrel *F. ardosiaceus*	0	0	0?	0	WD
Red-headed Falcon *F. chicquera*	0	0	0	0	WD
Merlin *F. columbarius*	2	3	1 (<5)	0	
Brown Hawk *F. berigora*	0	2	1 (<100)	1	SHN
New Zealand Falcon *F. novaezeelandiae*	1	2	0	0	
Oriental Hobby *F. severus*	1+	1	1 (<10)	0	
Little Falcon *F. longipennis*	1	2	0	0	SHN
Aplomado Falcon *F. femoralis*	1	1	0	1	SHN?
Lanner Falcon *F. biarmicus*	0	0	0	1	WD, ALT
Prairie Falcon *F. mexicanus*	1+	0	0	0	
Saker Falcon *F. cherrug*	2	2	0	0	
Gyrfalcon *F. rusticolus*	1+	3	0	0	
Peregrine Falcon *F. peregrinus*	2	3	0	0	SHN
Barbary Falcon *F. peregrinoides*	1	0	0	0	

(Bicolored Hawk [*Accipiter bicolor*] and Gray-bellied Hawk [*A. poliogaster*] were found to be migratory after analyses were completed)

LOCAL AND IRRUPTIVE MIGRANTS

Species	Migration Distance[a]	Water Crossing[b]	Flocking Tendency[c]	Insect Follower[d]	Comments[e]
Yellow-headed Vulture *Cathartes burrovianus*	0	0	0?	0	WD
California Condor *Gymnogyps californianus*	0	0	1 (<10)	0	ALT
Andean Condor *Vultur gryphus*	0	1	0?	0	ALT
Australian Black-shouldered Kite *Elanus notatus*	0	0	1 (<20)	1	SHN
Letter-winged Kite *E. scriptus*	1	0	1 (10s)	1?	
White-bellied Sea-Eagle *Haliaeetus leucogaster*	0	1	0?	0	
African Fish Eagle *H. vocifer*	0	0	0	0	WD

Appendix 1 *(continued)*

Species	Migration Distance[a]	Water Crossing[b]	Flocking Tendency[c]	Insect Follower[d]	Comments[e]
Himalayan Griffon *Gyps himalayensis*	0	0	0?	0	ALT
Brown Harrier Eagle *Circaetus cinereus*	0	0	0	0	
Smaller Banded Snake Eagle *C. cinerascens*	0	0	0	0	
Bateleur Eagle *Terathopius ecaudatus*	0	0	0	0	WD
Crested Serpent Eagle *Spilornis cheela*	0	0	0	0	
African Marsh Harrier *Circus ranivorus*	0	0	0	0	WD
Short-tailed Hawk *Buteo brachyurus*	0	0	0	0	
Madagascar Buzzard *B. brachypterus*	0	0	0	1	
Augur Buzzard *B. rufofuscus*	0	0	0	0	
Wedge-tailed Eagle *Aquila audax*	0	0	0	0	
Little Eagle *Hieraaetus morphnoides*	0	0	0	0	
Ayre's Hawk Eagle *H. dubius*	0	0	0	0	
Mountain Eagle *Spizaetus nipalensis*	0	0	0	0	ALT
Secretary Bird *Sagittarius serpentarius*	0	0	0	0	
Greater Kestrel *Falco rupiculoides*	0	0	0	0	
African Hobby *F. cuvieri*	0	0	1?	1	
Black Falcon *F. subniger*	1	0	0	0	

Note: The information in this appendix was gleaned from literature cited in this volume.

[a]Migration distance: 0 = short, <300 km one way; 1 = moderate–medium, 300–1,500 km; 2 = long distance, greater than 1,500 km.

[b]Water crossing: 0 = none; 1 = short, less than 25 km; 2 = moderate, 25–100 km; 3 = long, >100 km.

[c]Flocking: 0 = none; 1 = some or irregular flocking, usually small flocks of heterospecific flocks; 2 = regularly flocking, usually seen in flocks during migration (maximum flock sizes in parentheses).

[d]Insect Follower: 0 = not known to follow insect swarms or emergences; 1 = known to regularly or frequently follow insects (Isoptera, Orthoptera, Hymenoptera, etc.)

[e]Comments: SHN = north-south movements in the Southern Hemisphere known or strongly suspected; WD = wet-dry season movements; ITM = intratropical migrations known; ALT = altitudinal migrations known or suspected.

Appendix 2 Summary of Differential Migration Patterns among the Falconiformes

Species	Reference and Method[a]
SEASONAL TIMING OF MIGRATION: AUTUMN	
Males migrate before females	
Peregrine Falcon (immatures only)	Hunt, Rogers, and Stowe 1975, TB-M
European Sparrowhawk	Moritz and Vauk 1976, TB-M
Females migrate before males	
Northern Harrier	Broun 1949; Haugh 1972, MC
Sharp-shinned Hawk	Rosenfield and Evans 1980; Duncan 1982, TB-M
Cooper's Hawk	Hoffman 1985, TB-M
Merlin	Clark 1985b, TB-M
Adults migrate before immatures	
Black Kite	Schifferli 1967
Osprey	Newton 1979
Peregrine Falcon	Hunt, Rogers, and Stowe 1975; Ward and Berry 1972, TB-M
Steppe Buzzard	Broekhuysen and Siegfried 1970, SR
Golden Eagle (?)	Hoffman 1985, MC
Immatures migrate before adults	
Sharp-shinned Hawk	Broun 1949, MC; Mueller and Berger 1967b, TB-M
Goshawk	Mueller, Berger, and Allez 1977, TB-M
Cooper's Hawk	Broun 1949, MC; Hoffman 1985, TB-M
Northern Harrier (?)	Broun 1949; Haugh 1972, MC
European Sparrowhawk	Moritz and Vauk 1976, TB-M
Red-tailed Hawk	Haugh 1972, MC; Geller and Temple 1983, TB-M
American Kestrel (?)	Haugh 1972, MC
SEASONAL TIMING OF MIGRATION: SPRING	
Adults arrive or migrate earlier than immatures	
Broad-winged Hawk	Matray 1974 BG; Haugh 1972; Kerlinger and Gauthreaux 1985a, MC
Peregrine Falcon	Herbert and Herbert 1965, BG
Sharp-shinned Hawk	Haugh 1972, MC
Northern Harrier	Haugh 1972, MC
Red-tailed Hawk	Haugh 1972, MC
Goshawk	Haugh 1972, MC
Egyptian Vulture	Brown and Amadon 1968, BG?
Males arrive or migrate earlier than females	
Northern Harrier	Haugh 1972, MC
Merlin	Newton, Meek, and Little 1978, BG; Clark 1985b, TB-M
American Kestrel	Willoughby and Cade 1964; Smith, Wilson, and Frost 1972, BG
Gyrfalcon	Platt 1976, BG
Red Kite	Brown and Amadon 1968

Appendix 2 *(continued)*

Species	Reference and Method[a]
Pied Harrier	Brown and Amadon 1968
Marsh Harrier	Brown and Amadon 1968, BG
Montagu's Harrier	Robinson 1950, BG?
Northern Harrier	Hamerstrom 1969, BG
Bald Eagle	Harmata 1984, BG, TB-M
Females arrive or migrate earlier than males	
European Sparrowhawk (immatures)	Newton 1975, BG; Moritz and Vauk 1976, TB-M
Prairie Falcon	Enderson 1964, BG
American Kestrel (?)	Haugh 1972, MC
Peregrine Falcon	Mearns 1982, BG
No difference of arrival time or migration between sexes	
Peregrine Falcons (together)	Cade 1960, BG
American Kestrel (together)	Willoughby and Cade 1964, BG
Honey Buzzard (together)	Holstein 1944
Swallow-tailed Kite	Snyder 1974, BG

GEOGRAPHIC DISTRIBUTION DURING THE NON-BREEDING SEASON

Adults north of immatures (or closer to the breeding range)	
Steppe Eagle	Steyn 1973; Brooke et al. 1972; Smeenk 1974, BG?
European Sparrowhawk	Schelde 1960
Red-tailed Hawk	Brinker and Erdman 1985, TB-M; Kerlinger, unpublished data, WP
Gyrfalcon	Platt 1976, WP-SR
Goshawk	Mueller, Berger, and Allez 1977, TB-M
Bald Eagle (?)	Harmata 1984, WP-SR
No difference in range between adults and immatures	
Osprey	Osterlof 1977, TB
Males north of females (or closer to the breeding range)	
Pallid Harrier	Brown and Amadon 1968
Sharp-shinned Hawk	Rosenfield and Evans 1980, TB-M
American Kestrel	Cade 1955, WP-SR; Roest 1957, MUS; Willoughby and Cade 1964; Mills 1975, 1976, WP-SR; Enderson 1960, WP-SR
Females north of males (or closer to breeding range)	
Prairie Falcon	Enderson 1964, WP-SR, BG
European Sparrowhawk	Belopol'skij 1972, TB-M
Sharp-shinned Hawk	Clark 1985a, TB-M
Goshawk	Sulkava 1964; Haukioja and Haukioja 1970, TB-BG; Mueller, Berger, and Allez 1977, TB-M

[a]Method from which determination was made: MC = migration count; BG = study on breeding grounds; TB-M = trapping and banding during migration; WP-SR = skewed sex or age ratio at one location during the nonbreeding (winter) season; MUS = museum skins were the source of data; question mark next to species name indicates that the results were suggestive but not conclusive.

References

Able, K. P. 1970. A radar study of the altitude of noctural passerine migration. *Bird-Banding* 41:282–290.

———. 1972. Fall migration in coastal Louisiana and the evolution of migration patterns in the Gulf region. *Wilson Bull.* 84:231–242.

———. 1973. The role of weather variables and flight direction in determining the magnitude of noctural bird migration. *Ecology* 54:1031–1041.

———. 1977. The flight behaviour of individual passerine nocturnal migrants: A tracking radar study. *Anim. Behav.* 25:924–935.

———. 1980. Mechanisms of orientation, navigation, and homing. In *Animal migration, orientation, and navigation,* ed. S. A. Gauthreaux, Jr., pp. 283–373. New York: Academic Press.

Able, K. P., V. P. Bingman, P. Kerlinger, and W. Gergits. 1982. Field studies of avian nocturnal migratory orientation, 2. Experimental manipulation of orientation in White-throated Sparrows (*Zonotrichia albicollis*) released aloft. *Anim. Behav.* 30:768–773.

Agee, E. M., and P. J. Sheu. 1978. MCC and gull flight behavior. *Boundary Layer Meteorol.* 14:247–251.

Alerstam, T. 1978a. Analysis and a theory of visible bird migration. *Oikos* 30:273–349.

———. 1978b. A graphical illustration of pseudodrift. *Oikos* 30:409–412.

———. 1979a. Optimal use of wind by migrating birds: Combined drift and overcompensation. *J. Theor. Biol.* 79:341–353.

———. 1979b. Wind as a selective agent in bird migration. *Ornis Scand.* 10:76–93.

———. 1981. The course and timing of bird migration. In *Animal migration,* ed. D. J. Aidley, pp. 9–54. Society for Experimental Biology Seminar Series 13. New York: Cambridge University Press.

———. 1985. Strategies of migratory flight, illustrated by Arctic and Common Terns, *Sterna paradisaea* and *Sterna hirundo.* In *Migration: Mechanisms and adaptive significance,* ed. M. A. Rankin. *Contrib. Marine Sci. Suppl.* 27:580–603.

———. 1987. Radar observations of the stoop of the Peregrine Falcon *Falco peregrinus* and Goshawk *Accipiter gentilis. Ibis* 129:267–273.

Ali, S., and S. D. Ripley. 1978. *Handbook of the birds of India and Pakistan.* London: Oxford University Press.

Allen, R. P., and R. T. Peterson. 1936. The hawk migrations at Cape May Point, New Jersey. *Auk* 53:393–404.

Anderson, A. 1985. *North Sea bird club report for 1984*. Aberdeen: Offshore Petroleum Industry Training Board.

Andersson, M., and R. A. Norberg. 1981. Evolution of reversed sexual size dimorphism and role partitioning among predatory birds with a size scaling of flight performance. *Biol. J. Linnean Soc.* 15:105–130.

Ankney, C. D., and C. D. MacInnes. 1978. Nutrient reserves and reproductive performance of female Lesser Snow Geese. *Auk* 95:459–471.

AOU [American Ornithologists' Union]. 1983. *Check-list of North American birds*. 6th ed. Lawrence, KS: American Ornithologists' Union.

Baker, R. R. 1978. *The evolutionary ecology of migration*. New York: Holmes and Meiers.

———. 1981. *The mystery of migration*. New York: Viking Press.

Balcomb, R. 1977. The grouping of nocturnal passerine migrants. *Auk* 94:479–488.

Bates, M. R. 1975. Mt. Tom. In *Proceedings of the North American Hawk Migration Conference, 1974*. Syracuse, NY: Hawk Migration Association of North America.

Batschelet, E. 1981. *Circular statistics in biology*. New York: Academic Press.

Baudinette, R. V., and K. Schmidt-Nielsen. 1974. Energy cost of gliding flight in herring gulls. *Nature* 248:83–84.

Beaman, M., and C. Galea. 1974. The visible migration of raptors over the Maltese Islands. *Ibis* 116:419–431.

Bedard, J., and G. LaPointe. 1984. Banding returns, arrival times, and site fidelity in the Savannah Sparrow. *Wilson Bull.* 96:196–205.

Beebe, F. L. 1960. The marine Peregrines of the northwest Pacific coast. *Condor* 62:145–189.

Bellrose, F. C. 1967. Radar in orientation research. *Proc. Int. Ornithol. Congr.* 14:281–309.

Bellrose, F. C., and R. R. Graber. 1963. A radar study of the flight directions of nocturnal migrants. *Proc. Int. Ornithol. Congr.* 13:362–389.

Belopol'skij, L. O. 1972. Ecological peculiarities in *Accipiter nisus* (L.) migration. *Ekologiya* 3:58–63 (translated in *Soviet J. Ecol.* 3 [1972]: 138–174).

Bent, A. C. 1937. *Life histories of North American birds of prey*. Bulletin 167. Washington, DC: U.S. National Museum.

Berger, D. D., and H. C. Mueller. 1959. The bal-chatri: A trap for the birds of prey. *Bird-Banding* 30:18–26.

Bernis, F. 1973. Migración de Falconiformes y *Ciconia* spp. por Gibraltar, verano otoño 1972–1973: Primera parte. *Ardeola* 19:152–224.

———. 1980. *La migración de las aves en el Estrecho de Gibraltar*. Madrid: University of Madrid.

Bernstein, M. H., S. P. Thomas, and K. Schmidt-Nielsen. 1973. Power input during flight of the Fish Crow *Corvus ossifragus*. *J. Exp. Biol.* 58:401–410.

Berthold, P. 1975. Migration: Control and metabolic physiology. In *Avian biology,* ed. D. S. Farner and J. R. King, 5:77–128. New York: Academic Press.

Beske, A. E. 1982. Local and migratory movements of radio-tagged juvenile harriers. *Raptor Res.* 16:39–53.

Bildstein, K. L., W. S. Clark, D. L. Evans, M. Field, L. Soucy, and E. Henckel. 1984. Sex and age differences in fall migration of Northern Harriers. *J. Field Ornithol.* 55:143–150.

Binford, L. C. 1977. Fall migration of diurnal raptors at Point Diablo, California. *Western Birds* 10:1–16.

Bingman, V. P., K. P. Able, and P. Kerlinger. 1982. Wind drift, compensation, and the use of landmarks by nocturnal bird migrants. *Anim. Behav.* 30:49–53.

Blake, E. R. 1977. *Manual of Neotropical birds.* Chicago: University of Chicago Press.

Blem, C. 1980. The energetics of migration. In *Animal migration, orientation, and navigation,* ed. S. A. Gauthreaux, Jr., pp. 175–224. New York: Academic Press.

Bloch, R., and B. Bruderer. 1982. The air speed of migrating birds and its relationship to the wind. *Behav. Ecol. Sociobiol.* 11:19–25.

Borneman, J. C. 1976. California Condors soaring into opaque clouds. *Auk* 93:636.

Brinker, D. F., and T. C. Erdman. 1985. Characteristics of autumn Red-tailed Hawk migration through Wisconsin. In *Proceedings of the Fourth Hawk Migration Conference,* ed. M. Harwood, pp. 107–136. Rochester, NY: Hawk Migration Association of North America.

Broekhuysen, G. R., and W. R. Siegfried. 1970. Age and molt in the Steppe Buzzard in southern Africa. *Ostrich Suppl.* 8:223–237.

Broley, C. L. 1947. Migration and nesting of Florida Bald Eagles. *Wilson Bull.* 59: 3–20.

Brooke, R. K., J. H. Grobler, and M. P. S. Irwin. 1972. A study of the migratory eagles *Aquila nipalensis* and *A. pomarina* (Aves: Accipitridae) in southern Africa, with comparative notes on other large raptors. *Occas. Pap. National Museum of Rhodesia* 1972–B5(2):61–114.

Broun, M. 1935. The hawk migration during the fall of 1934, along the Kittatinny Ridge in Pennsylvania. *Auk* 52:233–248.

———. 1949. *Hawks aloft: The story of Hawk Mountain.* New York: Dodd, Mead.

Broun, M., and B. V. Goodwin. 1943. Flight-speeds of hawks and crows. *Auk* 60:487–492.

Brown, J. L. 1974. *The evolution of behavior.* New York: W. W. Norton.

Brown, L. H. 1970. *African birds of prey.* London: Collins.

———. 1976. *Eagles of the world.* London: David and Charles.

Brown, L. H., and D. Amadon. 1968. *Eagles, hawks and falcons of the world.* London: Country Life Books.

Browning, M. R. 1974. Comments on the winter distribution of the Swainson's Hawk (*Buteo swainsoni*) in North America. *Amer. Birds* 28:865–867.

Bruderer, B., and P. Steidinger. 1972. Methods in quantitative and qualitative analysis of bird migration with a tracking radar. *NASA Spec. Publ.* SP-262:151–167.

Buskirk, W. H. 1980. Influence of meteorological patterns and trans-Gulf migration on the calendars of latitudinal migrants. In *Migrant birds in the Neotropics*, ed. A. Keast and E. S. Morton, pp. 485–491. Washington, DC: Smithsonian Institution Press.

Cade, T. 1955. Experiment on winter territoriality of the American Kestrel, *Falco sparverius*. *Wilson Bull.* 67:5–17.

———. 1960. Ecology of the Peregrine and Gyrfalcon populations in Alaska. *Univ. Calif. Publ. Zool.* 63:151–290.

———. 1982. *The falcons of the world*. London: Collins.

Cameron, E. S. 1907. The birds of Custer and Dawson counties, Montana. *Auk* 24:241–270.

Cameron, R. A. D., L. Cornwallis, M. J. L. Percival, and A. R. E. Sinclair. 1967. The migration of raptors and storks through the near east in autumn. *Ibis* 109:489–501.

Caraco, T. 1981. Energy budgets, risk, and foraging preferences in dark-eyed juncos (*Junco hyemalis*). *Behav. Ecol. Sociobiol.* 8:213–217.

Casement, M. B. 1966. Migration across the Mediterranean observed by radar. *Ibis* 108:461–491.

Cherry, J. D. 1982. Fat deposition and the length of stopover of migrant White-crowned Sparrows. *Auk* 99:725–732.

Christensen, S. H., O. Lou, M. Mueller, and H. Wohlmuth. 1981. Spring migration of raptors in southern Israel and Sinai. *Sandgrouse* 3:1–42.

Cipriano, R. 1975. Sailplane measurements of planetary boundary layer structure. M.S. thesis, State University of New York, Albany.

Cipriano, R., and P. Kerlinger. 1985. Estimating the altitude of hawk migration. In *Proceedings of the Fourth Hawk Migration Conference*, ed. M. Harwood, pp. 67–73. Rochester, NY: Hawk Migration Association of North America.

Clark, R. J. 1971. Wing loading: A plea for consistency in usage. *Auk* 88:927–928.

Clark. W. S. 1981. A modified dho-gaza trap for use at a raptor banding station. *J. Wildl. Manage.* 45:1043–1044.

———. 1985a. The migrating Sharp-shinned Hawk at Cape May Point: Banding and recovery results. In *Proceedings of the Fourth Hawk Migration Conference*, ed., M. Harwood, pp. 137–148. Rochester, NY: Hawk Migration Association of North America.

———. 1985b. Migration of the Merlin along the coast of New Jersey. *Raptor Res.* 19:85–93.

Clark, W. S., and E. Gorney. 1986. Oil contamination of raptors migrating along the Red Sea. *Env. Poll.* 46:307–313.

Clark, W. S., and B. Wheeler. 1987. *Hawks: A field guide to the hawks of North America*. Peterson Field Guide series. Boston: Houghton Mifflin.

Cochran, W. W. 1972. A few days of fall migration of a Sharp-shinned Hawk. *Hawk Chalk* 11:39–44.

———. 1975. Following a migrating Peregrine from Wisconsin to Mexico. *Hawk Chalk* 14:28–37.

———. 1985. Ocean migration of Peregrine Falcons: Is the adult male pelagic? In *Proceedings of the Fourth Hawk Migration Conference*, ed. M. Harwood, pp. 223–237. Rochester, NY: Hawk Migration Association of North America.

———. 1987. Orientation and other migratory behaviours of a Swainson's Thrush followed for 1,500 km. *Anim. Behav.* 35:927–929.

Cochran, W. W., and R. D. Applegate. 1986. Speed of flapping flight of Merlins and Peregrine Falcons. *Condor* 88:397.

Cochran, W. W., G. G. Montgomery, and R. R. Graber. 1967. Migratory flights of *Hylocichla* thrushes in spring: A radiotelemetry study. *Living Bird* 6:213–225.

Cochran, W. W., and R. D. Lord, Jr. 1963. A radio-tracking system for wild animals. *J. Wildl. Manage.* 27:9–24.

Collopy, M. W., and J. R. Koplin. 1983. Diet, capture success, and mode of hunting by female American Kestrels in winter. *Condor* 85:369–371.

Cone, C. D., Jr. 1962a. Thermal soaring of birds. *Amer. Sci.* 50:180–209.

———. 1962b. The soaring flight of birds. *Sci. Amer.* 260:130–140.

Cox, G. W. 1968. The role of competition in the evolution of migration. *Evolution* 22:180–192.

———. 1985. The evolution of avian migration systems between temperate and tropical regions of the New World. *Amer. Nat.* 126:451–474.

Craighead, F. C., Jr., and T. C. Dunstan. 1976. Progress toward tracking migrating raptors by satellite. *Raptor Res.* 10:112–119.

Craighead, J. J., and F. C. Craighead. 1956. *Hawks, owls and wildlife*. Harrisburg, PA: Stackpole.

Cramp, S., and K. E. L. Simmons. 1980. *Handbook of the birds of Europe, the Middle East and North Africa: The birds of the western Palearctic*. Vol. 2. *Hawks to bustards*. Oxford: Oxford University Press.

Crook, J. H. 1965. The adaptive significance of avian social organizations. *Symp. Zool. Soc. Lond.* 14:181–218.

Curry-Lindahl, K. 1981. *Bird migration in Africa: Movements between six continents*. Vols. 1 and 2. New York: Academic Press.

Darrow, H. N. 1963. Direct autumn flight-line from Fire Island, Long Island, to the coast of southern New Jersey. *Kingbird* 13:4–12.

Dekker, D. 1970. Migrations of diurnal birds of prey in the Rocky Mountain foothills west of Cochrane, Alberta. *Blue Jay* 28:20–24.

Dementiev, G. P. 1951. Order Falconiformes (Accipitres): Diurnal raptors. In *Birds of the Soviet Union*, ed. G. P. Dementiev and N. A. Gladkov, pp. 126–136. Washington, DC: Israel Program for Scientific Translation.

Demong, N. J., and S. T. Emlen. 1978. Radar tracking of experimentally released migrant birds. *Bird-Banding* 49:342–359.

Dingle, H. 1980. Ecology and evolution of migration. In *Animal migration, orientation and navigation,* ed. S. A. Gauthreaux, Jr., pp. 2–101. New York: Academic Press.

Dingle, H., N. R. Blakeley, and E. R. Miller. 1980. Variation in body size and flight performance in milkweek bugs (*Oncopeltus*). *Evolution* 34:356–370.

Dobben, W. A. van. 1953. Bird migration in the Netherlands. *Ibis* 95: 212–234.

———. 1955. Nature and strength of the attraction exerted by leading lines. *Proc. Int. Ornithol. Congr.* 11:165–166.

Drost, R. 1938. Über den Einfluss von Verfrachtungen zur Herbstzugzeit auf den Sperber *Accipiter nisus* (L.). *Proc. Int. Ornithol. Congr.* 9:502–521.

Duncan, B. W. 1982. Sharp-shinned Hawks banded at Hawk Cliff, Ontario, 1971–1980. *Ontario Bird Banding* 15:24–38.

Dunne, P. J. 1977. Spring hawk movement along Raccoon Ridge, N.J.—1976. *Rec. N.J. Birds* 3:19–29.

———. 1978. Spring hawk migration along Raccoon Ridge, N.J.—1977. *Rec. N.J. Birds* 4:39–48.

Dunne, P. J., and W. S. Clark. 1977. Fall hawk movement at Cape May Point, N.J.—1976. *Rec. N.J. Birds* 3:114–124.

Dunne, P. J., and C. C. Sutton. 1986. Population trends in coastal raptor migrants over ten years of Cape May Point autumn counts. *Rec. N.J. Birds* 12:39–43.

Dunne, P. J., D. A. Sibley, and C. C. Sutton. 1988. *Hawks in flight: Flight identification of North American migrant raptors.* Boston: Houghton Mifflin.

Eastwood, E. 1967. *Radar ornithology.* London: Methuen.

Edelstam, C. 1972. The visible migration of birds at Ottenby, Sweden. *Fagelvarld,* suppl. 7.

Edscorn, J. B. 1974. Florida region. *Amer. Birds* 28:40–44.

Emlen, S. T. 1967. Migratory orientation in the Indigo Bunting, *Passerina cyanea.* 1. Evidence for use of celestial cues. *Auk* 84:309–342.

———. 1974. Problems in identifying species by radar signature analyses: Intraspecific variability. In *Proceedings of the Conference on the Bird/Aircraft Strike Hazard,* ed. S. A. Gauthreaux, Jr., pp. 509–524. Clemson, SC: Air Force Office of Scientific Research.

Enderson, J. H. 1960. A population study of the Sparrow Hawk in east-central Illinois. *Wilson Bull.* 72:222–231.

———. 1964. A study of the Prairie Falcon in the central Rocky Mountain region. *Auk* 81:332–352.

———. 1965. A breeding and migration survey of the Peregrine Falcon. *Wilson Bull.* 77:327–339.

Evans, D. L., and R. N. Rosenfield. 1985. Migration mortality of Sharp-shinned Hawks ringed at Duluth, Minnesota, USA. In *Conservation studies of birds of prey,* ed. I. Newton and R. D. Chancellor, pp. 311–316. ICBP

Technical Publication 5. Thessaloniki: International Council for Bird Preservation.

Evans, P. R. 1966. Migration and orientation of passerine night migrants in northeast England. *J. Zool. Lond.* 150:319–369.

———. 1970. Nocturnal songbird migration. *Nature* 228:1121.

Evans, P. R., and G. W. Lathbury. 1973. Raptor migration across the Straits of Gibraltar. *Ibis* 115:572–585.

Ferguson, A. L., and H. L. Ferguson. 1922. The fall migration of hawks as observed at Fishers Island, N.Y. *Auk* 39:488–496.

ffrench, R. 1980. *A guide to the birds of Trinidad and Tobago.* Newtown Square, PA: Harrowood Books.

Field, M. 1970. Hawk-banding on the northern shore of Lake Erie. *Ontario Bird Banding* 6:52–69.

Forbes, H. S., and H. B. Forbes. 1927. An autumn hawk flight. *Auk* 44:101–102.

Foy, R. W. 1983. A fellow in a hurry. *North Amer. Bird Bander* 8:108.

Friedman, H. 1950. *The birds of North and Middle America.* Bulletin 50. Washington, DC: U.S. National Museum.

Frisch, K. von. 1967. *The dance language and orientation of bees.* Cambridge: Harvard University Press.

Fuller, M. R., and J. R. Tester. 1973. An automated radio-tracking system for biotelemetry. *Raptor Res.* 7:105–106.

Galea, C., and B. Massa. 1985. Notes on raptor migration across the central Mediterranean. In *Conservation studies on raptors,* ed. I. Newton and R. D. Chancellor, pp. 257–262. ICBP Technical Publication 5. Thessaloniki: International Council for Bird Preservation.

Gallup, G. G. 1970. Chimpanzees: Self-recognition. *Science* 167:86–87.

Gauthreaux, S. A., Jr. 1970. Weather radar quantification of bird migration. *BioScience* 20:17–20.

———. 1971. A radar and direct visual study of passerine spring migration in southern Louisiana. *Auk* 88:343–365.

———. 1972. Behavioral responses of migrating birds to daylight and darkness: A radar and direct visual study. *Wilson Bull.* 84:136–148.

———. 1974. *Proceedings of the Conference on the Biological Aspects of Bird/Aircraft Collision Problem.* Clemson, SC: Air Force Office of Scientific Research.

———. 1978a. The ecological significance of social dominance. In *Perspectives in ethology,* ed. P. P. G. Bateson and P. H. Klopfer, pp. 17–54. New York: Plenum Press.

———. 1978b. The influence of global climatological factors on the evolution of bird migratory pathways. *Proc. Int. Ornithol. Congr.* 17:517–525.

———. 1978c. Importance of the daytime flights of nocturnal migrants: Redetermined migration following displacement. In *Animal migration, navigation, and homing,* ed. K. Schmidt-Koenig and W. T. Keeton, pp. 219–227. New York: Springer-Verlag.

———. 1982a. The ecology and evolution of avian migration systems. In

Avian biology, ed. J. R. King, D. S. Farner, and K. Parks, 6:93–168. New York: Academic Press.

———. 1982b. Age-dependent orientation in migratory birds. In *Avian navigation,* ed. F. Papi and H. G. Walraff, pp. 67–74. New York: Springer-Verlag.

———. 1985. Avian migration mobile research laboratory. In *Proceedings of the Fourth Hawk Migration Conference,* ed. M. Harwood, pp. 339–346. Rochester, NY: Hawk Migration Association of North America.

Gauthreaux, S. A., Jr., and K. P. Able. 1970. Wind and the direction of nocturnal songbird migration. *Nature* 228:476–477.

Geller, G. A., and S. A. Temple. 1983. Seasonal trends in body condition of juvenile Red-tailed Hawks during autumn migration. *Wilson Bull.* 95:492–495.

Gerrard, J. M., D. W. A. Whitfield, P. Gerrard, P. N. Gerrard, and W. J. Maher. 1978. Migratory movements and plumage of subadult Saskatchewan Bald Eagles. *Can. Field-Nat.* 92: 375–382.

Gessaman, J. A. 1979. Premigratory fat in the American Kestrel. *Wilson Bull.* 91:625–626.

———. 1980. An evaluation of heart rate as an indirect measure of daily energy metabolism of the American Kestrel. *Comp. Biochem. Physiol.* 65A:273–289.

Geyer von Schweppenburg, H. Frieh. 1963. Zur Terminologie und Theorie der Leitlinie. *J. Ornithol.* 104:191–204.

Gibo, D. L., and M. J. Pallett. 1979. Soaring flight of monarch butterflies *Danaus plexippus* (Lepidoptera: Danaidae), during the late summer migration in southern Ontario. *Can. J. Zool.* 57:1393–1401.

Glutz von Blotzheim, U. N., K. M. Bauer, and E. Bezzel. 1971. *Handbuch der Vögel Mitteleuropas.* Vol. 4. *Falconiformes.* Frankfurt am Main: Akademische Verlagsgesellschaft.

Grant, P. R. 1972. Convergent and divergent character displacement. *Biol. J. Linnean Soc.* 4:39–68.

Greenberg, R. 1980. Demographic aspects of long-distance migration. In *Migrant birds in the Neotropics,* ed. A. Keast and E. S. Morton, pp. 493–516. Washington, DC: Smithsonian Institution Press.

Greenewalt, C. H. 1962. Dimensional relationships of flying animals. *Smithsonian Misc. Coll.* 144(2):1–46.

———. 1975. The flight of birds. *Trans. Amer. Phil. Soc.* 65:1–66.

Griffin, C. R., J. M. Southern, and L. D. Frenzel. 1980. Origins and migratory movements of Bald Eagles wintering in Missouri. *J. Field-Ornithol.* 51:161–167.

Griffin, D. R. 1964. *Bird migration.* Garden City, NY: Doubleday.

———. 1973. Oriented bird migration in or between opaque cloud layers. *Proc. Amer. Phil. Soc.* 117:117–141.

———. 1984. *Animal thinking.* Cambridge: Harvard University Press.

Grigg, W. N. 1975. Whitefish Point, Michigan. In *Proceedings of the North*

American Hawk Migration Conference, 1974, pp. 27–28. Syracuse, NY: Hawk Migration Association of North America.

Grossman, M. L., and J. Hamlet. 1964. *Birds of prey of the world.* New York: Bonanza Books.

Gwinner, E. 1986. Circannual rhythms in the control of avian migration. *Adv. Study Behav.* 16:191–228.

Hall, F. F., Jr., J. G. Edinger, and W. D. Neff. 1975. Convective plumes in the planetary boundary layer, investigated with an acoustic echo sounder. *J. Appl. Meteorol.* 14:513–523.

Hamerstrom, F. 1969. A harrier population study. In *Peregrine Falcon populations: Their biology and decline,* ed. J. J. Hickey, pp. 367–385. Madison: University of Wisconsin Press.

Hamilton, W. J., III. 1962. Evidence concerning the function of nocturnal call notes of migratory birds. *Condor* 64:390–401.

Hankin, E. H. 1913. *Animal flight: A record of observation.* London. Iliffe.

Hardy, K. R., and I. Katz. 1969. Probing the clear atmosphere with high power high resolution radars. *Proc. IEEE* 57:468–480.

Hardy, K. R., and H. Ottersten. 1969. Radar investigations of convective patterns in the clear atmosphere. *J. Atmos. Sci.* 26:666–672.

Harmata, A. R. 1984. Bald Eagles in the San Luis Valley, Colorado: Their winter ecology and spring migration. Ph.D. dissertation, Montana State University.

Harmata, A. R., J. E. Toepfer, and J. M. Gerrard. 1985. Fall migration of Bald Eagles produced in northern Saskatchewan. *Blue Jay* 43:232–237.

Harrison, J. 1971. Wave soaring. *Sailplane and Gliding* 22:2–9, 92–100.

Haugh, J. R. 1972. A study of hawk migration in eastern North America. *Search* 2:1–60.

———. 1974. Large aggregations of soaring birds: The influence of weather and topography. In *Proceedings of the Conference on the Biological Aspects of the Bird/Aircraft Collision Problem,* ed. S. A. Gauthreaux, Jr., pp. 235–260. Clemson, SC: Air Force Office of Scientific Research.

Haugh, J. R., and T. J. Cade. 1966. The spring hawk migration around the southeastern shore of Lake Ontario. *Wilson Bull.* 78:88–110.

Haukioja, E., and M. Haukioja. 1970. Mortality rates of Finnish and Swedish Goshawks (*Accipiter gentilis*). *Finnish Game Res.* 31:13–20.

Heintzelman, D. S. 1975. *Autumn hawk flights: The migrations in eastern North America.* New Brunswick, NJ: Rutgers University Press.

———. 1986. *The migrations of hawks.* Bloomington: Indiana University Press.

Helbig, A. von, and V. Laske. 1986. Zeitlicher Verlauf und Zugrichtungen beim Wegzug des Stars (*Sturnus vulgaris*) in nordwestdeutschen Binnenland. Vogelwarte 33:169–191.

Hendriks, F. 1972. Dynamic soaring. Ph.D. dissertation, University of California at Los Angeles.

Heppner, F. H. 1974. Avian flight formations. *Bird-Banding* 45:160–169.

Herbert, R. A., and K. G. S. Herbert. 1965. Behavior of Peregrine Falcons in the New York City region. *Auk* 82:62–94.

Hilborn, R., and S. C. Stearns. 1982. On inference in ecology and evolutionary biology: The problem of multiple causes. *Acta Biotheor.* 31:145–164.

Hoffman, S. W. 1985. Autumn Cooper's Hawk migration through northern Utah and northeastern Nevada, 1977–1982. In *Proceedings of the Fourth Hawk Migration Conference,* ed. M. Harwood, pp. 149–166. Rochester, NY: Hawk Migration Association of North America.

Hoffman, S. W., and W. K. Potts. 1985. Fall migration of Golden Eagles in the Wellsville Mountains, northern Utah, 1976–1979. In *Proceedings of the Fourth Hawk Migration Conference,* ed. M. Harwood, pp. 207–218. Rochester, NY: Hawk Migration Association of North America.

Hofslund, P. B. 1966. Hawk migration over the western tip of Lake Superior. *Wilson Bull.* 78:79–87.

———. 1973. Do hawks feed during migration? *Raptor Res.* 7:13–14.

Holstein, V. 1944. *Hvepsevaagen* Pernis apivorus apivorus *(L.).* Copenhagen: Forlag.

Holt, J. B., and J. R. Frock. 1980. Twenty years of raptor banding on the Kittatinny Ridge, *Hawk Mountain News* 54:8–32.

Holthuizjen, A. M. A., and L. Oosterhuis. 1981. Migration patterns of female Sharp-shinned Hawks *(Accipiter striatus)* at Cape May Point, New Jersey. Technical report, Department of Fisheries and Wildlife Sciences. Blacksburg: Virginia Polytechnic Institute and State University.

———. 1985. Implications for migration counts from telemetry studies of Sharp-shinned Hawks *(Accipiter striatus)* at Cape May Point, New Jersey. In *Proceedings of the Fourth Hawk Migration Conference* ed. M. Harwood, pp. 305–312. Rochester, NY: Hawk Migration Association of North America.

Holthuizjen, A. M A., L. Oosterhuis, and M. R. Fuller. 1985. Habitat used by migrating Sharp-shinned Hawks at Cape May Point, New Jersey, USA. In *Conservation studies on raptors,* ed. I. Newton and R. D. Chancellor, pp. 317–327. ICBP Technical Publication 5. Thessaloniki: International Council for Bird Preservation.

Hopkins, D. A. 1975. The New England hawk watch. In *Proceedings of the North American Hawk Migration Conference, 1974,* pp. 137–146. Syracuse, NY: Hawk Migration Association of North America.

Hopkins, D. A., G. S. Mersereau, J. B. Mitchell, P. M. Roberts, L. J. Robinson, and W. A. Welch. 1979. *Motor-glider and cine-theodolite study of the 1979 fall Broad-winged Hawk migration in southern New England.* Greenwich, CT: Connecticut Audubon Council Hawk Migration Committee.

Houghton, E. W. 1964. Detection, recognition and identification of birds on radar. *Proc. Radar Meteor. Conf.* (Boulder), 11:14–21.

———. 1971. Spring migration at Gibraltar. Memorandum 2593. Malvern, Worcester, UK. Royal Radar Establishment.

———. 1974. Highlights of the NATO-Gibraltar bird migration radar study.

Privately distributed paper presented to Bird Strike Commission of Europe, Paris 1973. Paris: Royal Radar Establishment.

Houston, C. S. 1967. Recoveries of Red-tailed Hawks banded in Saskatchewan. *Blue Jay* 25:109–111.

———. 1968. Recoveries of Marsh Hawks banded in Saskatchewan. *Blue Jay* 26:12–13.

———. 1978. Recoveries of Saskatchewan-banded Great Horned Owls. *Can. Field-Nat.* 92:61–66.

———. 1981. Record longevity of Swainson's Hawks. *J. Field Ornithol.* 52:238.

Howell, T. R. 1953. Racial and sexual differences in migration in *Sphyrapicus varius*. *Auk* 70:118–126.

Hunt, W. G., R. R. Rogers, and D. J Stowe. 1975. Migratory and foraging behavior of Peregrine Falcons on the Texas coast. *Can. Field-Nat.* 89:111–123.

Hunt, W. G., F. P. Ward, and B. S. Johnson. 1981. The spring passage of Peregrine Falcons at Padre Island, Texas: A radiotelemetry study. Abstract of a paper presented at the Raptor Research Foundation 1981 Annual Meeting, Montreal.

Hussell, D. J. T. 1985. Analysis of hawk migration counts for monitoring population levels. In *Proceedings of the Fourth Hawk Migration Conference,* ed. M. Harwood, pp. 243–254. Rochester, NY: Hawk Migration Association of North America.

Janes, S. W. 1985. Habitat selection in raptorial birds. In *Habitat selection in birds,* ed. M. L. Cody, pp. 159–199. New York: Academic Press.

Johnston, D. W., and R. W. McFarlane. 1967. Migration and bioenergetics of flight in the Pacific Golden Plover. *Condor* 69:156–168.

Juillard, M. 1977. Observations sur l'hivernage et les dorloirs du Miland royal *Milvus milvus* (L.) dans le nord-ouest de la Suisse. *Nos Oiseaux* 34:41–57.

Kaimal, J. C., and J. A. Businger. 1970. Case studies of a convective plume and a dust devil. *J. Appl. Meteorol.* 9:612–620.

Katz, P. L. 1974. A long-term approach to foraging optimization. *Amer. Nat.* 108:758–782.

Keast, A., and E. S. Morton. 1980. *Migrant birds in the Neotropics.* Washington, DC: Smithsonian Institution Press.

Keeton, W. J. 1970. Comparative orientational and homing performances of single pigeons and small flocks. *Auk* 87:797–799.

Keith, L. B. 1964. Territoriality among wintering Snowy Owls. *Can. Field-Nat.* 78:17–24.

Kennedy, J. J. 1983. *Analyzing qualitative data: Introductory log-linear analysis for behavioral research.* New York: Praeger.

Kerlinger, P. 1982a. Aerodynamic performance and flight speed selection of migrating hawks. Ph.D. dissertation, State University of New York, Albany.

———. 1982b. The migration of Common Loons through eastern New York. *Condor* 84:97–100.

———. 1984. Flight behaviour of Sharp-shinned Hawks during migration. 2. Over water. *Anim. Behav.* 32:1029–1034.

———. 1985a. Water-crossing behavior of raptors during migration. *Wilson Bull.* 97:109–113.

———. 1985b. Daily rhythm of hawk migration, noonday lulls, and counting bias: A review. In *Proceedings of the Fourth Hawk Migration Conference,* ed. M. Harwood, pp. 259–266. Rochester, NY: Hawk Migration Association of North America.

———. 1985c. A theoretical approach to the function of flocking among soaring migrants. In *Proceedings of the Fourth Hawk Migration Conference,* ed. M. Harwood, pp. 41–50. Rochester, NY: Hawk Migration Association of North America.

Kerlinger, P., V. P. Bingman, and K. P. Able. 1985. Comparative flight behaviour of migrating hawks studied with tracking radar during migration in central New York. *Can. J. Zool.* 63:755–761.

Kerlinger, P., J. D. Cherry, and K. D. Powers. 1983. Records of migrant hawks from the North Atlantic Ocean. *Auk* 100:488–490.

Kerlinger, P., and S. A. Gauthreaux, Jr. 1984. Flight behaviour of Sharp-shinned Hawks during migration. 1. Over land. *Anim. Behav.* 32:1021–1028.

———. 1985a. Seasonal timing, geographic distribution, and flight behavior of Broad-winged Hawks during spring migration in south Texas: A radar and visual study. *Auk* 102:735–743.

———. 1985b. Flight behavior of raptors during spring migration in south Texas studied with radar and visual observations. *J. Field Ornithol.* 56:394–402.

Kerlinger, P., and M. R. Lein. 1986. Differences in winter range among age-sex classes of Snowy Owls *Nyctea scandiaca* in North America. *Ornis Scand.* 17:1–7.

———. 1988. Causes of mortality, fat condition, and weights of wintering Snowy Owls. *J. Field Ornithol.* 59:7–12.

Kerlinger, P., and F. R. Moore. 1989. Atmospheric structure and avian migration. *Current ornithology,* vol. 6, ed. D. Powers. New York: Plenum Press. In press.

Ketterson, E. D., and V. Nolan, Jr. 1976. Geographic variation and its climatic correlates in the sex ratio of eastern-wintering Dark-eyed Juncos (*Junco hyemalis hyemalis*). *Ecology* 57:679–693.

———. 1983. The evolution of differential bird migration. In *Current ornithology,* vol. 1, ed. R. F. Johnston, pp. 357–402. New York: Plenum Press.

King, J. R., and D. S. Farner. 1959. Premigratory changes in body weight and fat in wild and captive White-crowned Sparrows. *Condor* 61:314–324.

King, J. R., D. S. Farner, and L. R. Mewaldt. 1965. Seasonal sex and age ratios in populations of White-crowned Sparrows of the race *gambelii*. *Condor* 67:489–504.

Kipp, F. A. 1958. Zur Geschichte des Vogelzuges auf der Grundlage der Flügelanpassungen. *Vogelwarte* 19:233–242.

Kirkwood, J. K. 1985. Food requirements for deposition of energy reserves in raptors. In *Conservation studies on raptors,* ed. I. Newton and R. D. Chancellor, pp. 295–298. ICBP Technical Publication 5. Thessaloniki: International Council for Bird Preservation.

Klem, D., Jr., B. S. Hillegass, and D. A. Peters. 1985. Raptors killing raptors. *Wilson Bull.* 97:230–231.

Klem, D., Jr., B. S. Hillegass, D. A. Peters, J. A. Villa, and K. Kranick. 1985. Analysis of individual flight patterns of migrating raptors at a break in the Kittatinny Ridge: Lehigh Gap, Pennsylvania. In *Proceedings of the Fourth Hawk Migration Conference,* ed. M. Harwood, pp. 1–11. Rochester, NY: Hawk Migration Association of North America.

Knight, S. K., and R. L. Knight. 1983. Aspects of food finding by wintering Bald Eagles. *Auk* 100:477–484.

Kochenberger, R., and P. J. Dunne. 1985. The effects of varying observer numbers of raptor count totals at Cape May, New Jersey. In *Proceedings of the Fourth Hawk Migration Conference,* ed. M. Harwood, pp. 281–293. Rochester, NY: Hawk Migration Association of North America.

Kokshaysky, N. V. 1973. Tracing the wake of a flying bird. *Nature* 279:146–148.

Konrad, T. G. 1968. The alignment of clear-air convective cells. In *Proceedings of the International Conference on Cloud Physics* (Toronto), pp. 539–543. Boston: American Meteorological Society.

———. 1970. The dynamics of the convective process in clear air as seen by radar. *J. Atmos. Sci.* 27:1138–1147.

Konrad, T. G., and J. S. Brennan. 1971. Radar observations of the convective process in the clear air: A review. *Aero-Revue* 46:425–429.

Konrad, T. G., and R. A. Kropfli. 1968. Radar observations of clear-air convection over the sea. *Proc. Radar Meteorol. Conf.* 13:262–269.

Konrad, T. G., and F. L. Robison. 1973. Development and characteristics of free convection in the clear air as seen by radar. *J. Appl. Meteorol.* 12:1284–1294.

Koplin, J. R. 1973. Differential habitat use by sexes of American Kestrels wintering in northern California. *Raptor Res.* 7:39–42.

Kramer, G. 1957. Experiments in bird orientation. *Ibis* 101:399–416.

Krapu, G. L., G. C. Iverson, K. J. Reinecke, and C. M. Boise. 1985. Fat deposition and usage by arctic-nesting Sandhill Cranes during spring. *Auk* 102:362–368.

Kuettner, J. P. 1972. Cloudstreet waves. *Soaring* 37(6): 33–37.

Kuettner, J. P., and S. D. Soules. 1966. Organized convection as seen from space. *Amer. Meteorol. Soc.* 47:364–369.

Kuhn, T. 1962. *The structure of scientific revolutions.* Chicago: University of Chicago Press.

Kuroda, N. 1961a. A note on the pectoral muscle of birds. *Auk* 78:261–263.

———. 1961b. The over-sea crossings of land birds in the Pacific. *Yamashina's Inst. Ornithol. Zool.* 3:47–53.

Lack, D. 1954. Migration across the sea. *Ibis* 101:374–399.

———. 1960. The influence of weather on passerine migration: A review. *Auk* 77:171–209.

———. 1968. Bird migration and natural selection. *Oikos* 19:1–9.

Lack, D., and G. C. Varley. 1945. Detection of birds by radar. *Nature* 156:446.

Lack, D., and K. Williamson. 1959. Bird-migration terms. *Ibis* 102:255–256.

Lambert, A. 1983. *Annual report of the Great Lakes beached bird survey.* Long Point, Ont.: Long Point Bird Observatory.

Larkin, R. P. 1982. Spatial distribution of migrating birds and small-scale atmospheric motion. In *Avian navigation,* ed. F. Papi and H. G. Walraff, pp. 28–37. New York: Springer-Verlag.

Lengerke, J. von. 1908. Migration of hawks. *Auk* 25:315–316.

Lenschow, D. H. 1970. Airplane measurements of planetary boundary layer structure. *J. Appl. Meteorol.* 9:874–882.

Leshem, Y. 1985. Israel: An international axis of raptor migration. In *Conservation studies on raptors,* ed. I. Newton and R. D. Chancellor, pp. 243–250. ICBP Technical Publication 5. Thessaloniki: International Council for Bird Preservation.

———. 1987. Wings over Israel. *BBC Wildlife* 5:31–34.

———. n.d. *Following raptor migration from the ground, motorized glider, and radar at a junction of three continents.* ICBP volume 6. Eilat, Israel: ICBP. In press.

Lewontin, R. C. 1978. Adaptation. *Sci. Amer.* 239(3):212–230.

Lincoln, F. C. 1952. *Migration of birds.* Garden City, NY: Doubleday.

Lissaman, P. B. S., and C. A. Schollenberger. 1970. Formation flight of birds. *Science* 168:1003–1005.

Lorenz, K. 1981. *The foundations of ethology.* New York: Springer-Verlag.

Lowery, G. H. 1945. Trans-Gulf spring migration of birds and the coastal hiatus. *Wilson Bull.* 57:98–121.

Lundberg, A. 1979. Residency, migration and a compromise: Adaptations to nest-site scarcity and food specialization in three Fennoscandian owl species. *Oecologia* 41:273–281.

Lyons, W. A. 1972. The climatology and prediction of the Chicago lake breeze. *J. Appl. Meteorol.* 11:1259–1270.

McClure, H. E. 1974. Migration and survival of the birds of Asia. U.S. Army Medical Component, South East Asia Treaty Organization Medical Project, Bangkok, Thailand.

MacCready, P. B., Jr. 1976. Soaring bird aerodynamics—clues for hang gliding. *Ground Skimmer* 45:17–19.

———. 1984. Soaring aerodynamics of Frigate Birds *(sic). Soaring* 48(7):20–22.

McFarland, D. J. 1977. Decision making in animals. *Nature* 269:15–21.

McGahan, J. 1973a. Gliding flight of the Andean Condor in nature. *J. Exp. Biol.* 58:225–237.

———. 1973b. Flapping flight of the Andean Condor in nature. *J. Exp. Biol.* 58:239–353.

McLandress, M. R., and D. G. Raveling. 1981. Changes in diet and body composition of Canada Geese before spring migration. *Auk* 98:65–78.

MacRae, D. 1985. Over-water migration of raptors: A review of the literature. In *Proceedings of the Fourth Hawk Migration Conference,* ed. M. Harwood, pp. 75–98. Rochester, NY: Hawk Migration Association of North America.

Malkus, J. S. 1953. Aeroplane studies of trade-wind meteorology. *Weather, Lond.* 8:291–300.

———. 1956. Trade winds and trade wind clouds. *Oceanus* 4:8–12.

Malmberg, T. 1955. Topographical concentration of flight lines. *Proc. Int. Ornithol. Congr.* 11:161–164.

Manton, M. J. 1977. On the structure of convection. *Boundary-Layer Meteorol.* 12:491–503.

Marquiss, M., and I. Newton. 1982. Habitat preference in male and female Sparrowhawks. *Ibis* 124:324–328.

Marsh, R. L., and R. W. Storer. 1981. Correlation of flight-muscle size and body mass in Cooper's Hawks: A natural analogue of power training. *J. Exp. Biol.* 91:363–368.

Martin, S. J., and T. P. McEneany. 1984. Observations of migrating Swainson's Hawks in Wyoming. *Prairie Nat.* 16:62.

Masman, D., and M. Klaassen. 1987. Energy expenditure during free flight in trained and free-living European Kestrels (*Falco tinnunculus*). *Auk* 104:603–616.

Mayr, E. 1982. *The growth of biological thought: Diversity, evolution, and inheritance.* Cambridge, MA: Belknap.

Matray, P. F. 1974. Broad-winged Hawk nesting ecology. *Auk* 91:307–324.

Mead, C. J. 1973. Movements of British raptors. *Bird Study* 20:259–286.

Mearns, R. 1982. Winter occupation of breeding territory and winter diet of Peregrines in south Scotland. *Ornis Scand.* 13:79–83.

Mech, L. D. 1983. *Handbook of animal radio-tracking.* Minneapolis: University of Minnesota Press.

Melquist, W. E., and W. D. Carrier. 1978. Migration patterns of northern Idaho and eastern Washington Ospreys. *Bird-Banding* 49:234–236.

Mills, G. S. 1975. A winter population study of the American Kestrel in central Ohio. *Wilson Bull.* 87:241–247.

———. 1976. American Kestrel sex ratios and habitat selection. *Auk* 93:740–748.

Millsap, B. A., and J. R. Zook. 1983. Effects of weather on *Accipiter* migration in southern Nevada. *Raptor Res.* 18:10–15.

Mindell, D. P. 1983. Harlan's Hawk (*Buteo jamaicensis harlani*): A valid subspecies. *Auk* 100:161–169.

———. 1985. Biogeography of New World migrant raptors: A review. In *Proceedings of the Fourth Hawk Migration Conference,* ed. M. Harwood, pp. 369–385. Rochester, NY: Hawk Migration Association of North America.

Mises, R. von. 1959. *Theory of flight.* New York: Dover.

Moon, N. S., and L. W. Moon. 1985. Raptor counts at Braddock Bay, New York: The effects of observer coverage at auxiliary sites. In *Proceedings of the Fourth Hawk Migration Conference,* ed. M. Harwood, pp. 295–305. Rochester, NY: Hawk Migration Association of North America.

Moore, F. R. 1976. The dynamics of seasonal distribution of Great Lakes Herring Gulls. *Bird-Banding* 47:141–159.

———. 1984. Age-dependent variability in the migratory orientation of the Savannah Sparrow (*Passerculus sandwichensis*). *Auk* 101:875–880.

Moore, F. R., and P. Kerlinger. 1987. Stopover and fat deposition by North American Wood-Warblers (Parulinae) following spring migration over the Gulf of Mexico. *Oecologia* 74:47–54.

Moreau, R. E. 1953. Migration in the Mediterranean area. *Ibis* 95:329–364.

———. 1972. *The Palearctic-African bird migration systems.* New York: Academic Press.

Moritz, D., and G. Vauk. 1976. Der Zug des Sperbers (*Accipiter nisus*) auf Helgoland. *J. Ornithol.* 117:317–328.

Moynihan, M. H. 1962. The organization and probable evolution of some mixed species flocks of Neotropical birds. *Smithsonian Misc. Coll.* 143:1–140.

Mueller, H. C., and D. D. Berger. 1961. Weather and fall migration of hawks at Cedar Grove, Wisconsin. *Wilson Bull.* 73:171–193.

———. 1967a. Wind drift, leading lines, and diurnal migrations. *Wilson Bull.* 79:50–63.

———. 1967b. Fall migration of Sharp-shinned Hawks. *Wilson Bull.* 79:397–415.

———. 1969. Navigation by hawks migrating in spring. *Auk* 86:35–40.

———. 1970. Prey preferences in the Sharp-shinned Hawk: The role of sex, experience and motivation. *Auk* 87:452–457.

———. 1973. The daily rhythm of hawk migration at Cedar Grove, Wisconsin. *Auk* 90:591–596.

Mueller, H. C., D. D. Berger, and G. Allez. 1977. The periodic invasions of Goshawks. *Auk* 94:652–663.

———. 1979. Age and sex differences in size of Sharp-shinned Hawks. *Bird-Banding* 50:34–44.

———. 1981a. Age, sex, and seasonal differences in size of Cooper's Hawks. *J. Field Ornithol.* 52:112–126.

———. 1981b. Age and sex differences in wing loading and other aerodynamic characteristics of Sharp-shinned Hawks. *Wilson Bull.* 93:491–499.

Murray, B. G., Jr. 1964. A review of Sharp-shinned Hawk migration along the northeastern coast of the United States. *Wilson Bull.* 76:257–264.

———. 1969. Sharp-shinned Hawk migration along the northeastern coast of the United States. *Wilson Bull.* 81:119–120.

Murray, D. K. W. 1931. A North Sea bird log, 1928–1929. *Brit. Birds* 24:114–120.

Myers, J. P. 1981. A test of three hypotheses for latitudinal segregation of the sexes in winter. *Can. J. Zool.* 59:1527–1534.

Nagel, E. 1961. *The structure of science: Problems in the logic of scientific explanation.* New York: Harcourt, Brace and World.

Nagy, A. C. 1977. Population trend indices based on forty years of autumn counts at Hawk Mountain. In *World conference on birds of prey,* ed. R. D. Chancellor, pp. 243–253. Vienna: ICBP.

Nagy, K. A. 1980. CO_2 production in animals: Analysis of potential errors in the doubly labeled water method. *Amer. J. Physiol.* 238:R466-R473.

Newman, B. G. 1958. Soaring in gliding flight of the Black Vulture. *J. Exp. Biol.* 35:280–284.

Newton, I. 1975. Movements and mortality of British Sparrowhawks. *Bird Study* 22:35–43.

———. 1979. *Population ecology of raptors.* Vermillion, SD: Buteo Books.

———. 1986. *The Sparrowhawk.* Calton, Stoke-on-Trent: T. and A. D. Poyser.

Newton, I., E. Meek, and B. Little. 1978. Breeding ecology of the Merlin in Northumberland. *Brit. Birds* 71:376–398.

Nielsen, B. P., and S. Christensen. 1970. Observations on the autumn migration of raptors in the Lebanon. *Ornis Scand.* 1:65–73.

Nisbet, I. C. T. 1970. Autumn migration of the Blackpoll Warbler: Evidence for long flight provided by regional survey. *Bird-Banding* 41:207–240.

Oehme, H. 1977. On the aerodynamics of separated primaries in the avian wing. In *Scale effects in animal locomotion,* ed. T. J. Pedley, pp. 479–495. London: Academic Press.

Oke, T. R. 1978. *Boundary layer climates.* New York: John Wiley.

Olsson, O. 1958. Dispersal, migration, longevity, and death causes of *Strix aluco, Buteo buteo, Ardea cinerea* and *Larus argentatus. Acta Vertebr.* 1:91–189.

O'Malley, B. J., and R. M. Evans. 1982. Flock formation in White Pelicans. *Can. J. Zool.* 60:1024–1031.

Osterlof, S. 1977. Migration, wintering areas, and site tenacity of the European Osprey *Pandion h. haliaetus* (L). *Ornis Scand.* 8:61–78.

Paevskii, V. A. 1973. Atlas of bird migrations according to banding data at the Courland Spit. In *Bird migration: Ecological and physiological factors.* ed. B. E. Bykhovskii, pp. 1–124. New York: John Wiley.

Parrott, G. L. 1970. Aerodynamics of gliding flight of a Black Vulture *Coragyps atratus. J. Exp. Biol.* 53:363–374.

Pennycuick, C. J. 1960. Gliding flight of the Fulmar Petrel. *J. Exp. Biol.* 37:330–337.

———. 1968a. Power requirements for horizontal flight in the pigeon *Columba livia. J. Exp. Biol.* 49:527–555.

———. 1968b. A wind-tunnel study of gliding flight in the pigeon *Columba livia. J. Exp. Biol.* 49:509–526.

———. 1969. The mechanics of migration. *Ibis* 111:525–556.

———. 1971a. Gliding flight of the White-backed Vulture *Gyps africanus. J. Exp. Biol.* 55:13–38.

————. 1971b. Gliding flight of the dog-faced bat *Rousettus aegyptiacus* observed in a wind tunnel. *J. Exp. Biol.* 55:833–845.

————. 1972a. Soaring behaviour and performance of some East African birds observed from a motorglider. *Ibis* 114:178–218.

————. 1972b. *Animal flight.* London: Edward Arnold.

————. 1975. Mechanics of flight. In *Avian biology,* ed. D. S. Farner and J. R. King, 5:1–73. New York:Academic Press.

————. 1978. Fifteen testable predictions about bird flight. *Oikos* 30:165–176.

————. 1982a. The flight of petrels and albatrosses (Procellariiformes), observed in south Georgia and its vicinity. *Phil. Trans. Roy. Soc. Lond.,* ser. B, 300:75–106.

————. 1982b. The ornithodolite: An instrument for collecting large samples of bird speed measurements. *Phil. Trans. Roy. Soc. Lond.,* ser. B, 300:61–73.

————. 1983. Thermal soaring compared in three dissimilar tropical bird species *Fregata magnificens, Pelecanus occidentalis,* and *Coragyps atratus. J. Exp. Biol.* 102:307–325.

Pennycuick, C. J., T. Alerstam, and B. Larsson. 1979. Soaring migration of the Common Crane *Grus grus* observed by radar and from an aircraft. *Ornis Scand.* 10:241–251.

Perdeck, A. C. 1958. Two types of orientation in migrating starlings, *Sturnus vulgarus* L., and chaffinches, *Fringilla coelebs* L., as revealed by displacement experiments. *Ardea* 46:1–37.

Perkins, J. P. 1964. Seventeen flyways over the Great Lakes, part 1. *Audubon Mag.* 66:294–299.

Pielke, R. A. 1974. A comparison of three-dimensional and two-dimensional numerical predictions of sea breezes. *J. Atmos. Sci.* 31:1577–1585.

————. 1984. *Mesoscale meteorological modeling.* New York: Academic Press.

Pienkowski, M. W., P. R. Evans, and D. J. Townshend. 1985. Leap-frog and other migration patterns of waders: A critique of Alerstam and Hogstedt, and some alternative hypotheses. *Ornis Scand.* 16:61–70.

Platt, J. B. 1976. Gyrfalcon nest site selection and winter activity in the western Canadian Arctic. *Can. Field-Nat.* 90:338–345.

Poole, A. F., and B. Agler. 1987. Recoveries of Ospreys banded in the United States, 1914–1984. *J. Wildl. Manage.* 51:148–156.

Poole, E. L. 1938. Weights and wing areas in North American birds. *Auk* 55:511–517.

Porter, R., and I. Willis. 1968. The autumn migration of soaring birds at the Bosphorus. *Ibis* 110:520–536.

Porter, R. F., I. Willis, S. Christensen, and B. B. Nielsen. 1981. *Flight identification of European raptors.* Calton, Stoke-on-Trent: T. and A. D. Poyser.

Pulliam, H. R. 1973. On the advantages of flocking. *J. Theor. Biol.* 38:419–422.

Pulliam, H. R., and G. C. Millikan. 1982. Social organization in the non-reproductive season. In *Avian biology,* ed. D. S. Farner, J. R. King, and K. C. Parkes, 6:169–198. New York: Academic Press.

Pyke, G. H. 1981. Optimal travel speeds of animals. *Amer. Nat.* 118:475–487.

Pyke, G. H., H. R. Pulliam, and E. L. Charnov. 1977. Optimal foraging: A selective review of theory and tests. *Quart. Rev. Biol.* 52:137–154.

Rabenold, K. N., and P. P. Rabenold. 1985. Variation in altitudinal migration, winter segregation, and site tenacity in two subspecies of Dark-eyed Juncos in the southern Appalachians. *Auk* 102:805–819.

Rabenold, P. P. 1986. Family associations in communally roosting Black Vultures. *Auk* 103:32–41.

Rabol, J. 1974. Correlation between coastal and inland migratory movements. *Dansk. Orn. For. Tidsskr.* 68:5–14.

Rabol, J., and H. Noer. 1973. Spring migration in the Skylark (*Alauda arvensis*) in Denmark: Influence of environmental factors on the flocksize and the correlation between flocksize and migratory direction. *Vogelwarte* 27:50–65.

Raspet, A. 1960. Biophysics of bird flight. *Science* 132:191–200.

Raveling, D. G. 1976. Migration reversal: A regular phenomenon of Canada Geese. *Science* 193:153–154.

Rayner, J. M. V. 1977. The intermittent flight of birds. In *Scale effects in animal locomotion,* ed. T. J. Pedley, pp. 437–443. New York: Academic Press.

———. 1979a. A new approach to animal flight mechanics. *J. Exp. Biol.* 80:17–54.

———. 1979b. A vortex theory of animal flight. 1. The vortex wake of a hovering animal. *J. Fluid Mech.* 91:679–730.

———. 1979c. The vortex theory of animal flight. 2. The forward flight of birds. *J. Fluid. Mech.* 91:731–763.

———. 1985a. *Vertebrate flight: A bibliography to 1985.* Bristol: University of Bristol Press.

———. 1985b. Bounding and undulating flight in birds. *J. Theor. Biol.* 117:47–77.

Redig, P., G. E. Duke, and W. Jones. 1981. Recoveries and resightings of released rehabilitated raptors. *Raptor Res.* 15:97–107.

Reichmann, H. 1978. *Cross-country soaring.* Pacific Palisades, Ca: Thomson.

Richardson, W. J. 1972. Autumn migration and weather in eastern Canada. *Amer. Birds* 26:10–17.

———. 1975. Autumn hawk migration in Ontario studied with radar. In *Proceedings of the North American Hawk Migration Conference, 1974,* pp. 47–58. Syracuse, NY: Hawk Migration Association of North America.

———. 1976. Autumn migration over Puerto Rico and the western Atlantic: A radar study. *Ibis* 118:309–332.

———. 1978. Timing and amount of bird migration in relation to weather: A review. *Oikos* 30:224–272.

———. 1979. Southeastward shorebird migration over Nova Scotia and New Brunswick in autumn: A radar study. *Can. J. Zool.* 57:107–124.

Ricklefs, R. E. 1987. Community diversity: Relative roles of local and regional processes. *Science* 235:167–171.

Robbins, C. S. 1975. A history of North American hawkwatching. In *Proceedings of the North American Hawk Migration Conference, 1974*, pp. 29–40. Syracuse, NY: Hawk Migration Association of North America.

Roberts, T. S. 1907. A Lapland Longspur tragedy. *Auk* 24:369–377.

Robinson, W. 1950. Montagu's Harriers. *Bird Notes* 24:103–114.

Roest, A. I. 1957. Notes on the American Sparrow Hawk. *Auk* 74:1–19.

Rogers, W., and S. Leatherwood. 1981. Observations of feeding at sea by a Peregrine Falcon and an Osprey. *Condor* 83:89–90.

Roos, G. 1978. Counts of migrating birds and environmental monitoring: Long-term changes in the volume of autumn migration at Falsterbo 1942–1977. *Anser* 17:133–138.

———. 1985. Strackrakningar vid Falsterbo hosten 1984. [Visible bird migration in autumn 1984, with English summary]. *Anser* 24:1–28.

Rosenfield, R. N., and D. L. Evans. 1980. Migration incidence and sequence of age and sex classes of the Sharp-shinned Hawk. *Loon* 52:66–69.

Rowland, J. R. 1973. Intensive probing of the clear convective field by radar and instrumented drone aircraft. *J. Atmos. Sci.* 12:149–158.

Rowlett, R. A. 1980. Migrant Broad-winged Hawks in Tobago. *J. Hawk Migr. Assoc. North Amer.* 2:54.

Rudebeck, G. 1950. Studies on bird migration based on field studies in southern Sweden. *Fagelvarld*, suppl. 1:1–148.

———. 1950–1951. The choice of prey and modes of hunting of predatory birds with special reference to their selective effect. *Oikos* 2:65–88, 3:200–231.

Rusling, W. J. 1936. The study of the habits of diurnal migrants, as related to weather and land masses during the fall migration on the Atlantic coast, with particular reference to the hawk flights of the Cape Charles (Virginia) region. Unpublished manuscript, New York, National Audubon Society.

Safriel, U. 1968. Bird migration at Eilat, Israel. *Ibis* 110:283–320.

Salomonsen, F. 1955. The evolutionary significance of bird migration. *K. Dansk. Vidensk. Selsk. Biol. Medd.* 22:1–66.

Santilli, A. 1972. How to reach waves via thermals. *Soaring* 37(1): 24–30.

Sattler, G., and J. Bart. 1985a. Reliability of counts of migrating raptors: An experimental analysis. *J. Field Ornithol.* 55:415–423.

———. 1985b. A technique for evaluating observer efficiency in raptor migration counts. In *Proceedings of the Fourth Hawk Migration Conference*, ed. M. Harwood, pp. 275–280. Rochester, NY: Hawk Migration Association of North America.

Sauer, E. G. F. 1961. Further studies on the stellar orientation of nocturnally migrating birds. *Psychol. Forschung* 26:224–244.

Saunders, W. E. 1907. A migration disaster in western Ontario. *Auk* 24:108–110.

Savile, D. B. O. 1957. Adaptive evolution in the avian wing. *Evolution* 11:212–224.

Schelde, O. 1960. The migration of Danish Sparrowhawks (*Accipiter nisus* L.) *Dansk Orn. Foren. Tidsskr.* 54:88–102.

Schifferli, A. 1967. Vom Zug Schweizerischer und Deutscher Schwarzer Milane nach Ringfunden. *Orn. Beob.* 64:34–51.

Schmid, H., T. Steuri, and B. Bruderer. 1986. Zugverhalten von Mausebussard *Buteo buteo* und Sperber *Accipiter nisus* im Alpenraum. *Orn. Beob.* 83:111–134.

Schmidt-Nielsen, K. 1984. Scaling: Why is animal size so important? New York: Cambridge University Press.

Schnell, G. D., and J. J. Hellack. 1979. Bird flight speeds in nature: Optimized or a compromise? *Amer. Nat.* 113:53–66.

Schuz, E. 1972. *Vom Vogelzug: Grundriss der Vogelzugskunde.* 2d ed. Frankfurt am Main: Paul Schops.

Scorer, R. S. 1978. *Environmental aerodynamics.* New York: John Wiley.

Segal, M., R. T. McNider, R. A. Pielke, and D. S. McDougal. 1982. A numerical model simulation of the regional air pollution meterology of the greater Chesapeake Bay area—summer day case study. *Atmos Env.* 16:1381–1397.

Servheen, C. 1976. Bald Eagles soaring into opaque cloud. *Auk* 93:387.

Shelley, E., and S. Benz. 1985. Observations of aerial hunting, food carrying and crop size of migrant raptors. In *Conservation studies on raptors,* ed. I. Newton and R. D. Chancellor, pp. 299–301. ICBP Technical Publication 5. Thessaloniki: ICBP.

Sielman, M. S., L. A. Sheriff, and T. C. Williams. 1981. Nocturnal migration at Hawk Mountain, Pennsylvania: Preliminary results of a new method for studying nocturnal bird migration near mountains. *Amer. Birds* 35:906–909.

Smallwood, J. A. 1988. A mechanism of sexual segregation by habitat in American Kestrels (*Falco sparverius*) wintering in south-central Florida. *Auk* 105:36–46.

Smeenk, C. 1974. Comparative ecological studies on some east African birds of prey. *Ardea* 62:1–97.

Smith, D. G., C. R. Wilson, and H. H. Frost. 1972. The biology of the American Kestrel in central Utah. *Southwest. Nat.* 17:73–83.

Smith, N. G. 1973. Spectacular *Buteo* migration over Panama Canal Zone. *Amer. Birds* 27:3–5.

———. 1980. Hawk and vulture migrations in the Neotropics. In *Migrant birds in the Neotropics,* ed. A. Keast and E. S. Morton, pp. 50–61. Washington, DC: Smithsonian Institution Press.

———. 1985a. Dynamics of transisthmusian migration of raptors between Central and South America. In *Conservation studies of birds of prey,* ed. I. Newton and R. D. Chancellor, pp. 271–290. ICBP Technical Publication 5. Thessaloniki: International Council for Bird Preservation.

———. 1985b. Thermals, cloud streets, trade winds, and tropical storms: How migrating raptors make the most of atmospheric energy in Central

America. In *Proceedings of the Fourth Hawk Migration Conference,* ed. M. Harwood, pp. 51–65. Rochester, NY: Hawk Migration Association of North America.

Smith, N. G., D. L. Goldstein, and G. A. Bartholomew. 1986. Is long-distance migration possible for soaring hawks using only stored fat? *Auk* 103:607–611.

Snyder, N. F. R. 1974. Breeding biology of Swallow-tailed Kites in Florida. *Living Bird* 13:73–97.

Snyder, N. F. R., and J. W. Wiley. 1976. *Sexual size dimorphism in hawks and owls of North America.* Ornithological Monographs. Lawrence, KS: American Ornithologists' Union.

Southern, W. E. 1964. Additional observations on winter Bald Eagle populations: Including remarks on biotelemetry techniques and immature plumages. *Wilson Bull.* 76:121–137.

Spendelow, P. 1985. Starvation of a flock of Chimney Swifts on a very small Caribbean island. *Auk* 102:387–388.

Spillman, J. 1978. The use of wing tip sails to reduce vortex drag. *Aeronautical J.* 82:387–395.

Spofford, W. R. 1969. Hawk Mountain counts as population indices in northeastern America. In *Peregrine Falcon populations: Their biology and decline,* ed. J. J. Hickey, pp. 323–333. Madison: University of Wisconsin Press.

Stearns, E. I. 1949. The study of hawks in flight from a blimp. *Wilson Bull.* 61:110.

Stewart, P. A. 1977. Migratory movements and mortality rate of Turkey Vultures. *Bird-Banding* 48:122–124.

Steyn, P. 1973. Observations on the Tawny Eagle. *Ostrich* 44:1–22.

Stoddard, P. K., J. E. Marsden, and T. C. Williams. 1983. Computer simulation of autumnal bird migration over the western North Atlantic. *Anim. Behav.* 31:173–180.

Stone, W. 1937. *Bird studies at Old Cape May.* Philadelphia: Delaware Valley Ornithology Club.

Stong, C. L. 1974. Hang gliding, or sky surfing, with a high-performance low-speed wing. *Sci. Amer.* 231(6):138–143.

Storer, R. W. 1966. Sexual dimorphism and food habits in three North American accipiters. *Auk* 83:423–436.

Stuebe, M. M., and E. D. Ketterson. 1982. A study of fasting in Tree Sparrows (*Spizella arborea*) and Dark-eyed Juncos (*Junco hyemalis*): Ecological implications. *Auk* 99:299–308.

Stull, R. B. 1987. Atmospheric boundary-layer. In *The encyclopedia of physical science and technology,* vol. I, ed. R. A. Meyers. New York: Academic Press.

———. 1988. *An introduction to boundary-layer meteorology.* Dordrecht, Netherlands: Kluwer Academic Publications.

Stull, R., and E. Eloranta. 1984. Boundary layer experiment—1983. *Bull. Amer. Meteorol. Soc.* 65:450–456.

Sulkava, S. 1964. On the behaviour and food habits of the Sparrow Hawk (*Accipiter nisus*) during the nesting season. *Suomen Riista* 17:93–105.

Tabb, E. C. 1979. Winter recoveries in Guatemala and southern Mexico of Broad-winged Hawks banded in south Florida. *North Amer. Bird Bander* 4:60.

Terrill, S. B., and R. D. Ohmart. 1984. Facultative extension of fall migration by Yellow-rumped Warblers (*Dendroica coronata*). *Auk* 101:427–438.

Thake, M. A. 1980. Gregarious behaviour among migrating Honey Buzzards (*Pernis apivorus*). *Ibis* 122:500–505.

Thiollay, J. M. 1977. Distribution saisonnière des rapace diurnes en Afrique occidentale. *Oiseau* 47:25–85.

———. 1978. Les migrations de rapaces en Afrique occidentale: Adaptations écologiques aux fluctuations saisonnières de production des ecosystems. *Terre et Vie* 32:89–133.

———. 1980. Spring hawk migration in eastern Mexico. *Raptor Res.* 14:13–20.

Thomas, S. 1975. Hook Mountain, New York. In *Proceedings of the North American Hawk Migration Conference, 1974,* pp. 19–20. Syracuse, NY: Hawk Migration Association of North America.

Thompson, A. L. 1926. *Problems of bird migration.* Boston: Houghton Mifflin.

Tinbergen, N. 1951. *The study of instinct.* London: Oxford University Press.

Titus, K., and J. A. Mosher. 1982. The influence of seasonality and selected weather variables on autumn migration of three species of hawks through the central Appalachians. *Wilson Bull.* 94:176–184.

Torre-Bueno, J. R. 1976. Temperature regulation and heat dissipation during flight in birds. *J. Exp. Biol.* 65:471–482.

———. 1978. Evaporative cooling and water balance during flight in birds. *J. Exp. Biol.* 75:231–236.

Trowbridge, C. C. 1895. Hawk flights in Connecticut. *Auk* 12:259–271.

———. 1902. The relation of wind to bird migration. *Amer. Nat.* 36:735–753.

Tucker, V. A. 1966. Oxygen consumption of a flying bird. *Science* 154:150–151.

———. 1968. Respiratory physiology of House Sparrows in relation to high-altitude flight. *J. Exp. Biol.* 48:55–66.

———. 1970. The energetics of bird flight. *Sci. Amer.* 220:70–79.

———. 1972. Metabolism during flight in the Laughing Gull. *Amer. J. Physiol.* 222:237–245.

———. 1974. Energetics of natural avian flight. In *Publications of the Nuttall Ornithological Club, no. 15,* ed. R. A. Paynter, pp. 298–328. Boston: Nuttall Ornithological Club.

Tucker, V. A., and G. C. Parrott. 1970. Aerodynamics and gliding flight in a falcon and other birds. *J. Exp. Biol.* 52:345–367.

Tucker, V. A., and K. Schmidt-Nielsen. 1971. Flight speeds of birds in relation to energetics and wind directions. *Auk* 88:97–107.

Ulfstrand, S. 1958. The annual fluctuations in the migration of the Honey Buzzard (*Pernis apivorus*) over Falsterbo. (English summary.) *Fagelvarld* 17:118–124.

———. 1960. Some aspects on the directing and releasing influence of wind conditions on visible bird migration. *Proc. Int. Ornithol. Congr.* 12:730–736.

Ulfstrand, S., G. Roos, T. Alerstam, and L. Osterdahl. 1974. Visible bird migration at Falsterbo. *Fagelvarld,* suppl. 8:1–245.

Vagliano, C. 1985. The continental and island migration routes of the southeast Mediterranean: Problems and propositions. In *Conservation studies on raptors,* ed. I. Newton and R. D. Chancellor, pp. 263–269. ICBP Technical Publication 5. Thessaloniki: International Council for Bird Preservation.

Vaurie, C. 1965. *The birds of the Palearctic fauna.* Vol. 2. *Genus* Falco (*Part 1,* Falco peregrinus *and* Falco pelegrinoides). London: Witherby.

Videler, J. J., D. Weihs, and S. Daan. 1983. Intermittent gliding in the hunting flight of the Kestrel, *Falco tinunculus. J. Exp. Biol.* 102:1–12.

Walcott, C., and R. P. Green. 1974. Orientation of homing pigeons altered by a change in the direction of an applied magnetic field. *Science* 184:180–182.

Walker, J. 1985. The amateur scientist: A field formula for calculating the speed and flight efficiency of a soaring bird. *Sci. Amer.* 252(3):122–126.

Wallraff, H. G. 1978. Social interrelations involved in migration orientation of birds: Possible contributions of field studies. *Oikos* 30:401–404.

Walter, H. 1979. *Eleonora's Falcon: Adaptations to prey and habitat in a social raptor.* Chicago: University of Chicago Press.

Ward, F. P., and R. B. Berry. 1972. Autumn migration of Peregrine Falcons on Assateague Island, 1970–1971. *J. Wildl. Manage.* 36:484–492.

Ward, P., and A. Zahavi. 1973. The importance of certain assemblages of birds as "information centres" for food-finding. *Ibis* 115:517–534.

Ward-Smith, A. J. 1984. Aerodynamic and energetic considerations relating to undulating and bounding flight in birds. *J. Theor. Biol.* 111:407–417.

Warham, J. 1977. Wing loadings, wing shapes, and flight capabilities of Procellariiformes. *New Zealand J. Zool.* 4:73–83.

Warner, J. 1970. The microstructure of cumulus cloud. 3. The nature of the updraft. *J. Atmos. Sci.* 27:682–688.

Warner, J., and J. W. Telford. 1967. Convection below cloud base. *J. Atmos. Sci.* 24:374–381.

Weatherhead, P. J., and H. Greenwood. 1981. Age and condition bias of decoy-trapped birds. *J. Field Ornithol.* 52:10–15.

Webster, H. M. 1944. A survey of the Prairie Falcon in Colorado. *Auk* 61:609–616.

Wege, M. L., and D. G. Raveling. 1983. Factors influencing the timing, distance, and path of Canada Geese. *Wilson Bull.* 95:209–221.

———. 1984. Flight speed and directional responses to wind by migrating Canada Geese. *Auk* 101:342–348.

Weis-Fogh, T. 1975. Unusual mechanisms for the generation of lift in flying animals. *Sci. Amer.* 233(5):81–87.

Welch, A., L. Welch, and F. G. Irving. 1968. *New soaring pilot.* London: Murray.

Welch, W. A. 1975. Inflight hawk migration study. *J. Hawk Migr. Assoc. North Amer.* 1:14–22.

Welty, J. C. 1982. *The life of birds.* 3d ed. New York: Saunders.

White, C. M. 1968. Diagnosis and relationships of the North American tundra-inhabiting Peregrine Falcons. *Auk* 86:179–191.

Wilde, N. P., Stull, R. B., and E. W. Eloranta. 1985. The LCL zone and cumulus onset. *J. Climate Appl. Meteorol.* 24:640–657.

Williams, T. C., and J. M. Williams. 1978. An oceanic mass migration of land birds. *Sci. Amer.* 239(4):166–176.

———. 1980. A Peterson's guide to radar ornithology? *Amer. Birds* 34:738–741.

Williams, T. C., J. M. Williams, J. M. Teal, and J. W. Kanwisher. 1972. Tracking radar studies of bird migration. *NASA Spec. Publ.* 262:115–128.

Williamson, K. 1954. The migration of the Iceland Merlin. *Brit. Birds* 47:434–441.

———. 1955. Migrational drift. *Proc. Int. Ornithol. Congr.* 11:179–186.

Willis, E. O. 1963. Is the Zone-tailed Hawk a mimic of the Turkey Vulture? *Condor* 65:313–317.

Willoughby, E. J., and T. J. Cade. 1964. Breeding behavior of the American Kestrel (Sparrow Hawk). *Living Bird* 3:75–96.

Withers, P. C. 1979. Aerodynamics and hydrodynamics of the 'hovering' flight of Wilson's Storm Petrel. *J. Exp. Biol.* 80:83–91.

———. 1981. The aerodynamic performance of the wing in Red-shouldered Hawks *Buteo linearis* and a possible aeroelastic role of wing-tip slots. *Ibis* 123:239–247.

Withers, P. C., and P. L. Timko. 1977. The significance of ground effect to the aerodynamic cost of flight and energetics of the black skimmer *Rhynchops nigra. J. Exp. Biol.* 70:13–26.

Woffinden, N. D. 1986. Notes on the Swainson's Hawk in central Utah: Insectivory, premigratory aggregations, and kleptoparasitism. *Great Basin Nat.* 46:302–304.

Wood, C. J. 1973. The flight of albatrosses (a computer simulation). *Ibis* 115:244–256.

Woodcock, A. H. 1940. Convection and soaring over the open sea. *J. Marine Res.* 3:248–253.

———. 1975. Thermals over the sea and gull flight behavior. *Boundary-Layer Meteorol.* 9:63–68.

Woodcock, A. H., and J. Wyman. 1947. Convective motion in air over the sea. *Ann. N.Y. Acad. Sci.* 48:749–776.

Woodward, B. 1949. The motion in and around isolated thermals. *Quart. J. Roy. Meteorol. Soc.* 85:144–151.

World Meteorological Organization (WMO). 1978. *Handbook of meteorological forecasting for soaring flight.* Technical Note 158. Geneva: World
Meterological Organization.

Yapp, W. B. 1962. Some physical limitations on migration. *Ibis* 104:86–89.

Zar, J. H. 1984. *Biostatistical analysis.* 2d ed. Englewood Cliffs, NJ: Prentice-
Hall.

Zu-Aretz, S., and Y. Leshem. 1983. The sea—a trap for gliding birds. *Torgos*
50:16–17.

Index